Biophysics

A Physiological Approach

Specifically tailored to life science students, this textbook explains quantitative aspects of human biophysics with examples drawn from contemporary physiology, genetics, and nanobiology. It outlines important physical ideas, equations, and examples at the heart of contemporary physiology, along with the organization necessary to understand that knowledge.

The wide range of biophysical topics covered includes energetics, bond formation and dissociation, diffusion and directed transport, muscle and connective tissue physics, fluid flow, membrane structure, electrical properties and transport, pharmacokinetics, and system dynamics and stability. Enabling students to understand the uses of quantitation in modern biology, equations are presented in the context of their application, rather than derivation. They are each directed toward the understanding of a biological principle, with a particular emphasis on human biology.

Supplementary resources, including a range of test questions, are available at www.cambridge.org/dillon.

PATRICK F. DILLON is Professor in the Department of Physiology at Michigan State University. He has taught physiology for more than 30 years, ranging from high school to medical school level. He was awarded the Outstanding Faculty Award from Michigan State University in recognition of his teaching achievements.

Biophysics

A Physiological Approach

PATRICK F. DILLON

Michigan State University

CAMBRIDGE
UNIVERSITY PRESS

University Printing House, Cambridge CB2 8BS, United Kingdom

One Liberty Plaza, 20th Floor, New York, NY 10006, USA

477 Williamstown Road, Port Melbourne, VIC 3207, Australia

314-321, 3rd Floor, Plot 3, Splendor Forum, Jasola District Centre, New Delhi - 110025, India

79 Anson Road, #06-04/06, Singapore 079906

Cambridge University Press is part of the University of Cambridge.

It furthers the University's mission by disseminating knowledge in the pursuit of education, learning and research at the highest international levels of excellence.

www.cambridge.org
Information on this title: www.cambridge.org/9780521172165

© P. F. Dillon 2012

First published 2012
Reprinted 2016

A catalogue record for this publication is available from the British Library

Library of Congress Cataloging in Publication data
Dillon, Patrick F., 1951–
Biophysics : a physiological approach / Patrick F. Dillon.
    p. ; cm.
Includes index.
ISBN 978-1-107-00144-2 (hardback) – ISBN 978-0-521-17216-5 (pbk.)
I. Title.
[DNLM: 1. Biophysical Phenomena. QT 34]
LC classification not assigned
571.4–dc23

2011025051

ISBN 978-1-107-00144-2 Hardback
ISBN 978-0-521-17216-5 Paperback

Additional resources for this publication at www.cambridge.org/9781107001442

This book is dedicated to the two men who led me to Biophysics:

Fr. Donald Plocke, S. J., Ph.D.
  Professor and Former Chairman of Biology
  Boston College
  My undergraduate Biophysics professor, and the first one to suggest I might enjoy teaching.

Dr. Richard A. Murphy, Ph.D.
  Professor of Physiology
  University of Virginia
  Teacher, Mentor and Friend. Simply the best Ph.D. advisor, ever.

Thank you.

Contents

Acknowledgments

The first acknowledgments must go to the two scientist-teachers to whom this book is dedicated. My first exposure to biophysics came in an eponymous course taught by Fr. Donald Plocke during my junior year at Boston College. It focused mostly on protein and nucleic acid functions, and was the first course where I could regularly say, "That is so cool." Fr. Plocke, then chairman of the department, helped me far beyond any class work when I was looking to stay in Boston after graduation. He suggested that I apply for a high school teaching position through the BC department of education, telling me he thought I would make a good teacher. We're still waiting to find out if that is true.

Dr. Richard Murphy was my Ph.D. advisor in the Physiology department at the University of Virginia. In the late 1970s the UVA Physiology department was a thrilling place to be, especially in the Murphy lab. Dr. Murphy offered me a choice of biochemical or biophysical projects, and as no one else in the lab was doing biophysics, I went that way. His constant encouragement and support made all-night experiments the norm: you wanted him to enjoy the data you left on his desk at dawn as much as you did. Whenever he gave a talk, he gave credit to the student who had done the work on every single slide. We were all proud to be called "Murphettes" by our contemporaries in other labs. The most soft-spoken of men, his words had iron. He taught me how to think, how to write, and how it was always better to make a colleague than a competitor. He was always kind to my growing family. By the end of my time with Dr. Murphy, we would finish each other's sentences, with key points understood but unsaid, much to the confusion of the other people at lab meetings. By tradition, not by word, you got to call him Dick when you got your degree, and that was a big day. Dr. Richard Murphy, a great scientist, was the best Ph.D. advisor anyone ever had, and it was my great honor to have been his student. This book would not exist without him.

Being able to learn and relate the creative work of so many fine scientists was a privilege: there have been many, many bright people in biophysics. Everyone needs the help and support of their colleagues, and my help has come from my long-time friend and collaborator Bob Root-Bernstein, and the people in the lab, where one of the students once said, "You guys seem to have more fun than the other labs." Great contributions were made by Jill Fisher, Joe Clark, Pat Sears, Charlie Lieder and especially James Barger, who has helped develop both my biophysics course and this book, which his son Michael has been waiting to read. The comments of different reviewers proved very helpful in structuring the material. I thank all the supervisors I've had over the years: Marion Siegman, Marty Kushmerick, Harvey Sparks, John Chimoskey and Bill

Spielman for the encouragement and time freedom they've given me. I appreciate all the help and encouragement I've received from Katrina Halliday, Hans Zauner, Claire Eudall, Megan Waddington, Chris Miller and the rest of the staff at Cambridge University Press. They await their assignments in Biophysics: The Musical. My dear wife Maureen showed great patience, and will now get the computer back in the evening. Our children Caitlin, Brian, Mary and Laura will now have to ask something other than "What chapter are you working on now?" when they call home. And I've learned as much in teaching physiology and biophysics to more than 13,000 students at Michigan State University as I hope they have learned from me. I want to note that two biophysics students, Edwin Chen and Matt Moll, spent lots of time telling me how much a biophysics book for biology students was needed: conversations with these two initiated this book. Hopefully, after learning biophysics, many more of them will be able to say, "That is so cool."

The philosopher and mystic Meister Eckhart said that if the only prayer you ever said was "thank you," it would suffice. Great thanks to all of you.

Introduction

This book is designed for biological science majors with an interest in biophysics. It is particularly aimed at those students in medically oriented disciplines whose career goals include professional or graduate school in the medical sciences with the aim of linking biophysical principles to human physiological functions. In all parts of this book the general biophysical functions will have numerous physiology examples.

In general, pre-professional students have had a significant amount of mathematics during their education, but it may have been several years since they have taken calculus, typically the last formal mathematics education they have had. In contrast, they have had a great deal of current and relevant biological, and especially physiological, education. The presentations in this book take these factors into account, using many of the principles inherent in that level of mathematics education, without including many of the formal derivations of the formulas included in many biophysics texts. The equations used and the information derived from them fill the need of establishing the limits on physiological functions. When used properly, the calculations made by students will form the starting point from which they can then draw conclusions about physiological systems. Inherent in this is the way in which data is analyzed, as this analysis often presupposes how a system works, so that examples of how data can be manipulated are included.

As with many texts, we will start with the simplest systems and progress to the more complex. We will start by considering the environment around us, particularly the energy of that environment. All our molecules exist, except for brief periods during reactions, at equilibrium with the environment. The environmental energy, the product of the absolute temperature (T) times Boltzmann's constant (k), forms the background against which all biophysical functions occur. We will refer to the shorthand kT throughout the book, as it is the baseline energy toward which entropy will drive all of our systems. A number of our systems absorb energy directly from the environment. Among these are rhodopsin absorbance of visible wavelengths of light, and the damage done to DNA by ultraviolet light. The absorbance of UV light by melanin provides us with some protection from this damage. Some energy we absorb from the environment is not naturally generated, but is produced by man-made machines for medical and scientific purposes. Among these are the energies used for magnetic resonance imaging (MRI) and spectroscopy (MRS), and the imaging produced by ultrasound. All of these will be covered in the first chapter.

Molecules are not limited to absorbing and radiating the energy of the world around them. They also interact with other molecules, forming bonds of varying durations. Perhaps the most interesting aspect of biophysics is that it looks at things that happen, events that change on a timescale that we can follow. In the case of some bonds, especially covalent bonds, the lifetime of these bonds is so great that without the specific input of energy by an enzyme-linked system these bonds would never break in our lifetime. The energy in these bonds is so much greater than kT that they will never spontaneously break. As a result, covalent bond formation and rupture plays a very small role in biophysics, even if the structures they produce are enormously important. In contrast, hydrogen bonds and hydrophobic bonds have energies that are only slightly greater than kT, and thus are constantly being made and broken in a violent molecular environment. These bonds provide a tremendous range of different and interesting biophysical functions.

To understand the principles (and some examples) of how molecules recognize and bind to one another, we will look at the world of these molecules, or rather, their worlds. The intracellular and interstitial fluids are strikingly different, most notably in the differences produced by the presence (intracellular) and absence (interstitial) of very high protein concentrations, and by the different ion concentrations, particularly K^+ and Na^+, whose interaction with ubiquitous water molecules produces very different molecular worlds. While these form the bulk of the physiological world, there are other small but important environments, including the hydrophobic core of membranes and the high protein/Na^+ world of blood. Each of these produces unique conditions for molecular interaction.

The most profound developments in biology in the past 20 years have occurred in genetics. The elucidation of the human genome has given us the prospect of dramatic advances in medicine. For all its progress, however, most of the advances have remained descriptive. If there is an area in need of biophysical approaches, it is genetics. What processes control the entry and/or production of transcription factors at the membrane? Do these factors move to the nucleus by diffusion, or is direct transport using an ATP-dependent system involved? How do the factors bind to the genome, which must first unwind using considerable energy? The traditional calculations of dissociation constants assume molecular numbers approaching infinity, but at the chromosome for most genes there will be just two sites, one on each somatic chromosome. Instead of the fraction of the transcription factor that is bound, one has to consider the fraction of time the factor is bound, the retention time. And how long is the retention time relative to the reaction time, the time needed to initiate mRNA production? And the products of translation, the proteins, do not exist in a vacuum. The concentration of proteins inside the cytoplasm is far higher than can be achieved in a test tube, meaning that most proteins will be part of protein clusters. What determines protein–protein binding, and how does the formation of protein clusters alter protein activity? Needless to say, developing the potential of genetics will require extensive biophysical investigations.

The magnitude of binding constants is not limited to intracellular processes. Antibodies, for example, bind to their antigens with very high affinities. The binding is sufficiently long to trigger the non-specific immune response linked to the tail region of each antibody, resulting in the ultimate removal of the antigen, be it a cell or molecule. This system is sufficiently robust that the immune system clears most infections within weeks. In contrast, autoimmune diseases such as Type I diabetes take years to completely destroy their target cells. Why so long? If the antigen targets in autoimmune diseases are not the original antigens, but merely have some similarity, the binding will be weaker and have a shorter retention time, so that in most cases the immune response is not triggered.

But, binding duration is statistical, and occasionally that binding will be long enough to trigger a response, and slowly the cells will be destroyed.

Biological systems of course are not static: molecules and larger structures move from one place to another. In some cases, this movement is driven by concentration gradients, and Fick's diffusion equation is well known to most biology students. We will show how diffusion sets limitations on whether or not a molecule can participate in a physiological process. Can diffusion alone be sufficient to allow an ion to be part of the muscle contraction process? Or exocytosis? Or protein synthesis? What if the area or the diffusion coefficient changes, or more specifically, why do people die from emphysema or pneumonia?

For systems in which diffusion alone cannot support their activity, how will directed transport work? Muscles move entire cells, using alterations in a fundamental general process to produce the differences in skeletal, cardiac and smooth muscle contractions. Recent discoveries of intracellular transport using kinesin, dynein and non-polymerized myosin have answered questions that existed for decades before their discoveries. And for those white blood cells that must be able to respond in any direction, the transient nature of pseudopod formation and movement is also explored.

The systems producing movement must have cyclical binding between dynamic (ATP-dependent) and static structures, with alternating high affinity and low affinity binding constants. For forces and movements in a particular direction, there must be structures capable of bearing and transmitting those forces. Within cells, some structures can permanently perform this function, such as the Z-lines and dense bodies in muscle and the intermediate filaments in skin. Microtubules can bear internal loads in cells, but microtubules may also be restructured to respond to different forces that some cells, such as neurons, must respond to. The stress/strain characteristics of biological molecules show a range of behaviors, from linear responses through viscoelastic recovery to rupture when external force exceeds the molecule's ability to withstand that force. This also occurs in larger structures, such as the remodeling of bone and the rupturing of blood vessels leading to a stroke.

The flow of fluids also has biophysical properties, blood being the most obvious. Blood shows transitions from laminar to non-laminar flow, and the formed components of blood, red blood cells, white blood cells and platelets alter their flow patterns to minimize the physical stress they undergo. Changes in flow produced by atherosclerotic plaques produce non-laminar flow patterns that alter the movement of the formed elements as well as emboli that travel through the blood. Non-blood fluids also have distinct flow characteristics. Synovial fluid changes its physical characteristics as it lubricates and cushions joint movement. The draining of aqueous humor from the front of the eye is necessary to prevent glaucoma.

The physical separation of the intracellular fluid in the cytoplasm from the interstitial fluid by cell membranes produces special properties. The self-associative properties of phospholipids and cholesterol keep them separate from the hydrophilic environments adjacent to them. Membranes possess a degree of fluidity necessary to respond to forces applied to them without rupturing. This fluid nature extends to both the membrane as a whole and to the individual molecules of phospholipid and cholesterol attached by non-covalent bonds. This hydrophobic interface is then loaded with proteins whose myriad formations provide a wide range of functions that both alter and control the fluids around them. An important subset of these functions deals with membrane transport, using either the energy of ATP or that of concentration gradients for ion transport, hydrophilic molecule transport or protein transport, as well as the resting, graded and action potentials that regulate so many physiological functions.

All biology students are familiar with the basics of the different membrane electrical functions. They know well that K^+ is more permeable than Na^+ at rest, and that this reflects a difference in the number of open channels, as if all channels are the same. But, both the ionic interactions with the environment around them and the channels themselves have biophysical differences, providing a more diverse control of the electrical events. Even the resting membrane potential itself produces important physiological functions altering the entry of ions through channels and the binding of agonists to their receptors. We will spend a significant amount of time discussing membrane behavior.

We include a consideration of compartment analysis, methods that are used to model metabolic fluxes and pharmacokinetics. While these systems provide information on how these systems as a whole behave, they are limited by the statistical variations between individuals. These variations, reflecting the inherent risks associated with any pharmaceutical treatment, mean that a small, predictable number of individuals, the identities of whom cannot be known in advance, will not follow the majority pattern, often with disastrous results. These outliers recall the variations in the energy level of an ensemble of molecules discussed in the first chapter.

The interactions of different components of the body produce complex cases. The stability of physiological systems involves control at both the cellular level and the whole body level. There are also transitory metastable states, such as those that occur in enzyme–substrate complexes, which will be considered. Among the most interesting physiological phenomena with biophysical control points are those associated with positive feedback reactions, such as the rupturing of the ovarian follicle, the clotting of blood and, most profoundly, life–death transitions. These systems produce an irreversible state change, and can be modeled using catastrophe theory. Fractal behavior appears in many physiological systems and in some cases devolves into chaotic behavior. These non-linear states may provide systemic stability, preventing pathological state transitions. Regulation by homeostasis may only be a part of systemic control in which multiple inputs associated with allometric control produce a wide window of parameter variability consistent with a healthy state.

This book is not intended to cover the breadth of all of biophysics. Many interesting elements, such as the flight of birds, or the behavior of protein under non-physiological conditions, have been omitted, as have many of the formal, mathematical derivations of the equations presented. These and other elements are presented in many other, fine biophysics books, and the interested reader is referred to those texts. I hope you find this work focusing on those biophysical processes relevant to human physiology useful and enlightening.

1 The energy around us

1.1 Forms of energy

We are all subject to the laws of physics. Every process, living or not, obeys the laws of thermodynamics. Biophysical systems in living organisms must have a constant input of energy to remain alive, but will reach thermal equilibrium after death. Sufficiently small subsystems within an organism will be at thermal equilibrium, even if the organism as a whole is not at thermal equilibrium with its environment. Biophysical systems can neither create nor destroy energy, but they can manipulate energy by doing work or altering the internal energy of the system. Biophysical processes removed from equilibrium will produce an increase in entropy. When biophysical systems, including physiological ones, have an increase in energy produced by ordering the local environment, there must be a greater decrease in the energy of the universe. The difference in these energies is the change in entropy. These principles of the zeroth, first and second laws of thermodynamics appear all the time in discussing biophysics. Understanding these general principles will make understanding energy absorbance, bond formation, ion diffusion, fluid flow, muscle contraction and dozens of other processes possible.

Everyone has a basic idea of muscle contraction or blood flow. These are biophysical, and physiological, processes. A process is a transition between state functions. State functions are thermodynamic quantities, and are therefore, in the absence of external energy input, at equilibrium. Processes describe the quantitative transition between state functions or, more simply, different states. The internal energy of a system is the sum of the different states comprising that system. In the world of chemistry, there are multiple ways in which the internal energy of a system can be subdivided, including the enthalpy and the Helmholtz free energy. In discussing the energy of a physiological system, the Gibbs free energy is most relevant.

At its most fundamental level, the Gibbs free energy is

$$G = H - TS \qquad (1.1)$$

where H is the enthalpy, T is the absolute temperature and S is the entropy. The exact value of S cannot be known. The change in the Gibbs free energy dG is more commonly used:

$$\mathrm{d}G = -S\mathrm{d}T + V\mathrm{d}p + F\mathrm{d}l + \sum_{i=1}^{m} \mu_i \mathrm{d}n_i + \psi \mathrm{d}q \qquad (1.2)$$

where V is the volume, dp is the change in pressure, F is the mechanical force, dl is the change in length, μ_i is the chemical potential, dn_i is the change in the number of molecules, ψ is the electric potential and dq is the change in electric charge. This daunting equation is more complex than the molar Gibbs free energy of reaction, familiar to many students from their biochemistry classes:

$$\Delta G = -RT \ln K_{eq} \tag{1.3}$$

where R is the molar gas constant and K_{eq} is the equilibrium constant.

The difference between the equations is illustrative. The smaller free energy of reaction equation is a molecular subset of the larger Gibbs free energy equation, the reaction equation being derived by assuming that temperature, pressure, length and charge are all constant, reducing the free energy equation to a statement only relating to the chemical potential of the system. In practical terms, this is the goal of experimental science: to hold all variables constant except the one we are interested in measuring. Studies of thermal regulation focus only on dT; respiration depends on dp; muscle contraction measures dl; and electrophysiology depends on dq. All of these elements are always present, but using logic and control conditions we try to minimize the effect of outside forces that would alter our results. These forces manifest themselves as system "noise." Noise is nothing but the unintended input of one of those other elements, whether biological, such as the heating of muscle during contraction, or mechanical, such as a faulty recorder switch.

Other processes that have little influence on normal physiological systems come into play when the body is exposed to unusual energetic input, such as magnetic fields during magnetic resonance imaging (MRI). In this case, the term BdM would have to be added to the free energy equation, with B being the external magnetic field and dM the change in magnetization. Outside of the magnetic resonance magnet, this term is negligible because B is so small that there is no significant change in this term. When you analyze an MRI, consider the other terms of the free energy equation: does the temperature change during the input of radio-frequency pulses used to alter the magnetization? Does the pressure change (unlikely)? Does the person move (dl)? Does the person have a cardiac pacemaker whose performance could be altered by the magnetic field? And, importantly, is there a difference in chemical potential in different areas, such as the more hydrophobic white matter and more hydrophilic gray matter of the brain, the hydrogen nuclei of which respond differently in the magnetic field? It is the interaction between the magnetic field and the chemical potential that is used to produce contrast in the MRI.

The energy of biophysical systems, then, takes on many forms. Even as we focus on individual elements, it is important to bear in mind that other elements can sometimes play a role, even unanticipated ones. The best scientists have so much familiarity with their equipment they know that when they make an unusual finding, the *sine qua non* for all new knowledge, that finding is not due to the limitations of their equipment, but because they have uncovered something novel. True creativity, true genius, requires both technical expertise and inspired insight.

1.2   Ambient energy

We live in a world where our body temperature is 98.6° F, 37° C or 310 K. We regulate this temperature closely, no matter the temperature around us. With the exception of a few molecules near our body surfaces, the constant temperature of the body produces an average energy that the molecules in the body are exposed to. This energy is the absolute temperature T (in K) times the Boltzmann constant, k, 1.38×10^{-23} J/K·molecule. Except for those occasions when a molecule is involved in a reaction, molecules will be in equilibrium with the energy of the environment around us, E_o,

$$E_o = kT = 310\,\text{K} \cdot 1.38 \times 10^{-23}\,\text{J/K} \cdot \text{molecule} = 4.28 \times 10^{-21}\,\text{J/molecule}. \quad (1.4)$$

This is the equilibrium energy of a single molecule. When we deal with an ensemble of molecules, we measure molecules on the molar scale, using Avogadro's number, N_A, to convert Boltzmann's constant to the gas constant, R,

$$R = N_A \cdot k = 6.02 \times 10^{23}\,\text{molecule/mol} \cdot 1.38 \times 10^{-23}\,\text{J/K} \cdot \text{molecule}$$
$$= 8.31\,\text{J/K} \cdot \text{mol} \quad (1.5)$$

and the molar equilibrium energy to

$$E_m = RT = 8.31\,\text{J/K} \cdot \text{mol} \cdot 310\,\text{K} = 2.58\,\text{kJ/mol}. \quad (1.6)$$

Many of the traditional physical chemical measurements of molecule activity use the kT scale. We will find the RT scale useful when dealing with cellular energy levels, as they are routinely measured in kJ/mol, so that it is important to be familiar with both scales.

Like all physical systems, biological systems will go to the lowest energy state, maximizing entropy. No matter which scale is used, kT or RT, the environmental energy is the lowest energy state that living physiological systems will go toward. This energy is elevated above global thermal equilibrium (GTE), the energy of the world around us. We exist at a steady-state minimum removed from global thermal equilibrium (Figure 1.1).

All systems will try to maximize entropy, always tending to the lowest energy state possible. All living systems on the earth will tend toward the energy of the environment around them, global thermal equilibrium. All beings will reach GTE when they die. Until then, living beings must have a constant input of energy in order to counter entropy.

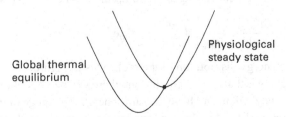

Global thermal equilibrium

Physiological steady state

Figure 1.1   All living systems exist at a steady state whose minimum energy position (●) has a higher energy than the global equilibrium it is part of. To maintain this steady state, there must be a constant input of energy to offset entropy.

The • in Figure 1.1 represents the steady-state point at which the vectors for energy input and entropy exactly balance one another. When death occurs, there is no longer energy input, and the • will slide to the GTE minimum, and the being will be at equilibrium with the earth. (One could make further nestings as the earth is in a steady state relative to the sun, the sun is in a steady state relative to the galaxy, and the galaxy is in a steady state relative to the universe.) Within any substructure in the body, the molecules will be at equilibrium with the environment around them, at 2.58 kJ/mol, unless they are involved in a reaction, such as absorbance of light by melanin or 11-cis-retinal. Melanin will spontaneously return to thermal equilibrium but, as we will see below, the new all-trans state of retinal does not spontaneously return to the 11-cis state without an enzymatic reaction.

Every ensemble of molecules at equilibrium will have the same average energy, but each individual molecule within the ensemble will not have the same energy. The Boltzmann function of energy distribution shows the number (n_i) of molecules that have a given energy level (E_i), according to the relationship

$$n_i = Ce^{-E_i/kT} \tag{1.7}$$

where C is a normalization constant. Because the exponent is negative, there will be fewer molecules with a given energy as E_i increases. Each molecule in the ensemble will be subject to local conditions, such as collisions with other molecules, which will constantly change its velocity and thus its energy. Maxwell developed the equation of velocity distribution:

$$\frac{dn(v)}{n_o dv} = \frac{4}{\sqrt{\pi}} \left(\frac{m}{2kT} \right)^{3/2} v^2 e^{-\frac{mv^2}{2kT}} \tag{1.8}$$

in which the fraction of molecules (dn/n_o) is within a particular velocity range (dv). The velocity distribution is nearly symmetrical, as shown in the dashed line of Figure 1.2. Look at the exponential term: $mv^2/2$ represents the kinetic energy of the molecule, divided by kT. Since the molecules are at equilibrium, there is no potential energy, only kinetic energy, and the energy of a molecule E_m is

$$E_m = \frac{mv^2}{2}. \tag{1.9}$$

The Maxwell velocity distribution equation can be modified, multiplying the non-exponential part by $2/m \cdot m/2$ and converting $mv^2/2$ to E_m to give the energy distribution:

$$\frac{dn(E)}{n_o dE} = \frac{8}{m\sqrt{\pi}} \left(\frac{m}{2kT} \right)^{3/2} E_m e^{-\frac{E_m}{kT}}. \tag{1.10}$$

This allows us to plot the energy distribution of the molecules in Figure 1.2.

The important concept here is that even at equilibrium there will be a distribution of the energy of the molecules. They will not all have the same energy. The range of the energy distribution will be determined by the physical conditions surrounding the molecule. While its mass will not change, the molecular interactions with its surroundings will affect its velocity, and thus alter its particular energy. As can be seen in Figure 1.2, the

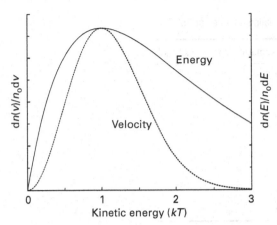

Figure 1.2 The energy and velocity distributions of molecules at equilibrium. The abscissa is plotted in units of kT relative to the ambient energy. The velocity distribution will have its peak when the $mv^2/2kT = 1$. The energy distribution will have its peak when $E_m/kT = 1$.

energy distribution will not be a normal distribution, with the most common energy occurring at a lower energy value than the average energy value.

1.3 Molecular energy

The energy associated with each atom and each bond is not continuous, but quantal, based on the electron shells occupied in the electron cloud around the nucleus. For a given atom, there would be a quantal energy distribution, with the lowest energy configuration being the most common, as Boltzmann demonstrated. Within each quantum domain small variations in thermal excitation exist. In a molecule, however, not every bond will be at its lowest energy: instead, the molecule as a whole will seek its lowest overall energy, out of the many possible configurations of attractions and repulsions that will alter bond angles and the energy in the bonds. The larger the molecule, the greater the number of potential configurations and energy levels that are possible. Because of this, the energy distributions of a molecule will appear to be continuous, as seen in Figure 1.2, but if magnified sufficiently the digital nature of molecular configurations would be revealed. For those special atoms which respond to a particular electromagnetic frequency of radiation, they may have an electron raised to a higher electron orbital. This occurs in fluorescent and phosphorescent systems and for molecules in which a particular bond is sensitive to a particular electromagnetic frequency.

The thermal energy of molecules must be distinguished from the chemical energy of molecules. All molecules, regardless of chemical structure, will see the same thermal energy kT. The different chemical structure of molecules means that different amounts of energy are trapped within the chemical bonds of different molecules. The organic part of a molecule will have its atoms connected by covalent bonds. Each of these bonds has an energy associated with it: the dissociation bond energy necessary to break the bond.

Table 1.1 Average bond dissociation energies at 25 °C

Bond	Dissociation energy (kJ/mol)
C—C	344
C≡C	615
C—H	415
C—N	292
C—O	350
C≡O	725
N—H	391
O—O	143
O—H	463
O_2	498

Source: Tinoco *et al.*, 1995

Some of the most common covalent bond dissociation energies are listed in Table 1.1. The ambient energy in the human body is 2.58 kJ/mol. This energy is far below that of any covalent bond in the body. This means that it is statistically unlikely that any covalent bond would spontaneously break due to the random thermal fluctuations around it.

1.4   Molecular energy absorbance

Despite the thermal stability of covalent bonds in physiological systems, some of these bonds are sensitive to energy input from external sources. When energy is absorbed by a molecule, it will either release the energy as heat, returning to its original configuration, or trap some of the energy within the molecule by altering its structure, as shown in Figure 1.3. In the first case, the molecule can absorb heat from the environment without changing its chemical structure, as will occur when there is a local temperature increase. The molecule will have a higher energy. If the increase in energy is above kT (i.e., the entire environment has not increased its temperature), the molecule will come to thermal equilibrium with the environment around it, and return to its original energy state. This scenario is shown in the upper part of Figure 1.3. The absorbance of radiant energy by protein in the skin, for instance, would be an example of this. This is the most common type of energy absorbance in physiological molecules.

In the second case, shown in the lower part of Figure 1.3, a molecule will absorb energy, alter the electrons of the bonds of the molecule, and change its chemical structure. The new structure, on the right, will have its own energy minimum. It may be possible for the molecule to revert to its original structure, but this will be determined by the height of the energy barrier between the two states. The greater the height of the energy barrier, the less likely a molecule will spontaneously revert due to random energy fluctuations in its environment. If the energy barrier is less than kT, then spontaneous reversion will occur. The equilibrium between the two states of the molecule will be determined by the relative basal energy states. The higher of the two minima will have fewer elements at

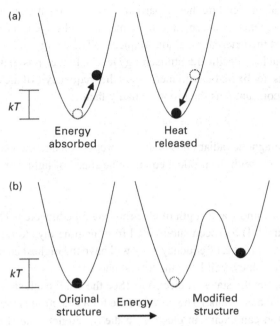

(a)

kT

Energy
absorbed

Heat
released

(b)

kT

Original
structure

Energy

Modified
structure

Figure 1.3   Energy absorbance within molecules. The molecule may absorb energy and radiate heat
(a), or alter its chemical structure (b). The effect of kT on any state is measured from its local
minimum relative to the lowest local energy barrier.

equilibrium than the lower minimum. If the two minima are equal, then at equilibrium
there will be an equal number of both states. The absorption of energy may be so great
that no reversion to the original state can occur. This is the case when a molecule absorbs
large amounts of heat, destroying its three-dimensional structure, as occurs in the thermal
denaturation of a protein. The energy barrier between the original and the new state is so
great that no enzyme is capable of lowering the energy barrier sufficiently to return the
molecule to its original configuration. When foods are cooked by increasing their
temperature, whether by radiant heat, conduction (heating by contact) or by microwave
radiation increasing the friction between water molecules and thus the internal heat, the
molecules cannot revert to their original structure, as seen in the translucent to opaque
conversion of egg whites. There is an energy barrier between the new and original states
that greatly exceeds kT.

1.5   Molecular transduction

Between the cases in which the molecule that changes its structure can spontaneously
revert to its original conformation and in which it is denatured so that no return is
possible, there are particular alterations in physiological systems that can revert with
the assistance of enzymes. In these cases, a particular bond can absorb energy from the
surrounding environment and alter its structure. Unlike the case in which all parts of the

molecule see a higher local temperature, here a particular bond is sensitive to a particular wavelength of electromagnetic radiation, due to a match of the electron oscillation frequency of the bond and the external radiation frequency. The energy of an individual bond is not continuous, but has specific quantum energy levels. Planck postulated that the energy ε of the quantum is not fixed, but will increase as the frequency v of the oscillation increases, with Planck's constant h as the proportionality factor:

$$\varepsilon = hv. \tag{1.11}$$

The frequency of electromagnetic radiation is inversely proportional to the wavelength λ of the electromagnetic wave, with the product equal to the speed of light c in a vacuum:

$$c = \lambda v. \tag{1.12}$$

Thus, the energy, frequency and wavelength of electrons are all connected. The electromagnetic spectrum (Figure 1.4) has been subdivided from gamma rays to radio waves. Since gamma waves have the highest frequency, they will have the highest energy. Radio waves, with the lowest frequency, will have the lowest energy.

Quantum mechanics limits the states of electrons. (See the Pauli exclusion principle, not covered here, for the details of this.) The ground state for an electron is the S_0 singlet state, from which a photon can excite an electron to the S_1 singlet state (Figure 1.5).

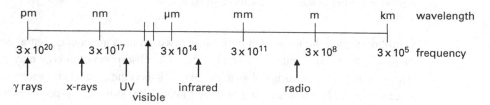

Figure 1.4   Electromagnetic spectrum. Wave energy is directly proportional to frequency and inversely proportional to wavelength.

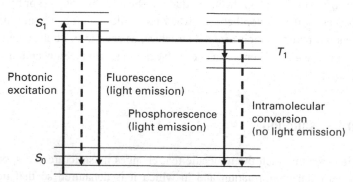

Figure 1.5   At the single atom level, the different quantum states (S_0, S_1, T_1) have thermal variations at each quantum level, indicated by the horizontal parallel lines. Fluorescence and phosphorescence do not play a role in human physiology. Intramolecular conversion occurs in phototransduction, and the destructive effects of ultraviolet and x-rays.

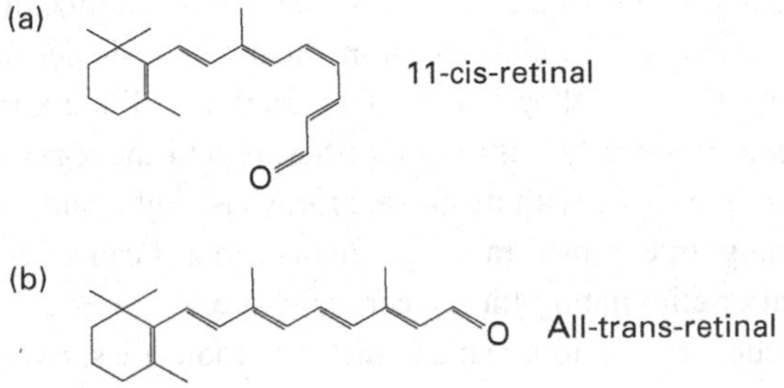

(a)

11-cis-retinal

(b)

All-trans-retinal

Figure 1.6   Structural forms of retinal: (a) 11-cis-retinal and (b) all-trans-retinal.

The electron can spontaneously release this energy as a photon of light: this is fluorescence. Or, there may be an energetic transfer to an adjacent triplet state T_1, from which a photon of light can be released at a different frequency: this is phosphorescence. Fluorescent radiation (10^{-9} to 10^{-5} s) is faster than phosphorescent radiation ($> 10^{-5}$ s) due to the greater stability of the T_1 state (Glaser, 2001). These mechanisms do not play a direct role in physiological systems, although innumerable laboratory methods use both in the study of biological molecules.

The third path, indicated by the dashed lines in Figure 1.5, shows an energy release after quantum absorption of energy without light emission from either the S_1 or T_1 state. In this case, a portion of the energy remains within the molecule and alters its structure. The retinal portion of rhodopsin is an example of this, as shown in Figure 1.6.

Upon exposure to light, with a maximum absorbance at 500 nm, the 11-cis form of retinal absorbs sufficient energy to be converted to the all-trans form of retinal. The all-trans form partially dissociates from the opsin portion of rhodopsin, triggering the cascade that produces phototransduction. The all-trans form does not spontaneously revert entirely to the 11-cis form, indicating that there is an energy barrier between the forms that exceeds ambient energy. An enzyme, retinal isomerase, is responsible for the conversion back to the 11-cis form. Since at equilibrium the 11-cis form predominates, it must have a lower energy minimum than the all-trans form.

DNA is particularly sensitive to UV-B radiation. Upon exposure to this wavelength, 280–315 nm, thymine–thymine and thymine–cytosine pyrimidine bridges may form. These mutations in the DNA can be corrected enzymatically, so that from an energetic perspective this change is analogous to the changes in retinal. In both cases, the product of the energy absorbance has lost its physiological function, the ability to absorb light or to pass on genetic information, respectively. But in both cases, there has been evolutionary development of enzymatic activity that can return the molecules to their original, functional state.

1.6   Ionizing radiation

DNA damage can also occur from exposure to x-rays. X-rays have wavelengths in the nanometer range, shorter than the visible and UV-B wavelengths in the cases above.

X-rays produce ionizing radiation, in which molecules are altered by ionization of one of their electrons. The usefulness of x-ray radiation for medical diagnosis lies in the ability of the x-rays to interact with tissue: if there were no interactions, all the x-rays coming from the x-ray source would uniformly pass through tissue, and the x-ray film would have no contrast. Damage to the tissue with diagnostic x-rays is slight, and in most cases can be repaired enzymatically. In contrast, radiation therapy to kill cancer cells is much more intense, with the goal of eliminating the cancerous tissue.

X-rays are highly energetic, and can ionize many different molecules. Every molecule has an energy of ionization: in the case of water, that energy is 1200 kJ/mol, 100–1000 times less than the energy of x-rays. The collision of an x-ray with a water molecule will result in the ionization of the water molecule. The ionization of water leads to the formation of the destructive radicals that are a product of ionizing radiation. There are multiple reaction sequences generated by this initial reaction. An example of one of these leading to a chain reaction of destructive reactions is as follows:

$$H_2O + h\nu \rightarrow H_2O^+ + e^- \qquad \text{Ionization of water} \qquad (1.13)$$

$$H_2O^+ + e^- \rightarrow H_2O^* \qquad \text{Excited water molecule} \qquad (1.14)$$

$$H_2O^* \rightarrow H^\bullet + OH^\bullet \qquad \text{Dissociation to radicals} \qquad (1.15)$$

$$R_1H + OH^\bullet \rightarrow R_1^\bullet + H_2O \quad \text{Generation of organic radical } R_1 \qquad (1.16)$$

$$R_1^\bullet + O_2 \rightarrow R_1O_2^\bullet \qquad \text{Generation of oxygen radical } R_1 \qquad (1.17)$$

which leads to the chain reaction:

$$R_1O_2^\bullet + R_2 \rightarrow R_1O_2H + R_2^\bullet \text{ Covalent change in } R_1\text{, generation} \atop \text{of radical } R_2 \qquad (1.18)$$

$$R_2^\bullet + O_2 \rightarrow R_2O_2^\bullet \ldots \qquad \text{Generation of oxygen radical } R_2\text{, etc.} \quad (1.19)$$

The presence of oxygen extends the destructive power of radical formation. Each organic radical has its covalent structure permanently altered, often in a manner that eliminates the normal function of that molecule. The chain reaction nature of radical formation means many molecules will be destroyed and the cell potentially killed. In the case of radiation therapy on a cancerous growth, this is the desired outcome; in the case of healthy tissue, it is not. Antioxidant molecules such as Vitamin C can stop this chain reaction. Non-lethal removal of radicals results in the production of H_2O, H_2, and hydrogen peroxide, H_2O_2. H_2O_2 is also a destructive molecule, but the enzyme catalase in peroxisomes converts hydrogen peroxide to H_2O and O_2 and stops the cycle of destruction.

X-rays are of course not the only imaging technology. Computerized tomography (CT) images produce two-dimensional images using Radon transforms of multiple x-ray scans. Closely related to CT is positron emission tomography scanning, or PET. PET scans use a metabolic tracer radio-labeled with a positron-emitting nuclide.

The radiotracer is injected in the body and is absorbed by the tissue of interest, often a cancerous tumor. When the positron is emitted, it collides with an electron, mutually annihilating both, and emitting two photons 180 degrees apart. The paired photons are detected, and an image is produced. Since the image is related to the metabolite (fluorodeoxyglucose is commonly used), PET scans give information on the functional state of the tissue. They are often collected simultaneously with CT scans, which have better spatial resolution.

1.7 Magnetic resonance

Other imaging technologies, such as magnetic resonance imaging (MRI) and ultrasound, use different methods of energy absorbance. MRI takes advantage of a different property of molecules, the spin associated with the nucleus. The nucleus is charged due to the presence of protons, and a spinning charge generates a magnetic field. Different atomic nuclei (H1, F19, P32, C13) have different spin rates, and therefore have different magnetic moments. When an external magnetic field is applied to the nuclei of a tissue, two spin states exist: one in which some magnetic moments line up with the applied field (the low energy spin state), and one in which some magnetic moments line up against the applied field (the high energy state) as shown in Figure 1.7.

There will always be more spins in the low energy state than in the high energy state, creating a net magnetic moment along the z-axis in the direction of the magnetic field. The stronger the applied magnetic field, the greater the energy difference and the stronger the potential signal. When a radio frequency signal is applied that matches the spin frequency of the protons, the energy states are equilized. When the radio field is turned off, the spins will re-equilibrate, releasing a radio frequency signal that can be detected. This is T1 relaxation. In addition to the energy difference between the states, the magnetic moments, spread out in the shape of a cone when no radio frequency energy

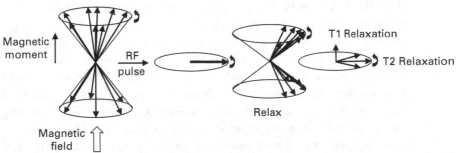

Figure 1.7 Magnetic moments precess, or spin, around the nucleus. When atoms are placed in a magnetic field, the magnetic moments of the atoms will align with or against the field, with the greater number aligned with the field. An applied radio frequency pulse will equalize the magnetic moments. When the radio frequency field is removed, the magnetic moments will re-equilibrate using T1 and T2 relaxations, and generate a signal that produces MR images and MR spectra. The smaller cones on the right show the signals in the middle of relaxation.

(a)   (b)

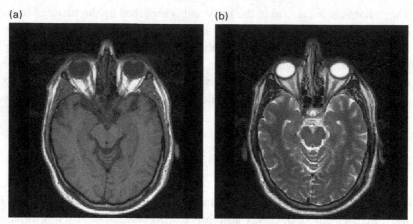

Figure 1.8   T1 (a) and T2 (b) weighted MR images of the brain show the eyeballs with the optic nerves, medulla, vermis, and temporal lobes with hippocampal regions. (Courtesy of MR-TIP.com, used with permission).

is applied, will be lined up together at a single angle of the cone in the x–y plane when the radio waves are applied. When the applied field is turned off, the aligned magnetic moments will spread out in the plane perpendicular (in the x–y plane) to the external field. This is T2 relaxation. T1 and T2 relaxation energy emissions can be collected and, using two- and three-dimensional reconstruction programs also based on the Radon transform, can have images produced from H1 nuclei (Deans, 1985). Since the hydrogen atoms of water and fats have slightly different nuclear environments, they will have different spins and different relaxation characteristics. These differences produce contrast between tissue with different amounts of fat and water, as in white and gray matter in the brain. Depending on the collection parameters, the image produced can be T1 or T2 weighted. Examples of these different images of the same brain are shown in Figure 1.8. There is a reversal of the image intensity of the water/fat regions in the T1 and T2 images.

Functional MRI (fMRI) is a differential imaging technique based on local blood flow. Blood flow is related to local oxygen consumption, and therefore the levels of oxyhemoglobin and deoxyhemoglobin, the ratio of which will change slightly in metabolically active tissues. This difference can be detected, especially in the brain. Metabolically active tissue will produce increased signals, allowing a correlation between a particular activity and a local brain region. While not as sensitive as MRI, and therefore not as spatially localized, there is an increasing use of fMRI in the diagnosis of neurological pathology, as well as to improve understanding of normal neural function. An example of fMRI imaging is shown in Figure 1.9.

A few other nuclei have properties that can produce MR signals, but the natural concentration of the molecules with these nuclei is too small for diagnostic imaging. The next strongest magnetic moment after H1 is in F19. Specific compounds are labeled with F19, and the metabolism of these compounds can be followed over time.

Three nuclei, H1, P32 and C13, can be used to follow metabolism of particular chemicals in physiological tissue. Spectra showing different compounds, their peak

(a) (b)

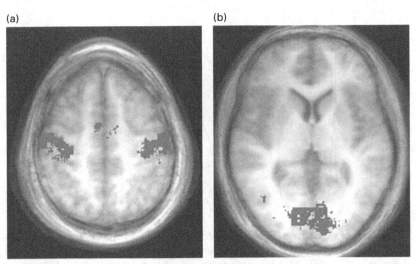

Figure 1.9   fMRI images of the brain. (a) Brain activity associated with motor activity (finger snapping). (b) Brain activity associated with visual activity (viewing a checkerboard pattern) (Reproduced from Slade *et al.*, 2009).

size proportional to their concentration, can be generated. H1, for example, can be used to detect lactate and creatine. P32 is used to measure tissue energy use. C13, while not naturally sufficient for measurement, is used to radiolabel compounds, usually related to carbohydrate metabolism.

P32 is the most abundant of the P atoms and is present in sufficient concentrations that many molecules can be measured in their native amounts. The energy-using reactions involving ATP and phosphocreatine (PCr) can be monitored in many tissues. In these cases, the relaxation of the MR signal, which has amplitude and time information, is Fourier transformed into a spectrum with amplitude and frequency information. Because the physical environment around each different P nucleus, such as the three phosphates of ATP, has a slightly different magnetic moment, these peaks of the spectrum will occur at slightly different frequencies. An example of P32 spectra is shown in Figure 1.10, in which changes in inorganic phosphate and phosphocreatine are illustrated in an exercising human muscle. In addition to changes in the concentration of the different molecules, changes in the physical environment of a tissue can cause some peaks to shift position, producing an internal measurement of a particular factor. For example, the inorganic phosphate peak is sensitive to the local H^+ concentration, so that the position of the P_i peak is used to measure the pH. Also, ATP molecules bind to Mg^{2+} inside cells. In many tissues, the position of the βATP peak will be proportional to the free Mg^{2+} concentration.

For a few reactions, such as overall tissue ATPase rate and creatine kinase reaction, magnetic resonance can be used to measure their in vivo kinetics. Saturation transfer of one peak in the spectrum negates the signal from that nucleus, such as the γ-ATP phosphate or the phosphate of phosphocreatine (Brown, 1980). When the atom with the negated signal is transferred to another molecule, that rate of the transfer can be measured.

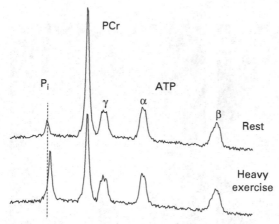

Figure 1.10 Magnetic resonance spectra of human skeletal muscle at rest and during heavy exercise. Peaks are P$_i$, inorganic phosphate; PCr, phosphocreatine; γ-, α- and β-ATP. Note the increase in P$_i$ and the decrease in PCr, while ATP does not significantly change. There is also a shift to the right of the P$_i$ peak, indicating a decrease in pH. The curve in the baseline is due to the signal from bound calcium phosphate in bone.

Although this method uses signal negation rather than addition, the concept is similar to that of radio labeling a molecule to measure enzyme kinetics. For the reaction

$$A \underset{k_1}{\overset{k_2}{\longleftrightarrow}} B \tag{1.20}$$

such as the reaction between ATP and PCr catalyzed by the creatine kinase reaction, there will be forward and backward rate constants k_1 and k_2. The changes in magnetization of A and B following the normal applied radio frequency are

$$\frac{d}{dt} M_z^A = -\frac{M_z^A - M_o^A}{T_1^A} - k_1 M_z^A + k_2 M_z^B \tag{1.21}$$

$$\frac{d}{dt} M_z^B = -\frac{M_z^B - M_o^B}{T_1^B} + k_1 M_z^A - k_2 M_z^B \tag{1.22}$$

where M_z is the magnetization after the radio signal is applied, M_o is the magnetization without the radio signal, and T_1 is the relaxation in the z-direction. The terms with the rate constants k_1 and k_2 account for the changes in magnetization caused by the transfer of magnetization during the enzymatic reaction. The reaction rate must be on the timescale of T1 to be measured using MR saturation transfer. At equilibrium,

$$k_1 M_o^A = k_2 M_o^B \tag{1.23}$$

with the forward and backward fluxes being equal. When the B resonance is specifically saturated,

$$M_z^B = 0 \tag{1.24}$$

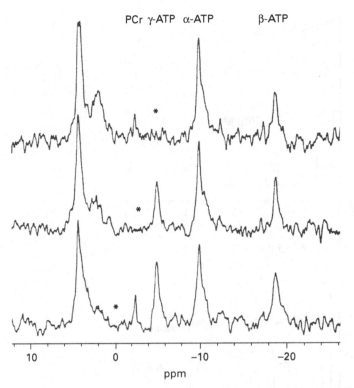

Figure 1.11 Saturation transfer between PCr and γ-ATP in the pig carotid artery. The asterisks (*) are the frequencies at which the radio frequency saturation signal was applied. (Reproduced from Clark and Dillon, 1995.)

and the change in the magnetization of A will be reduced to

$$\frac{\mathrm{d}}{\mathrm{d}t} M_z^A = -\frac{M_z^A - M_o^A}{T_1^A} - k_1 M_z^A \tag{1.25}$$

decreasing the magnetization of A by the amount $k_2 M_z^B$. By varying the saturation duration, the values of k_1 and k_2 can be measured. In Figure 1.11, the control spectrum in the lower figure shows the unsaturated PCr and γ-ATP peaks in a pig carotid artery, with a control saturation at the asterisk. The upper two spectra show the specific saturations of the PCr and γ-ATP peaks at the asterisks, totally negating their signals. Note that when ATP is saturated in the upper spectrum, the PCr peak is decreased compared with the control, and that when the PCr peak is saturated in the middle spectrum, the ATP peak is decreased. These changes are due to the activity of creatine kinase in this tissue.

There is not sufficient naturally occurring C13 to produce a spectrum. Compounds can be C13-enriched, however, and the metabolism of the original compound can be followed as new, C13-enriched compounds are produced downstream of the original molecule. Figure 1.12 shows an example of a C13 spectrum, with the new peaks

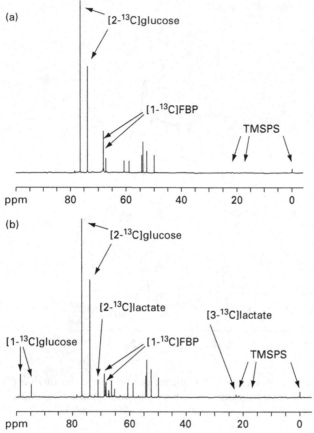

Figure 1.12 ^{13}C-nuclear magnetic resonance spectra showing the metabolic fates of [1-^{13}C]fructose 1,6-bisphosphate (FBP) and [2-^{13}C]glucose in pig cerebral microvessels. (a) ^{13}C-NMR spectrum of the solution before incubation with PCMV, showing the positions of the ^{13}C-labeled substrates [2-^{13}C]glucose and [1-^{13}C]FBP. Peaks at 0, 17.6, and 21.7 ppm represent 3-(trimethylsilyl)-1-propanesulfonic acid (TMSPS). (b) ^{13}C-NMR spectrum of the solution after incubation with PCMV. Major new resonances corresponding to [2-^{13}C]lactate derived from [2-^{13}C]glucose (71.1 ppm) and [1-^{13}C]glucose derived from [1-^{13}C]FBP (β, 98.6; α, 94.8 ppm) are present. The small peak at 22.7 ppm represents [3-^{13}C]lactate derived from [1-^{13}C]FBP. (Redrawn from Lloyd and Hardin, 1999, with permission.)

appearing following the introduction of C13-labeled [2-C]glucose and fructose [1-C] 1,6-bisphosphate into a solution bathing blood vessels. The appearance of lactate and altered glucose ([1-C]glucose) can be followed.

1.8   Sound

There is a range of sound waves that can be detected by the human ear. Sound waves, unlike electromagnetic radiation, require a medium for the waves to pass through.

The frequency of a sound wave is its pitch, and the amplitude of the sound wave is its loudness. Sound waves produce cyclic compressions of the medium, whether gas, liquid or solid. All of these media are part of the sound transduction system of the human ear: the sound waves arrive through the air, pass though the solid structures of the tympanic membrane, bones of the middle ear, and the oval window, the liquid of the inner ear, and the solid basilar membrane before being transduced into electrical signals.

The relation between the speed of sound in a medium c_m, the frequency of the sound wave v, and the wavelength of the sound wave λ is

$$c_m = v\lambda. \tag{1.26}$$

The speed of sound is not constant. It is faster in solid than liquids, and faster in liquids than in air. In addition, the physical properties of the medium will change what the speed of sound is in that medium, such as the density of the material or the humidity of the air. The relationship between the speed of sound, the stiffness of the medium K, and the density of the medium δ is

$$c_m = \sqrt{K/\delta}. \tag{1.27}$$

The stiffness of the medium is in part determined by how much physical change is retained by the material as it absorbs sound energy, a significant property of solids but unimportant in fluid and air. In the ear, this aspect of hearing is manifested by changes in the bones of the middle ear as people age and the slight thickening of the oval window end of the basilar membrane over time, resulting in the gradual loss of high frequency hearing with age.

Sound pressure can be measured on an absolute scale in units of pascals. This is not, however, how differences in the sound pressure that humans can detect is normally presented. The different sound levels are compared with an arbitrary standard: the lowest sound detectable by the human ear at 1 kHz. This sound pressure, p_o, is 20 μPa, and is defined as 0 decibels. The decibel units use a logarithmic scale. The relationship between sound intensity L and pressure p is

$$L = 10 \ \log\frac{p^2}{p_o^2} = 20 \ \log\frac{p}{p_o}. \tag{1.28}$$

This is the decibel scale. For a sound equal to p_o, the ratio p/p_o is 1 and the log is 0, so $L = 0$. This sound has zero decibels. Because there is a squared relationship between sound intensity and sound pressure, a tenfold change in sound pressure produces a sound intensity change of 20 decibels.

When sound waves enter the ear canal, they produce vibrations of the tympanic membrane, or eardrum. The tympanic membrane separates the outer ear from the middle ear. The first of the three bones of the middle ear, the malleus, spans the tympanic membrane and receives its vibrations from it. The oscillations of the tympanic membrane have a displacement of about 100 nm up to a frequency of 1 kHz, and decrease to less than 10 nm at 10 kHz. The vibrations are conducted through the middle ear bones, the malleus, incus and stapes to the oval window, the membrane that separates the middle ear from the inner ear. The stapes vibrates against the oval window with displacements that parallel those of the tympanic membrane, but with a displacement about tenfold lower, the movement being 10 nm up to 1 kHz and falling to less than 1 nm at 10 kHz. The transmission system thus acts as a low pass filter, with the signal decreasing as the frequency increases. The oval window is only one-twentieth of the area of the tympanic membrane, and that combined with the lever action of the ear bones produces a 30-fold peak amplification of sound pressure in the middle ear at 1–3 kHz. Dysfunctional ear bones are a major cause of conductive hearing loss in the elderly, and hearing aids amplify sounds to activate the oval window directly. A buildup of fluid in the middle ear caused by blockage of the Eustachian tube dampens the ear bone vibrations and reduces sound transmission. The use of tympanic membrane ear tubes to drain the middle ear is often effective in relieving this problem, especially in children.

When the oval window vibrates, the sound waves move through the perilymph, the fluid of the inner ear. The sound waves pass the structures of the Organ of Corti: the basilar membrane, the hair cells whose cell bodies sit on the basilar membrane, and the tectorial membrane, into which the hairs of the hair cells are imbedded. The basilar membrane is sensitive to the sound waves and will vibrate. The tectorial membrane is much stiffer, and is not significantly displaced by the sound waves compared with the significant movement of the basilar membrane. The differential displacement of the two membranes causes strain in the hair cells proportional to the amplitude of the sound wave. The movement of the hair cells opens ion channels, initiating the electrical signal that will ultimately be interpreted as sound by the brain.

The basilar membrane changes its structure as it winds from its base at the oval window to the apex of the cochlea over 2.5 turns and a length of 3–3.5 mm. At the base it has a width of about 0.1 mm, increasing to a width of about 0.5 mm at its apex. The structure determines at which frequency the sound waves will cause the local vibrations along the basilar membrane. High frequency sounds, 10 kHz, have the peak displacement within 0.3 mm of the oval window; 1 kHz waves peak at 1.5–2 mm; and 400 Hz waves peak near 3 mm. The positional, differential activation of the Organ of Corti is interpreted as different frequency sounds. Thus, the inner ear can distinguish both the amplitude and frequency of the sound waves that reach it. The combination of all the sound transmission structures in the ear results in a frequency range of 20–20 000 Hz. The peak of the range occurs at 1–3 kHz (Figure 1.13). There is often a loss of high frequency sound detection with age. Since the highest frequency sounds will vibrate the smallest structures, any thickening of the sound transmission system, such as inflammation of the base of the basilar membrane, would result in loss of high frequency sound detection.

Ultrasound imaging uses sound waves, not electromagnetic radiation, with frequencies of 1–5 MHz penetrating tissue, a higher frequency range than that of human hearing (20 Hz to 20 kHz). As the ultrasound waves pass through tissue, they reach boundaries between tissues of different densities. When this occurs, some of the waves are reflected back, while some continue on to the next boundary, where again some waves are reflected back. The image is constructed based on the differential reflections, so that the areas of different densities appear with different intensities. Current developments in this field

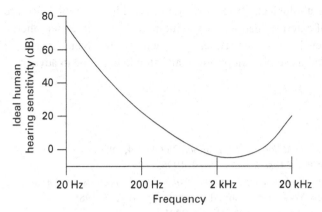

Figure 1.13   Frequency dependence of ideal human hearing sensitivity. The standard of human hearing is set at 0 decibels at 1 kHz. Humans have their most sensitive hearing at about 3 kHz. The limitations on hearing are set by the sensitivity of the tympanic membrane, the ear bones, and the vibrations of the basilar membrane.

allow three-dimensional construction of images, as well as Doppler images, in which the flow of blood allows imaging of blood vessels.

The ultrasound waves are absorbed by different molecules, and the energy dissipation is in the form of heat. In all types of imaging the applied energy must interact with molecules to produce the image. This will cause an increase in heat during image production. The side effect of increased local temperature has to be considered whenever an image is produced, regardless of the imaging technology. In the development of each imaging modality, the increase in temperature is measured to determine the safety of the procedure. Routine procedures only use energies that produce insignificant changes in temperature.

The destructive properties of ultrasound are medically applied in the fracturing of kidney stones, a process called lithotripsy. Kidney stones will shatter when specific wavelengths of sound impart a force on a stone. The force F applied to the stone is

$$F = \frac{W}{c} \tag{1.29}$$

where W is is the irradiating power and c is the wave velocity. The stones, which come in multiple crystal forms, absorb energy in the frequency range of 0.1–1 MHz (Schroeder et al., 2009). The power is determined by the amplitude of the wave. Any system exposed to an irradiating frequency of sufficient amplitude with which it has a matching resonance is in danger of damage or destruction.

In summary, physiological systems exist in a steady state removed from global thermal equilibrium; the energy of these systems has an average ambient energy of 2.58 kJ/mol, with the energy of individual molecules spread across a Boltzmann distribution; these systems can absorb energy over a wide range of forms of the electromagnetic spectrum; most energy absorbance is radiated as heat with no molecular transformation, but in some cases energy absorbance leads to molecular transformation, transformations that may be physiological or pathological; all sensory systems and imaging modalities require specific interaction of external energy of a specific frequency with either inherent or added molecules. The development and refinement of energy absorbance systems in the body has already fulfilled its significant promise, and should continue to advance.

References

Brown T R. *Philos Trans R Soc Lond B Biol Sci.* **289**:441–4, 1980.

Clark J F and Dillon P F. *J Vasc Res.* **32**:24–30, 1995.

Deans, S R. In *The Radon Transform in Mathematical Analysis of Physical Systems*, ed. Mickens, R E. New York: Von Nostrand Reinhold, pp. 81–133, 1985.

Glaser, R. *Biophysics*. Berlin: Springer-Verlag, 2001.

Lloyd P G and Hardin C D. *Amer J Physiol.* **277**:C1250–62, 1999.

MR-TIP.com, Brain Tranversal T1 001 and Brain Transversal T2 002. http://www.tr-tip.com

Schroeder A, Kost J and Barenholz Y. *Chem Phys Lipids.* **162**:1–16, 2009.

Slade J M, Carlson J J, Forbes S C, *et al. Hum Brain Mapp.* **30**:749–56, 2009. www.interscience.wiley.com

Tinoco Jr I, *et al. Physical Chemistry: Principles and Applications in the Biological Sciences*, 3rd edn. Upper Saddle River, NJ: Prentice Hall, 1995.

2 Molecular contacts

In this chapter, we will consider the connections that molecules make. In most cases, these contacts will be with other molecules. In a few important cases, however, ion–molecular complexes are formed in physiological systems. In some instances, the contacts formed are virtually permanent, while in others the transitory nature of the connections is far shorter than the timescale of any physiological process. In all cases, a set of parameters controls molecular contacts. These factors include the concentrations of the molecules (and ions, in those cases), the dissociation constant between the binding pair, the energy barrier between the pair, and the energy of the environment.

2.1 Dissociation constants

Every chemistry text, including those of physical chemistry and biochemistry, has a section on dissociation. This concept is of fundamental importance to interactions of all pairs of factors, be they ions or molecules. When salts are placed in solution, they will dissociate to some degree. Strong acids, such as hydrochloric acid, HCl, will completely dissociate:

$$HCl \rightarrow H^+ + Cl^-. \tag{2.1}$$

Note that the arrow is single headed. HCl present in the lumen of the stomach is completely split into the H^+ and Cl^- forms. There is no measurable HCl present in the combined form. The pH of the stomach is 1–2, a condition at which proteins will denature and cells will die. This is an important function of the stomach lumen: to destroy pathogens that may have been ingested. In the cells of the body, the pH is about 7.0, and in the plasma it is 7.4 under normal conditions.

In contrast to HCl, phosphate is a weak acid at physiological pH, in equilibrium between two intermediate forms:

$$H_2PO_4^- \leftrightarrow HPO_4^{2-} + H^+. \tag{2.2}$$

Not all the hydrogens are dissociated from the oxygens. The completely dissociated form, PO_4^{3-}, would only occur at very high pH (greater than 11), and the fully protonated form, H_3PO_4, would only occur at very low pH (less than 3). Since the cells of the body cannot function at these extreme pH conditions, only the partially dissociated forms of phosphate occur in physiological systems.

The conditions under which dissociation occurs are governed by the dissociation constant. When two substances A and B can combine to form a complex AB, that complex can also separate into the individual components A and B:

$$AB \leftrightarrow A + B. \tag{2.3}$$

The separation of AB into A and B will continue until the system reaches equilibrium. At equilibrium, there will be a ratio between the concentrations of the components [A], [B] and [AB]:

$$K_D = \frac{[A][B]}{[AB]} \tag{2.4}$$

with K_D the dissociation constant for the complex AB. (If the ratio is inverted, the association constant K_A is determined. While both K_D and K_A are used in the literature, K_D is more commonly used in physiological systems.)

The relationship between K_D and the concentrations of A and B is illustrated in Figure 2.1. If the K_D is much greater than the concentration of A or B, e.g., when $K_D/A_{total} \gg 1$, very little AB complex will form. When the concentrations of A and B exceed K_D a substantial fraction of AB will form. Figure 2.1 also shows that when there is a large concentration difference between A and B, the lower concentration places an upper limit on the amount of complex that can form. In the figure, A is held constant and B is altered in the three lines. This illustrates that when two elements can combine a substantial increase in the concentration of one element will result in increased complex formation. This is the situation that occurs, for example, when an increase in intracellular calcium binding to a constant amount of troponin leads to activation of muscle contraction.

For a dissociation in which one of the factors is H^+, the K_D equation and its rearrangement solving for H^+ are

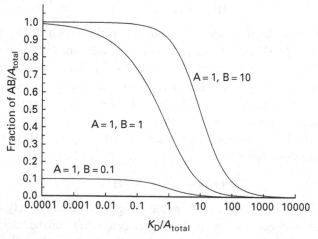

Figure 2.1   Effect of K_D on complex formation. A is held constant and B is changed tenfold on the three lines. When K_D is lower than the concentrations of A and B, the complex AB will form. When K_D is higher, less complex will form.

$$K_D = \frac{[H^+][A^-]}{[HA]}$$ (2.5)

$$[H^+] = K_D \frac{[HA]}{[A^-]}.$$ (2.6)

Taking the negative logarithm of both sides of the lower equation yields

$$-\log[H^+] = -\log K_D - \log \frac{[HA]}{[A^-]}.$$ (2.7)

Recognizing the left side as the definition of pH, the right side of the equation can be rearranged to give the Henderson–Hasselbalch equation routinely used in physiological measurements of pH:

$$pH = pK_D + \log \frac{[A^-]}{[HA]}.$$ (2.8)

Both K_D and pK_D values are used in physiological systems. A familiarity with both formats is necessary, as they convey the same information in different forms.

There are several assumptions made when dissociation constants are measured. First, it is assumed that the different species, A and B, are completely independent of one another and of all other molecules or ions. That is, there are no interactions altering the binding of interest: the calculated concentrations of A and B are the true concentrations of A and B. When this is not the case, the term activity is used, where the activity α_i of an individual molecule or ion is a fraction of the concentration c_i of that molecule or ion:

$$\alpha_i = f_i c_i$$ (2.9)

with f_i the coefficient of activity, equal to 1 in an ideal solution in which the activity and concentration are the same. This usually occurs in dilute solutions. In measurements of chemical potential μ, where μ^o is the standard chemical potential, the relation to concentration and activity is often shown as

$$\mu = \mu^o + RT \ln c \cong \mu^o + RT \ln \alpha.$$ (2.10)

In the cases where this relation applies, there are presumably reasonable assumptions that the concentration and activity are the same. If, however, there is significant binding to other molecules not involved in a particular event, the activity may be considerably lower than the concentration. Inside cells, protein concentrations may exceed 40 mg/ml. In a laboratory test tube, it is exceedingly difficult to produce protein solutions greater than 5 mg/ml. This alone (and there is considerable other evidence, as we will see later) indicates that proteins form complexes that greatly increase the amount of protein that can exist inside cells. If one of these proteins is an enzyme involved in the binding of a substrate, the activity of the protein may be substantially less than the concentration of the protein, altering conclusions about the molar reactivity of the enzyme.

The second assumption made is that there are so many individual molecules A and B in a solution that they can be treated as an ensemble; that is, they can be treated as a group rather than as individual molecules. In solution chemistry this is a reasonable assumption. Take, for example, a 1 nM solution in one liter. This 10^{-9} M solution, multiplied by Avogadro's number 6.023×10^{23} molecules/mole, would have 6.023×10^{14} molecules in a liter. A cell, being much smaller than a liter, will have fewer molecules. For a cell that is 10 μm wide, 10 μm thick and 100 μm long, the volume would be 10^{-11} liters. In this cell, a 1 nM concentration represents 6023 molecules, enough so that statistically they can be treated as an ensemble. There are few chemicals inside cells with functional concentrations of less than 1 nM, so the assumption of a group behavior for soluble chemicals is usually a good one.

2.2   Promotor sites and autoimmune diseases

There is, however, a very important set of binding sites for which an ensemble is not a good assumption. The binding sites for transcription factors on promotor sites of genes are much less in number. For a somatic gene, present only on two chromosomes, the maximum number of sites is two, one on each chromosome. For a sex-linked gene only present on the X or Y chromosome, the binding site number would be one in males. Even in the cases where there are multiple transcription factor binding sites, the number is small enough that the assumption of ensemble behavior, rather than digital behavior, cannot be correct. (In the case of multiple, consecutive sites in the same promotor region, the independence of each site must also be questioned.) For these cases, a different type of analysis is needed.

Consider the binding of transcription factor A to promotor site B. The association/dissociation reactions will be

$$A + B \xrightarrow{k_1} AB \tag{2.11}$$

$$AB \xrightarrow[k_2]{} A + B \tag{2.12}$$

where k_1 and k_2 will be the rate constants of the reactions. Since B is limited, the key element in the first reaction is A locating the limited number of B sites. This process will be covered in the next chapter on diffusion and directed movement. Here, we are concerned with the second reaction, the dissociation of AB. Since B is so much less than A, the concentration of A is essentially constant. Conversely, B will have its concentration as B or AB changed drastically whenever A binds or dissociates. In the sex linked case, with one site, that site will either have A bound (AB) or not (B). The concept of fractional binding is meaningless here. In this digital world, you cannot have 42% of the sites bound. For two sites, the potential states are 100% bound (both AB), 50% bound (one AB, one B) or 0% bound (both B). Since the binding of transcription factors to promotor regions is a key element in biology, this has to be dealt with quantitatively using different methods than those of ensemble solution chemistry.

The rate constant k_2 is related to the molar change in AB. At the molecular level, it will be associated with the rate at which A separates from the binding site B. The more likely the separation, the larger (faster) the rate constant; the more stable the AB complex, the smaller (slower) the rate constant. The dissociation of AB changes the concentration of AB such that

$$-\frac{[AB]}{dt} = k_2[AB] \tag{2.13}$$

which can be integrated (Moore, 1972) to yield

$$\ln[AB] = -k_2t + \ln[AB]_o \tag{2.14}$$

where $[AB]_o$ is the initial concentration of AB. Converting the above equation into its exponential form yields

$$[AB] = [AB]_o e^{-k_2 t}. \tag{2.15}$$

If the dissociation of A from B is presumably independent of any other AB events at different locations, then the probability that A and B are still bound at time t can be treated as an exponential process

$$\frac{[AB]}{[AB]_o} = e^{-k_2 t} \tag{2.16}$$

where [AB] are the total number of occupied AB sites, t is time, and k_2 is the molecular rate constant identical with k_2 above, with units of 1/time. For conditions in which there is a known average rate of dissociation of the transcription factor from the promotor site, that is, there is already an estimate for k_2, a slightly more complex analysis using the Poisson distribution, describing random, discrete, rare events occurring in a defined time interval, would be appropriate. If in this case the dissociation of one factor altered the rate of adjacent factors, the non-homogeneous Poisson distribution would be appropriate. As the molecular details of a defined dissociation rate needed to use the Poisson distribution are not known until k_2 has been determined, the simpler exponential process is the proper starting point.

Using the exponential analysis, the dissociation of A from a limited number of B sites is analogous to radioactive decay. In radioactive material, a limited number of atoms can radiodecay, losing energy in the form of photons (gamma) or subatomic particles (alpha, beta), and reverting to the lower energy level of most of the atoms in the material. In the dissociation of transcription factor A from the binding site B, consider the set of all promotor sites. Only a limited number of the total number of promotor sites for all transcription factors will have A bound. The dissociation ("decay") of A from the AB sites, the sites reverting to the general "non-A bound" state of all the other promotor sites, will be a random exponential process.

Given the exponential behavior of transcription factor binding, and the limited number of sites for any transcription factor, the concept of the fraction of either transcription factor bound or of binding sites bound does not convey meaningful information. Instead, when there are only a few molecular sites for any binding, the molecular rate constant k_2 is the key factor, which we will term k_{AB} for the description of the dissociation of AB. The units of k_{AB} are 1/time. Its inverse will be the time constant τ_{AB}. τ_{AB} will relate how long A will

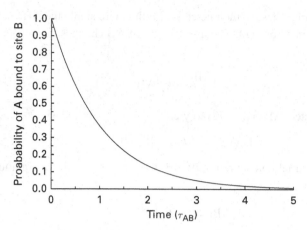

Figure 2.2   Probability that transcription factor A will remain bound to genome site B as a function of the dissociation time constant τ_{AB}. The probability will approach zero, but will not reach zero, always remaining a positive probability.

be bound to B. Since the separation is a random exponential process, τ_{AB} will not define how long A will be bound, but the time at which there is a 37% (1/e) chance that A is still bound (Figure 2.2). By the time five time constants have passed, there is only a very small probability that the original A will still be bound. Despite the limitations on the exact duration of A binding, knowing τ_{AB} does let us estimate the retention time for A. The retention time tells how long a binding element is in position to trigger a response. In the case of a transcription factor, it is the time A is present on the promotor. The complement to the retention time is the reaction time. The reaction time is the time taken for the bound element to evoke a response, in this case the initiation of transcription. If the retention time is less than the reaction time, it is unlikely that the response will be triggered. If the retention time is greater than the reaction time, there is a high probability that the event will be triggered. Thus, different transcription factors activating the same gene may have different retention times and different reaction times, and therefore may have different efficiencies in triggering mRNA production.

It may seem trivial to note that the retention time must be greater than the reaction time to trigger a response, as it must. But, like the distribution of molecular energy or the time required for dissociation, the duration of retention or of reaction initiation is not a fixed value at the molecular level. This is a non-linear, probabilistic world. Take the exponential equation above: if τ_{AB} were 1 ms, then 1 ms after observing factor A bound to site B, there is only a 37% chance that A is still bound to B. After 5 ms, there is only a 0.67% chance, or 1/150, that A is still bound. While A is unlikely to still be bound after 5 ms, on average, for every 150 bindings of A to site B, one of those bindings will endure for 5 ms. Thus, even if an event will rarely occur, it will occasionally occur, and can produce a response. The reaction time will also have a time distribution in a manner similar to the retention time. Transcription occurs when the retention time is long enough that the reaction starts.

The idea of retention time and reaction time is not limited to genetics. This concept is important in understanding autoimmune disease as well. In a common infection, such as a cold, the immune system will respond and eliminate the antigenic trigger within a few weeks. Autoimmune diseases take much longer to eliminate the self-antigen that triggers the disease. For example, Type I or juvenile diabetes is considered an autoimmune disease, with the immune system gradually destroying the beta cells of the pancreas over a period of several years. This is considerably longer than the normal time course of an immune response. It is thought that the prolonged time course of autoimmune diseases arises from weak binding of the antibodies or T cells to their target antigen. The original target of the immune system may have been a combination of antigens: the complex formed by several proteins, or a protein and a bacterium, for example, present a unique antigen that the body responds to with a specific immune response. When that complex is eliminated, the antibodies or T cells that recognized and eliminated the complex can still bind to the individual elements of the complex, but much more weakly; that is, they will bind with a higher dissociation constant. Alternatively, the autoimmune target may have a similar structure to that of an immune target, so that the antibodies that bound to the immune target may bind to the autoimmune target, but again with a weaker binding.

Antibodies eliminate antigens by binding with their variable region to the antigen and then triggering a non-specific response with their constant region. These responses include activation of natural killer cells, the complement system, or phagocytotic cells. The antibody must bind long enough, that is, have a long enough retention time, to activate the non-specific response, the reaction time. For a normal immune response, the binding is very strong, and the retention time will be sufficient to activate the response and eliminate the antigen. Thus, normal infections are gone within weeks. If the binding is weak, however, the antibody (or T cell) may not have a long enough retention time to trigger the immune response and remove the antigen. The duration of binding, however, is not a set time, but reflects the probability of dissociation. As discussed above, occasionally even those dissociations that normally occur rapidly will occasionally not occur for an extended time, allowing reactions that need longer activation times to go forward. This may be the case in autoimmune diseases: only occasionally will the binding of the antibody or T cell be of sufficient duration to eliminate the antigen. But over the course of years, this rare activity will be enough to do the job, and cells like the beta cells of the pancreas will be eliminated, with Type I diabetes the result.

Since the exact target(s) of the immune system in Type I diabetes has (have) not been identified, this analysis represents reasonable speculation, not proven pathology. There is evidence, however, of antigenic responses to molecular complexes in a related area. Insulin and glucagon are both made in the pancreas by the beta cells and the alpha cells, respectively. They have complementary functions, as insulin reduces the plasma glucose concentration, while glucagon increases the plasma glucose concentration. It has been hypothesized that molecules with complementary functions may form molecular complexes (Root-Bernstein and Dillon, 1997). This complex formation has been demonstrated for insulin and glucagon (Root-Bernstein and Dobblestein, 2001; Dillon et al., 2006) and is shown in Figure 2.3.

The peaks in Figure 2.3 come from capillary electrophoresis (CE), which separates molecules based on their charge-to-mass ratio. Samples are placed at one end of a capillary tube and driven to the other end by exposure to an electric field, which drives

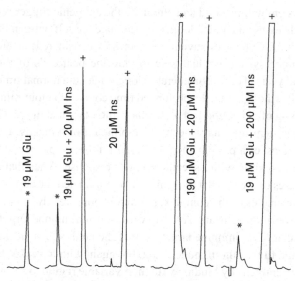

Figure 2.3   Electropherograms of insulin and glucagon. The asterisk indicates the glucagon peak. The + indicates the insulin peak. When present together, the molecules bind with a reduction in the glucagon peak and increase in the insulin peak. (Reproduced from Dillon *et al.*, 2006, with permission from Elsevier.)

a carrier buffer forward. Molecules will separate based on their electrophoretic mobility, with smaller molecules whose charge is opposite to the driving voltage moving more slowly, while those with the same charge as the driving voltage moving faster. For a given charge on a molecule, the smaller the molecule the greater the charge density, and the stronger the electric field effect. CE can be used to measure molecular binding (Dillon *et al.*, 2000; Dillon *et al.*, 2006). Figure 2.3 shows CE runs of insulin and glucagon, proteins already known to bind to each other (Root-Bernstein and Dobblestein, 2001). When insulin is present, glucagon shows a concentration-dependent decrease in size; when glucagon is present, insulin shows a concentration-dependent increase in size. This can only occur if the molecules are binding to one another in solution. Using deviations in the Beer–Lambert law to infer insulin-glucagon binding, Root-Bernstein and Dobblestein estimated the K_D of insulin-glucagon binding to be 1 μM.

Antibodies to insulin have been shown initially to develop against the insulin–glucagon complex. Antibodies against the complex can then bind to insulin and label it for destruction. This process has interesting biophysical aspects. The in vivo concentrations of insulin and glucagon are in the high picomolar range, and their dissociation constant is about 1 μM. Because the dissociation constant is so much greater than the protein concentrations, there will be negligible endogenous formation of the insulin–glucagon complex, as shown in Figure 2.3, and therefore no immune response. During insulin injections, however, nearly 1 mM insulin is injected, enough to form complexes with glucagon near the injection site. Local damage to the injection site draws macrophages that recognize the complex as a foreign antigen and produce an immune response against it. The antibody products of this response then attack insulin even when it is not complexed with glucagon, producing an immune response to a molecule that is no longer part of an immune complex (Root-Bernstein and Dobblestein, 2001). This can lead to the rejection of a molecule, insulin, that is present in the body and which the body has had since birth.

The initial determination of insulin-glucagon binding used one of the most common methods for measuring molecular interactions. The basis of this measurement involves deviations from the Beer–Lambert law, also called Beer's law. This law relates the transmission and absorbance of electromagnetic radiation as it passes through a sample chamber containing a liquid or a gas. The radiation can be visible light, but the principle applies to other radiation, such as ultraviolet or infrared radiation, as well. The light intensity I_o that enters a sample chamber interacts with the molecules in the chamber, reducing the light intensity that exits the chamber to an intensity I_t. This decrease is dependent on the concentration of the substance in the chamber. If the substance is homogenously distributed in the chamber, the decrease in intensity $-dI/I$ is

$$-\frac{dI}{I} = \alpha c \, dx \tag{2.17}$$

where α is a constant, c is the concentration, and dx is the interval through which the light passes. Over the length l of the chamber, integration of this equation yields

$$-\int_{I_o}^{I_t} \frac{dI}{I} = \alpha c \int_0^l dx \tag{2.18}$$

which is a logarithmic relation:

$$\ln \frac{I_o}{I_t} = \alpha c l, \text{ or its exponential form } I_t = I_o e^{-\alpha c l}. \tag{2.19}$$

(This is the same equation form used above in calculating the dissociation of AB.) When used to measure solutions with a spectrophotometer in the laboratory, the total length of the chamber is standardized to 1 cm with a cross-sectional area of $1 \, cm^2$, so that the decrease in intensity is measured through a sample volume of $1 \, cm^3$, or 1 ml. The ratio I_t/I_o is the transmittance of the sample, which will be zero if the sample absorbs all the light, and is related to the logarithm of the concentration. The absorbance of the sample, the amount of light absorbed by the molecules, is defined as

$$A = -\log T = \log \frac{I_o}{I_t} = \frac{\alpha c l}{2.303} = \varepsilon c l \tag{2.20}$$

where $\varepsilon = \alpha/2.303$ converts the natural logarithm ln to the base 10 log. Thus, the Beer–Lambert law says that the absorbance (A) is directly proportional to the concentration ($\varepsilon c l$). If you increase the concentration over a sufficiently small range, the increase in absorbance is linearly related to the concentration. If a sample contains two types of molecules A and B, the absorbance of the sample will be the sum of the two molecular concentrations measured separately. If there is deviation from the Beer–Lambert law (i.e., the absorbance of the combined solution is not the sum of the two separate solutions), then it is inferred that the two molecules are interacting to form AB. This is among the most common ways in which molecular interaction is shown. The deviation in

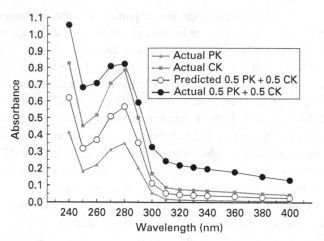

Figure 2.4   Application of the Beer–Lambert law to pyruvate kinase and creatine kinase. The frequency spectra of PK and CK were determined separately (small symbols, dashed lines). A solution with one-half the concentration of each protein was made. The Beer–Lambert law predicts an absorbance intermediate between the control absorbances. The actual absorbance (closed circles) is much higher than the predicted absorbance (open circles).

absorbance will vary with concentration, but since the absolute absorbance of the combined complex AB is usually not known, this method is said to be semi-quantitative. This approach was used in the measurement of the interaction of insulin and glucagon discussed above. Deviation from expected absorbance was also used to show the interaction of pyruvate kinase and creatine kinase in Figure 2.4 (Sears and Dillon, 1999). In this figure, it is apparent that the absorbance when both proteins are present is far greater than predicted, more than fivefold greater at wavelengths above 300 nm. While the use of spectrophotometry is a common method for measuring the concentration of solutes, the potential for deviant data based on the absorbance of molecular complexes cannot be ignored. The more solutes that are present in the solution, the more potential for complex formation, and the more potential for error. In the next section, we will cover various methods for measuring the binding of molecules.

2.3   Methods of measuring dissociation constants

The determination of K_D is made by varying the concentrations of the elements A and B and measuring the change in some parameter that will vary as a function of AB complex formation. There are two hypothetical methods of doing these measurements. There are four elements in the K_D equation: K_D, A, B, and AB. Since K_D is the desired value, the other three elements have to be measured independently of one another to exactly determine K_D. Alternatively, the measurement of some physical parameter can be used to infer one of the elements and, by measuring the change in that parameter as the total concentration of A and B are changed, can create a range of measures from which K_D can then be determined by extrapolation.

For systems in which one of the binding elements is an ion, the formation of the complex will reduce the free concentration of the ion, and changes in the measurement of electrical conductance can be used to infer complex formation. While knowledge of the conductance properties of the ion alone may be relatively easy to measure, the conductance of the complex is more difficult to know. By changing both the relative concentrations of A and B as well as the total ionic concentrations of A and B, changes in ionic conductance can be used to estimate the formation of the AB complex and therefore the K_D. The exposure of a weak acid, which can release a H^+ ion, to a high electric field can produce the field dissociation effect (Moore, 1972). The higher the electric field, the greater the dissociation of the H^+, and the larger the conductance of the solution.

Electric fields can also be used to measure the dissociation constants of bimolecular binding. When two molecules bind or dissociate, the change in conductance will be very small in most cases, not sufficient to determine a dissociation constant. It is, however, possible to use the electric field method described in the previous section, capillary electrophoresis, to measure K_D. The use of CE to measure K_D employs the following principle: if two molecules form a complex when no electric field is applied, and separate in a high electric field, then there must be an electric field where they are 50% bound and 50% separated. This would allow the measurement of a dissociation constant at that electric field, the K_e. By measuring the K_e at several electric fields and extrapolating to zero electric field, the dissociation constant at zero electric field, which is the K_D, can be determined. Figure 2.5 shows this effect. Capillary electropherograms of norepinephrine (NE) and ascorbic acid (Asc; Vitamin C) are shown in separate runs at the top of the figure, each showing a separate peak. When the same concentrations of NE and Asc are run together, a third peak appears (Dillon et al., 2000). This peak is the NE–Asc complex. When the concentration of Asc is increased and NE held constant, the complex peak increases and the free NE peak decreases, producing the curves in the lower part of the figure, each at a different electric field with a different K_e. The K_D for NE–Asc binding is calculated by extrapolating $\log K_e$ values to zero electric field and determining the $\log K_D$. This method can be used with both small molecules and proteins (Dillon et al., 2006). It has the advantage of needing very little material for the measurement, as CE requires only nanoliter injection volumes. Because of heating produced by high currents, this method is not appropriate for solutions with high ionic strength.

Other methods of measuring K_D use a variety of different techniques. The key element has to include a change in the amount of complex AB formation as one of the binding factors, A or B, is altered while the other is held constant. Measurement of a solution's osmotic pressure is among the most straightforward. If two molecules do not bind, an increase in the concentration of either or both will result in a linear increase in osmotic pressure. If there is binding, however, there will not be a linear increase, as complex formation will result in a lower total number of osmotic elements, and the osmotic pressure will be lower than that predicted by the concentration increases. This method assumes that there is no autobinding of one of the elements, for example BB complex formation. In studying proteins where self-association is a common characteristic, this complicating factor must be taken into consideration. Other techniques to measure K_D include changes in fluorescence as one molecule binds to another (Rand et al., 1993) and changes in the

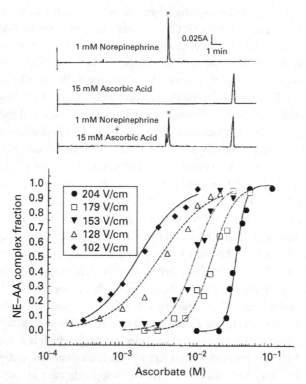

Figure 2.5   The upper panel shows three electropherograms with norepinephrine (NE; free NE has an *),
ascorbic acid (AA), and a combination of the two. The peak to the left of free NE is the NE–
ascorbate complex. The lower panel shows NE–ascorbate complex formation over a range of
electric fields. Higher concentrations of ascorbate are needed at higher electric fields to overcome
the dissociating effect of the electric field. The NE–Asc log K_D can be calculated by extrapolating
the electric field log K_e values to zero electric field. (Reproduced from Dillon *et al.*, 2000, with
permission from Elsevier.)

kinetics of enzymatic activity (McGee and Wagner, 2003). The fluorescence of hexokinase
(HK) decreases when glucose binds to the protein. The K_D of glucose binding was
determined by saturating HK with glucose to establish the maximum change in fluorescence, then measuring the glucose concentration at which one half of the fluorescence
change occurs. This assumes that there is a linear relation between binding and fluorescence, a reasonable assumption in a sufficiently dilute system. The K_D for glycosaminoglycan binding antithrombin was obtained by measuring the GAG concentration that gave
half-maximal antithrombin activity on water transfer reactions. This required using conditions in which the K_D is approximately equal to the $K_{1/2}$, the concentration at which the
reaction rate was one-half of maximal. While the assumptions in these methods may
produce measured K_D values different from the true K_D values, the variations in well-
controlled experiments using reasonable assumptions are likely to be small, yielding useful
K_D values. Any system in which A, B and AB can be measured can be used to measure K_D.
Every technique will have its own advantages and limitations: understanding those limitations is critical in using any method appropriately.

2.4   Metal–molecular coordination bonds

All bonds that have a significant lifetime on a physiological timescale must have an energy in excess of kT. In the case of physiologically important metal ions, there is a large range in the duration of the binding between the metal ion and the molecule it binds. Some of these bonds, once formed, are never broken, while others have a transient lifetime appropriate for control mechanisms. The coordination bonds formed between the metal and the organic molecule depend on multiple factors. Critical elements in bond formation include the affinity between the two elements, the structural elements of the molecule, the molecular reactions the molecule is involved in and the relative concentrations of the metal and molecule.

The coordination bond between a metal and a molecule involves the electrostatic attraction of the metal with dipoles or ionic side groups of the molecule. Coordination bonds have elements of ionic bonds and covalent bonds, but cannot be fully described by either. Most coordination bonds have energies significantly higher than "weak" bonds such as hydrogen bonds, but lower energy than covalent bonds. In some cases, coordination bonds hold metals permanently within molecular structures, such as the iron atom within the heme group. In other cases, the metal is freely exchanged, as in the case of calcium binding to calmodulin. The duration of a coordination bond is a function of how many valences attach the metal to the molecule; how much strain the coordination bonds make on the internal structure of the molecule; what the concentration of the metal ion is in the surrounding solution; and the interaction of the metal and molecule with water.

The heme complex is shown in Figure 2.6. The iron atom has coordination bonds with four nitrogen atoms in the plane of the heme moiety. The right-side triangle shows the attachment of a fifth coordination bond between the iron atom and the histidine of an adjacent protein. The left-side triangle shows the bound oxygen molecule. This complex has sufficient binding energy that the iron is never dislodged from the molecular complex while in the body. The most common form of heme, heme b present in the hemoglobin of red blood cells, is transformed during the production of bilirubin after hemolysis of the red blood cells. There is a shift in the absorption of light when this occurs, giving rise to the color of feces and urine. Iron remains attached to the ring structure as long as it is present in the body.

In Figure 2.6, note the planar nature of the heme–Fe complex. There is no significant alteration of the bond angles of the heme when the iron is attached. Molecules will always tend toward their lowest free energy state, so that modification of bond angles when a coordination bond is formed will usually increase the free energy of the molecule, and reduce the stability of the coordination bond, as if the addition of the metal ion loads a spring, making it more likely to expel the metal. In the case of heme–Fe binding, this strain is minimal, enhancing the stability of the complex.

Zinc is present in multiple protein molecules, providing structural integrity by forming coordination bonds with negatively charged amino acid side groups of aspartate, glutamate, histidine and cysteine with four coordination bonds commonly formed between the metal and the protein (Maret, 2005). Zinc binding to some proteins is very tight: the time

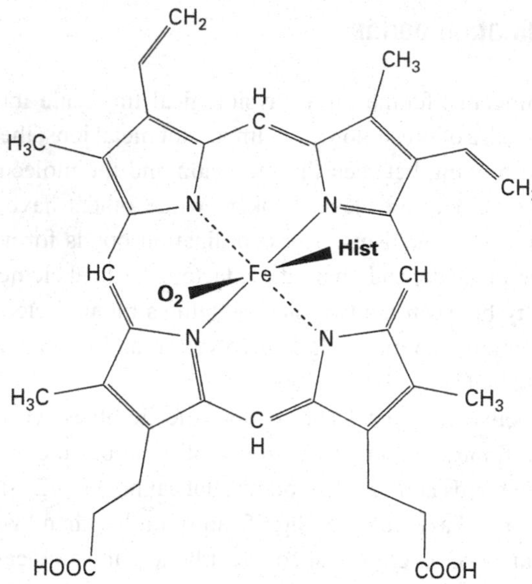

Figure 2.6   The structure of heme.

constant for zinc exchange in carbonic anhydrase, for example, is on the order of years. Zinc controls the redox state of many proteins (nitric oxide synthase, protein kinase C, heat shock protein) and protein–protein complex formation (insulin hexamer). Some zinc interactions involve proteins with rapid turnover rates, however, as in the case of the zinc finger proteins that control gene expression. There must be a labile zinc pool available for inclusion in newly manufactured proteins. The protein metallothionein (MT) has 20 cysteines, and can reversibly bind seven zinc ions. Zinc binds to metallothionein at four cysteine sites, providing considerable thermodynamic stability. The protein structure of MT, however, is kinetically very active, allowing rapid exchange of zinc between the MT and the environment around it (Romero-Isart and Vasák, 2002).

There are multiple forms of zinc finger complexes (Krishna *et al.*, 2003). The zinc atom forms a structural domain within a protein. Many proteins with zinc finger domains act as transcription factors, although they have multiple other functions in different proteins. The name zinc finger derives from its original discovery, in which the zinc region of the protein grips a DNA domain (Klug and Schwabe, 1995). A common form of complexation of zinc with the protein involves four binding sites: two cysteines and two histadines. The four binding sites provide sufficient structural stability that the zinc atom does not easily dissociate from the protein, in a manner similar to the Fe-stability of heme, but zinc fingers can also have kinetic flexibility, as in metallothionein, which makes their zinc binding more labile. Heavy metals such as cadmium can substitute in one of the zinc binding sites (Figure 2.7). Cadmium binding to the transcription factor for the sodium-dependent glucose transporter (SGLT) reduces mRNA production, the mechanism for the reduction of glucose transport in the kidney during cadmium exposure (Kothinti *et al.*, 2010). Unlike the permanent nature of iron binding in hemoglobin within

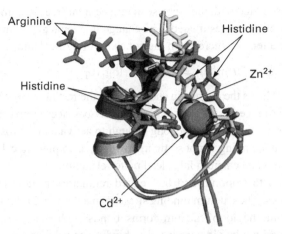

Figure 2.7 Zinc finger domain of the transcription factor for the SGLT mRNA with cadmium in the zinc binding site. Cadmium binding down regulates mRNA production. (Reproduced from Kothinti *et al.*, 2010, with permission from American Chemical Society.) Histidines and cysteines form the metal binding pocket in different zinc finger proteins.

a red blood cell, there is intracellular turnover of transcription factors, including those containing zinc. The proteolytic digestion of zinc-containing transcription factors results in the release of zinc, which can then be reconstituted into other proteins, either transiently in MT or in a new protein.

The use of calcium as a signaling ion requires that it bind and dissociate on the timescale of physiological functions. Molecules like calmodulin and troponin regulate important cellular activities. There is a substantial calcium gradient across the cell membrane. The extracellular calcium concentration is 2.5 mM, although a substantial fraction of this is bound to negatively charged metabolites and buffers such as lactate and sulfate as well as to albumin. The free, ionized concentration of plasma calcium is approximately 1.2 mM.

Intracellular calcium has a much lower concentration. In cells at rest the ionized calcium concentration is 10–100 nM. Increases in calcium lead to the activation of many cellular processes, such as muscle contraction and exocytosis. The increase in calcium during activation occurs through the opening of membrane calcium channels or the release of calcium from intracellular stores, such as the sarcoplasmic reticulum and, to a lesser extent, the mitochondria. The very large free calcium gradient, more than 10 000-fold, means that the opening of calcium channels leads to a rapid increase in available calcium for intracellular processes. In most cases, calcium-triggered systems are initally activated at 0.1–1 µM, and are fully activated when the calcium approaches the 10 µM range.

Calcium binds to some proteins with binding energies of hundreds of kJ/mol, energies far in excess of the local thermal energy. Calcium would be unlikely to dissociate from these sites, which would not be useful for regulatory processes. Conversely, in measurement of the energy of calcium binding to human cardiac troponin C, the regulatory site for cardiac muscle contraction, the decrease in entropy (increase in molecular energy) when calcium binds to its functional site is 8 kJ/mol (Spyracopoulos *et al.*, 2001).

The increase in energy associated with the increase in intracellular free calcium leading to the activation of contraction can be estimated. The change in chemical potential μ in kJ/mol that would occur for a tenfold increase in the concentration of calcium is

$$\mu = RT\ln[10] = 2.303RT\ \log[10] \tag{2.21}$$

or a value of 5.93 kJ/mol. This is the change in energy above the resting energy value. The increase in concentration produces a sufficient increase in energy that calcium is driven into its binding site in troponin and remains there long enough to activate the myosin ATPase and initiate contraction. This binding is not so tight, however, as to prevent calcium from dissociating from troponin when calcium falls, leading to relaxation.

The binding of calcium to troponin and the related regulatory protein calmodulin occurs at four EF-hand sites. The calcium binding sites are in a loop region of the protein between α-helices. Within the loop calcium forms bonds with negatively charged aspartate and glutamate carboxylic side groups. The binding of calcium causes considerable rearrangement of the protein helices compared with the non-bound, apoprotein state. This rearrangement moves the helices away from their lowest energy configuration, and makes the dissociation of calcium more likely, especially when the surrounding environment has a low calcium concentration. The range of calmodulin-regulated processes is very large, including nuclear division, smooth muscle contraction, nitric oxide synthase, and many more. As with troponin, calmodulin acts as the intermediary between an enzyme and the calcium ion. Calcium itself does not bind to the enzyme, but the complex of Ca^{2+}-calmodulin-enzyme leads to increased enzymatic activity. In the case where EF-hand-containing proteins are extracellular, the millimolar calcium concentration ensures that the calcium is a permanent part of the protein structure, rather than having a regulatory role (Busch *et al.*, 2000).

Magnesium forms an important complex with ATP inside cells. This was discussed briefly in Chapter 1 in discussing 31P-NMR. Mg^{2+} forms a complex in which all three phosphates, or rather their surrounding oxygens, form coordination bonds with magnesium. When Mg^{2+} is titrated against ATP, all three phosphates show a chemical shift in the NMR spectrum, indicative of binding to Mg^{2+}. By inference, we can conclude that the Mg^{2+} binds most tightly to the β phosphate of ATP, as that phosphate has the greatest chemical shift when Mg^{2+} binds. Calcium, also a divalent cation, can also bind to ATP. Comparison of the two cations binding to ATP is illustrative. Both bind with the same equilibrium constant, but under non-equilibrium, comparative ion-loading conditions, the Ca–ATP complex has a higher concentration because the rate constant for its formation is 100 times faster than that for Mg–ATP formation (Glaser, 2001). We can see this represented graphically in Figure 2.8. Since both complexes have the same equilibrium constant, the energy difference between the free and complexed ATP states, the lowest points of the energy wells, must be the same. Since Ca^{2+} has the faster rate constant for formation (and because the K_{eq} is the ratio of the forward and backward rate constants, the Ca dissociation rate constant must be faster as well), the energy barrier between the free and complexed states must be smaller for Ca–ATP. Thus, equilibrium constants alone, although the ratio of rate constants, tell us nothing about what those rate constants are. At least one of the rate constants must be determined independently.

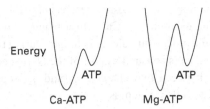

Figure 2.8   Relative free energies associated with Ca^{2+} and Mg^{2+} binding with ATP.

Why then do we use 31P-NMR to measure free Mg^{2+} in cells, rather than free Ca^{2+}? The explanation lies in the concentrations. Free, ionized calcium occurs in the micro-molar range inside cells, while both ATP and free magnesium occur in the millimolar range. Any Ca–ATP complex formation would be inconsequential compared with Mg–ATP complex concentrations. For this reason fluorescent dyes, aequorin and fura for example, are added to cells to measure the intracellular free calcium concentration. Any quantification of binding has to consider equilibrium constants, association and dissociation rate constants, concentrations relative to the K_{eq}, and the total concentrations when multiple, competing bindings are possible. Extracellularly, the similar concentrations of calcium and magnesium will have different consequences, where increased plasma magnesium is used to reduce contractions of blood vessels and the uterus by reducing calcium entry into smooth muscle cells. This mechanism will be discussed in the section on membrane transport.

2.5   Hydrogen bonds

Hydrogen bonds play an important role in the dynamics of physiological systems. They have a much shorter lifetime than covalent bonds, which are never spontaneously broken under physiological conditions. Their lower energy makes them susceptible to random thermal energy, and in the case of hydrogen bonds between water molecules they have a duration of about 10^{-11} seconds. Some structures, like DNA, have many hydrogen bonds which are mutually supportive, giving greater long-term stability to the entire structure, while still enabling the system to respond rapidly to changing demands. Hydrogen bonds, or H-bonds, are primarily electrostatic in nature, resulting from the dipole nature of molecules. We will discuss electrostatics and the formation of dipoles before covering H-bonds in water, DNA, and in ion–water cluster formation.

Electrostatics is based on the interactions between charged molecules or ions. The unit of charge is based on the charge on a single positron (or the negative of the electron charge). This charge e is

$$e = 1.602 \times 10^{-19} \text{ C} \tag{2.22}$$

and this value is multiplied by Avogadro's number N to get Faraday's constant F.

$$eN = (1.602 \times 10^{-19})(6.022 \times 10^{23}) = 9.648 \times 10^4 \text{ C/mol} = F. \tag{2.23}$$

At the molecular level, units of e are routinely used to designate the charge on a molecule or ion, such as the charge of 2+ on the calcium ion, Ca^{2+}. In a macroscopic system, Faraday's constant is used to designate the molar equivalent of charge. During the section on membrane potential, we will see Faraday's constant as part of the Nernst equation.

Coulomb's law describes the force between two charges, q_1 and q_2, separated by distance r from one another. The force F in this system is

$$F = \frac{q_1 q_2}{4\pi\varepsilon_0 \varepsilon r^2}. \tag{2.24}$$

ε_0 is a conversion factor, called the permittivity of free space, which connects electrical and mechanical factors. It has a value of 8.854×10^{-12} C/V·m. In a vacuum, it is the only correction factor needed. There is an additional correction factor ε, the dielectric constant, based on the physical characteristics of the system in which a real substance, the medium of the system, is between the two charges. In physiological systems, water is the dominant substance between ions in solution. Its dielectric constant at 37°C is 74. It is dimensionless; that is, it has no units. If the two charges have the same sign, then F will have a positive value and will be a repulsive force. If the two charges have opposite signs, the value of F will be negative and will be an attractive force. The energy of a mechanical system is the integral of the force. The electrostatic binding energy E of two charges at a given distance r is

$$E = \int \frac{q_1 q_2}{4\pi\varepsilon_0 \varepsilon r^2} \, dr = -\frac{q_1 q_2}{4\pi\varepsilon_0 \varepsilon r}. \tag{2.25}$$

If two charges are the same, the binding energy will be negative, and if they are opposite, the binding energy will be positive.

In a molecular dipole, there is an internal separation of charge within the molecule. The molecule overall may be neutral, but different parts of the molecule may have positive and negative charges. Dipoles may be permanent dipoles or induced dipoles. Permanent dipoles have a separation of charge at all times. Water, in which the oxygen always has a negative charge and the hydrogens always have a positive charge, is a permanent dipole. Other molecules may become dipoles when exposed to an electric field. Proteins in membranes that respond to electrical signals are induced dipoles. The dipole moment μ is the product of the charge q, in its $q+$ or $q-$ form, and the distance r between the separated charges:

$$\mu = qr. \tag{2.26}$$

The distance r in fact can be quite small relative to the size of the molecule. In water the dipole moment distance is 3.9 pm, compared with the hydrogen Bohr radius of 53 pm. Since the dipole must be within the sphere of both atoms, the centers of the positive and negative charges of the water dipole must be near the interface of the hydrogen and oxygen atoms. Qualitatively, the electrons shared by the atoms must be shifted slightly toward the oxygen nucleus and away from the hydrogen nucleus, creating the dipole. The dipole nature of water is a key element in its ability to act as a solvent. The charges on water's atoms enhance its interactions with hydrophilic molecules.

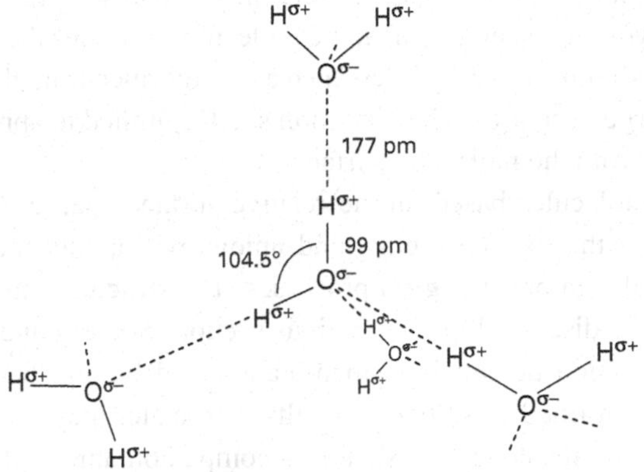

Figure 2.9   Hydrogen bonding in water. Covalent bonds between oxygen and hydrogen are connected by solid lines. Hydrogen bonds (dashed lines) are formed between the positive hydrogen dipole ($\sigma+$) and the negative oxygen dipole ($\sigma-$).

The dipole nature of water produces attraction between the positive hydrogen of one water molecule and the negative oxygen of an adjacent water molecule. This attraction forms the hydrogen bond. Because each water molecule has two hydrogens, water forms a continuous network of hydrogen bonds (Figure 2.9). The energy of water hydrogen bonds is an area of continuing research interest. In a vacuum, the energy of the water H-bond is about 25 kJ/mol, and in pure water the H-bonds have an energy of 18.4 kJ/mol (Markovitch and Agmon, 2007). This is much higher than the ambient energy in the body of 2.58 kJ/mol. With these energies water would form structures with a long duration, not easily disrupted by random thermal collisions of molecules. Physiological solutions, however, contain significant amounts of solutes, which interact with water and reduce the energy of water hydrogen bonds to 5 kJ/mol (Tinoco *et al.*, 1995). At this energy, random thermal collisions will occasionally rupture water H-bonds. This aspect is also important in the solvation of other molecules: if the energy between the water molecules were very high, their energetic preference for water–water binding would limit binding to other molecules, and limit the solubility of those molecules. The energy of water H-bonds exists in an intermediate state: significantly stronger water binding energy would make life very different.

Water will form temperature dependent clusters, from several hundred molecules per cluster at the melting point to several dozen per cluster at the boiling point. The frequency of oscillation of water hydrogen bonds is 5×10^{-13} Hz, and the duration of a given cluster is 10^{-10} to 10^{-11} seconds (Glaser, 2001). This means that each H-bond will have several thousand oscillations before it breaks and the components form a new H-bond with other water molecules. Dozens of water molecules together will form transient, flickering clusters.

In addition to water–water H-bond formation, water will also form hydrogen bonds with other, solute molecules. Oxygen and nitrogen in these molecules have negatively charged dipoles, and will attract the hydrogen dipole of water. Molecules are more soluble in water at higher temperature because the smaller cluster size creates more sites for potential interactions with other molecules: the greater number of interactions between a solute and a solvent increases the solute solubility in that solvent. The surface of molecules will have a hydration shell in which the molecule structures the water in its immediate vicinity. This hydration structure is lessened as the distance from the molecule increases and the local charge decreases. The hydration shell is limited to approximately two water diameters away from the molecular surface.

Ions will attract water molecules based on the relative surface charge on the ionic surface. Cations will attract the oxygen dipole, and anions will attract the hydrogen dipole. Cations play key roles in physiological processes. The structural importance of calcium and magnesium was discussed in the previous section. Sodium and potassium fluxes control graded and action potentials in membranes, and lithium has important clinical uses. When cations formed in solution as salts dissociate, they donate one or more electrons from their outermost electron shell to a companion anion, like chlorine. As a result, cations in solution have a smaller radius than their crystal radius if all the outermost electrons have been removed. The charge on the ions will be distributed over their "surface." The electron cloud of an atom of course does not have a defined surface, but has a probability function defining the position and velocity of the electrons. The quantum physics of this area is beyond the scope of this book. It is instructive, however, to calculate the charge density of such a theoretical surface.

Table 2.1 shows this information for five physiologically important cations in order of increasing molecular weight. Note that the charge density is not a direct function of mass. Na^+ and K^+ have the lowest charge densities, in a range where an important distinction occurs. As was discussed above, hydrogen bonds form between the negative dipole of the oxygen atom in water and the positive dipole of the hydrogen of an adjacent water molecule. The small size of the dipole moment relative to the atomic radii places the center of both the positive and negative charges near the H–O interface. Calculation of the surface charges on water is a complex and ongoing research field. Experimental evidence using NMR can estimate the residence time t that a water molecule associates with its nearest water neighbor, and the residence time t_i that a water molecule associates with an adjacent ion (Glaser, 2001). When the ratio of $t_i/t > 1$, the ion is restricting the movement of nearby water molecules, effectively structuring the water. If the ratio is < 1, then the ion is disrupting the local water structure. Na^+ is a structure-making ion, and K^+ is a structure-breaking ion. The oxygens of water are more strongly attracted to the $+$ charge on Na^+ than to the $+$ charge on the hydrogen dipole, resulting in a cluster of water molecules forming around Na^+. No such cluster surrounds K^+, as the charge density on K^+ must be lower than the effective charge density on the hydrogen dipole, so that the oxygen dipole preferentially forms hydrogen bonds with other water molecules rather than with K^+. The other cations in Table 2.1 all have charge densities greater than Na^+, so they will also be structure-making ions. Each will have a shell of water molecules surrounding it.

Hydrogen bonds also contribute to the structure of organic molecules. These bonds may be part of the internal structure of a molecule, or contribute to the intermolecular binding within a cluster of molecules. As with water, internal dipoles produce the electrostatic attractions leading to H-bond formation. For small molecules, these dipoles may contribute significantly to the overall dipole moment of the molecule. For large molecules, particularly proteins, the local dipole that is part of a H-bond may only be a

Table 2.1 Ionic radii of physiologically important cations

Cation	Charge (e)	Radius (nm)	Ionic surface area (nm^2)	Charge density (e/nm^2)
Li	1	0.076	0.0726	13.77
Na	1	0.102	0.1307	7.65
Mg	2	0.072	0.0651	30.72
K	1	0.138	0.2393	4.18
Ca	2	0.100	0.1257	15.92

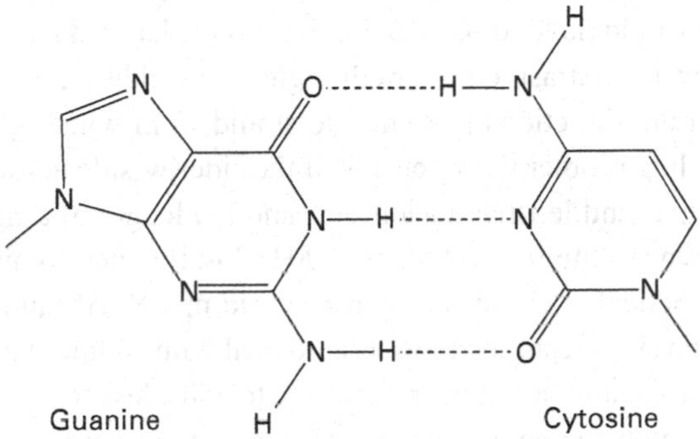

Guanine   Cytosine

Figure 2.10 Hydrogen bond formation between guanine and cytosine in DNA. There are two hydrogen bonds between the base pair adenine and thymine.

minor component of the overall dipole nature of the molecule. In most cases, the positive dipole contributing to a molecular dipole is the hydrogen of either a –NH or –OH side group, although others are possible. The negative dipole is often an =N— or =O group. Figure 2.10 shows H-bond formation between guanine and cytosine bases of DNA. Each bond has an energy on the same order of magnitude as those of water H-bonds, but the large number of H-bonds, two or three for every base pair, gives double-stranded DNA tremendous stability. These molecules do not undergo the same rapid restructuring seen in the flickering clusters of water. Helicases use the energy of ATP to rupture the H-bonds of DNA as well as contributing to the separation of the strands, energy use not seen in the cycling of water H-bonds.

DNA structure presents one of the most long-standing and important problems in biophysics. Immediately after the introduction of the double helix structure, the physical difficulty of unwinding a DNA helix was apparent. The angular momentum required for the thousands of revolutions required to separate the two strands had no energetic source, as well as the difficulties present in unwinding a very long molecule in a confined space (Delmonte and Mann, 2003). There is no question that the pairing of the bases, adenine with thymine and guanosine with cytosine, is correct, and allows for the replication of genetic material. While the elegance of the double helix made it attractive, dismantling a double helix over the length of a chromosome is so energetically daunting that alternatives must exist (Root-Bernstein, 2008). The unwinding problem has been addressed in two different ways: a series of enzymes that cut short segments of DNA, unwind them, and then repair the ends; or a side-by-side structure that has both right-hand and left-hand segments of DNA, resulting in a "warped zipper" that eliminates the need to unwind DNA over long distances. There is no consensus in favor of either model, even after more than 50 years. Evidence using scanning tunneling microscopy shows that double-stranded DNA is asymmetric along its long axis, having a major axis of 2.0 to 2.2 nm and a minor axis of 1.2 to 1.6 nm (Delmonte, 1997). This evidence is consistent with the side-by-side warped zipper models of DNA; helical DNA would be symmetrical with a width of 2.2 nm.

A side-by-side model of DNA presents an interesting energetic possibility. DNA polymerase III adds nucleotides to the growing end of a DNA chain. For a side-by-side model to be correct, bases would have to be added in both right-handed and left-handed directions as the chain grew, in contrast to the strictly right-hand addition in a full double helical model. The alternating directions prevent one strand from winding around the other, obviating the unwinding problem. Proponents of the side-by-side model acknowledge that the number of right- and left-hand additions, and the length of a right- or left-hand sequence, are unknown (Delmonte and Mann, 2003), but if either dominates, there would still be a low-angle helical pitch, resulting in a hybrid model. At ratios of greater than 4:1, there would still be significant local twisting to deal with. Without knowing the details of the total system, is the amount of energy needed to add a less frequent left-hand base reasonable compared with the energy needed to add a more common right-hand base? From the Boltzmann distribution in Chapter 1, the number of molecules with a given energy relative to RT can be calculated. Also, the average energy of the added bases must be at RT. Given these constraints, it is possible to calculate what the different energies would be for the addition of a right-hand and left-hand base that would result in a 4:1 right to left ratio (or any other ratio of interest). In humans with an ambient energy of 2.58 kJ/mol, a necessary energy of 1.86 kJ/mol for the addition of a right-hand base (an energy present in 0.485 of the bases, based on the Boltzmann equation) and an energy of 5.44 kJ/mol for the addition of a left-hand base (present in 0.121 of the bases) would produce a ratio of four right-hand bases to one left-hand base in the DNA molecule. The higher energy in the left-hand addition would be needed to have DNA–PIII in position to form the covalent bond using the energy of ATP, the hydrolysis of which would yield enough energy to form the bond. This energy estimate can only be an approximation, as the energy requirement of different bases, as well as the four possibilities of bases addition (R on R, R on L, L on R, and L on L), will probably be slightly different. Still, the energy requirements for left-handed additions are reasonable.

Intermolecular H-bonds also contribute to the formation of complementary molecular clusters. Ascorbate (Vitamin C) and norepinephrine (NE) form a bimolecular complex, mentioned in the methods section above, with four H-bonds contributing to the stability of the complex. There is also a π–π bond formed by the overlap of ring structures in this complex, a type of molecular binding covered in the next section. The complex of ascorbate and NE protects the catecholamine from oxidation. Interestingly, as the complex approaches a membrane, the electric field produced by the membrane will cause dissociation of the complex, making NE available to bind to its receptor on the membrane. This mechanism will be discussed in the section on membrane function.

2.6   Non-bonded molecular interactions

Significant interactions occur between molecules even in the absence of direct bond formation. The motions of the electron around any atom produce instantaneous dipoles that in turn induce the formation of other attraction dipoles in neighboring atoms. The closer the two atoms are, the stronger the instantaneous dipole attraction, and the stronger the attractive energy between the two atoms. The attraction occurs because the instantaneous position of the electron cloud of one atom is positioned between the two positively charged nuclei, with the instantaneous position of the second electron located on the far side of the second atom, producing a net $+ - + -$ effect. Named for Fritz London, who derived the equations describing this phenomenon, London attraction is related to the distance between two atoms i and j and the types of atoms involved. The energy E_L between the atoms at distance r_{ij} is

$$E_L = \frac{A_{ij}}{r_{ij}{}^6} \qquad (2.27)$$

where A_{ij} is a constant specific for the pair of atoms i and j. In Figure 2.11, the energy in London attraction decreases, that is, becomes more attractive, as the distance between the atoms decreases, as a 6th power function. As the atoms get very close, however, there is steric repulsion of the atoms due to the quantum mechanical limitations on the number of

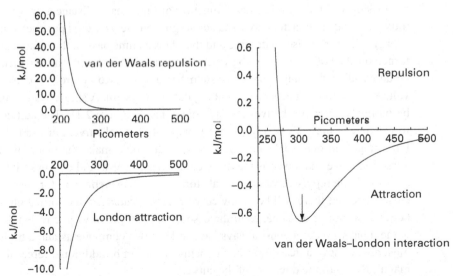

Figure 2.11   Non-bonded molecular interactions. At distances less than the atomic radius, the energy of molecular interaction increases rapidly as van der Waals repulsion occurs between electrons trying to occupy the same orbitals. London attraction occurs as electronic motion produces instantaneous dipoles between neighboring atoms. A balance point occurs (arrow) where the attraction and repulsion forces are equal. For these diagrams, the numerical values are for two neighboring, but not bonded, oxygen atoms.

electrons allowable in the outermost shell. This van der Waals repulsion, also shown in Figure 2.11, has an energy E_{vdW} related to the 12th power of the distance

$$E_{vdW} = \frac{B_{ij}}{r_{ij}^{12}} \tag{2.28}$$

where B_{ij} is again a specific constant (different from A_{ij}) related to the atoms. Both London attraction and van der Waals repulsion occur at the same time, so that a balance point must exist between attraction and repulsion. This will be the distance, referred to as the van der Waals radius, indicated by the arrow in the combined L–vdW graph in Figure 2.11, at which there is the minimum energy between the two atoms.

Force is the slope vector of the energy-position relation, so that any movement away from this minimum energy position will produce a force driving the atoms back to the minimum distance. Because A_{ij} and B_{ij} are specific for any atom pair, this distance will also be specific. Two non-bonded oxygen atoms, for example, have a minimum energy at 3.04 Å or 304 pm. The distance between two hydrogen bonded oxygens is 276 pm. Hydrogen bonding decreases the distance between two neighboring oxygen atoms below the van der Waals radius that would occur in the absence of a hydrogen bond. (Covalently bonded oxygen atoms are about 130 pm apart.)

Figure 2.11 demonstrates a common useful phenomenon. The arrow indicates a change in direction of a relation: it reverses its direction as the distance increases. This biphasic relation occurs because the graph is a sum of two separate, monophasic relations: the increase in London attraction as distance decreases and the increase in van der Waals repulsion as distance decreases. Whenever a biphasic relation appears, with rising and falling phases as a function of time, mass, distance, or whatever, there must be at least two factors involved converging on the measured parameter, in this case, energy. If a system rises with time and then falls as time passes, the system is just being turned on, then off. There must be separate systems that first turn it on and then turn it off. Understanding that any biphasic system has at least two control systems provides a valuable clue to the mechanisms that regulate that system. A single system must always be monophasic, as in the two left-side diagrams of Figure 2.11. The addition of other forces may or may not alter the general shape of the total curve. For example, there are long-range van der Waals attraction forces that would make the slope of the London attraction curve shallower as the distance increased, but would not make it biphasic. In contrast, if the repulsive van der Waals forces can be overcome by forcing the electrons closer together, there will be a new curve reversal leading to the formation of covalent bonds at a shorter distance, as in the case of the 130 pm oxygen covalent bond. Thus, while a biphasic curve must always have at least two components and a triphasic curve must always have at least three components, there can be additional forces that alter the magnitude if not the direction of the curve.

As you can see from the London–van der Waals interaction curve, the energy between non-bonded molecules at its minimum is much less than the human ambient energy of 2.58 kJ/mol. Molecules with these associations will constantly have the connection broken by random thermal motions of molecules. There are, however, environments in which these interactions will produce significant, longer term structures. The fatty acid tails of phospholipids consist of non-polar $-CH_2-$ residues, with occasional $-CH=CH-$ residues in unsaturated fatty acids. These chains associate using non-bonded attraction forces. The great number of interactions produces an additive force that keeps them together. These attraction forces are sometimes referred to as "hydrophobic," although they occur whether water is present or not. The association of the fatty acid tails in membranes is not so strong that thermal energy cannot rearrange them. This mobility leads to the fluidity associated with the molecules in membranes: unless anchored to structures outside the membrane, molecules diffuse laterally, limited only by temperature, size and the steric hindrance produced by nearby molecules.

Cholesterol plays a key role in the stability of biological membranes. The phospholipids that make up the backbone of membranes have a molecular makeup similar to detergents. All children know that when they make soap bubbles in the air, touching the bubble causes it to burst immediately. This occurs because the fatty acid tails form a semi-crystalline structure that absorbs the mechanical energy of the touch, transmits it through the bubble's membrane structure until a weak point is reached and the bubble bursts. If our membranes had only phospholipids, they would be just as vulnerable as soap bubbles. The presence of cholesterol in our membranes provides the mechanical buffering that prevents our membranes from rupturing. Cholesterol has a steroid four-ring structure in a planar, oblong shape, with a hydroxyl group at one end and a hydrocarbon chain at the other end (Figure 2.12). The hydroxyl group will associate with the phosphate head groups of phospholipids. The steroid rings and hydrocarbon chain will associate with the fatty acid tails of phospholipids with the same London attraction forces that hold the fatty acids together (Figure 2.12). A membrane will, like all physical structure, always assume its lowest energy configuration. (That is why soap bubbles always form spheres: the sphere has the lowest surface-to-volume ratio, a physical shape that minimizes free energy.) In having its lowest energy, the cholesterol will maximize its

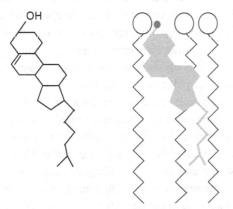

Figure 2.12   Cholesterol structure (left) and cholesterol imbedded with phospholipids in a membrane. Imbedded cholesterol is shown with the gray shading. The small ring at the top of the cholesterol with the darker shading is the hydroxyl group. The ovals represent the phosphatidyl groups of phospholipids, and the alternating black segments the $-CH_2$ groups. The middle fatty acid tail is passing behind the cholesterol.

contacts with neighboring fatty acids, meaning it will have the longer sides of its oblong shape in contact with the fatty acids. When mechanical energy is applied, the membrane will stretch. Without cholesterol, this stretch cannot be very great before adjacent phospholipids would separate and the membrane would rupture. With cholesterol present, the cholesterol can reorient its contacts holding the fatty acid chains together, allowing greater stretch to the membrane. The reorientation can be seen as either reducing the number of contacts between cholesterol and the fatty acid chains, or stretching bond angles to maintain contact. Both will only occur if there is an input of mechanical energy, and the weak strength of the London attractions relative to ambient energy means that both reduced contacts and bond stretching will occur to some extent. The ability of the membrane to reorient its structure in the presence of cholesterol will keep the membrane intact until the mechanical force is removed. The membrane will then return to its original, lowest energy state. The hydrophilic head groups of a membrane will have a negatively charged phosphatidyl group and an additional side group which may have a positive charge (choline) or be neutral (serine), so that the different combinations of attraction and repulsion will have a complex relationship. The presence of negatively charged neuraminic acid (sialic acid) residues on the extracellular side of membranes will further alter electrostatic interactions. In red blood cells, the phosphatidylserine residues are in the inner surface, and the neuraminic acids are on the outside, so that the net charge on both sides is negative. Electrostatic repulsion in these cases would not contribute to membrane stability, which will reside in the hydrophobic tails. The interaction of the hydroxyl group of cholesterol with the hydrophilic components of a membrane provides an additional anchor point helping to bridge the strain on the hydrophobic core of the membrane.

Cholesterol cannot of course prevent a membrane from rupturing under all conditions. Mechanical deformation or physical pressure can be great enough to rupture any biological membrane. Rupturing is an irreversible state transition. We will deal with the general principles of state transitions in a later section. Here, we can give a further example of the effect of cholesterol. Red blood cells (RBCs) have a lifetime of 120 days. They contain no organelles, but are mostly bags of hemoglobin with a few additional enzymes, structural proteins and ion pumps (carbonic anhydrase, spectrin, membrane-bound glycolytic enzymes, and the Na–K ATPase). There is no nucleus, nor any mechanism for cellular repair. Over time, cholesterol will gradually decrease in the RBC membrane, making the membrane progressively stiffer as time passes. Twice each minute an 8 μm red blood cell must squeeze through our 7 μm capillaries, deforming each time and then returning to its biconcave disc shape as it enters a venule. As they age, RBCs' stiffer membranes are increasingly less able to conform to the stresses they see in the capillaries. The hemolysis of RBCs takes place primarily in the spleen and liver where phagocytes destroy aged RBCs. The loss of cholesterol from RBC membranes will make them more susceptible to rupture both in capillaries and through immune system mediated lysis.

Non-covalent bonds can also occur between aromatic ring structures, such as the purine and pyrimidine bases of DNA. These ring structures have delocalized π-electrons produced by conjugated bond systems in the rings. When the rings are parallel to one another, π–π interactions occur between the p-orbitals. The energy of these bonds is stronger than other non-bonded interactions, and is approximately that of hydrogen bonds in the physiological environment, 5 kJ/mol. While still much weaker than covalent or ionic bonds, when many of these bonds occur in supramolecular structures, like DNA, they provide substantial structural stability. Non-covalent π–π interactions can also contribute to the complementarity of smaller molecules. The ascorbate–norepinephrine complex discussed earlier has a π–π bond between the rings of the two molecules (Figure 2.13). The binding energy of Asc–NE is about 24 kJ/mol, consistent with the four hydrogen bonds and one π–π bond linking the molecules.

The measurement of the binding energy could be measured using the CE data above, and is shown in Figure 2.14. The relation between association energy E_A and K_D is

$$E_A = -2.303RT\log K_D. \tag{2.29}$$

The log K_e values from Figure 2.5 were plotted to calculate the K_D. The parallel y-axis in Figure 2.14 shows the association energy calculated from the K_D. The model of four

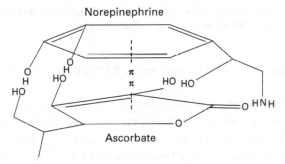

Figure 2.13   Model of norepinephrine–ascorbate binding. The four hydrogen bonds and one π–π bond produce at binding energy about ten times greater than ambient energy. The ascorbate binding protects the oxidation sites on NE, preventing it from being oxidized. The bond angles are not quantitative.

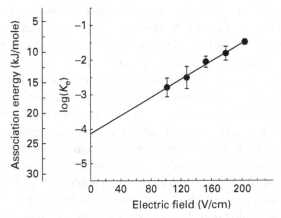

Figure 2.14   Log K_e values from Figure 2.5 are plotted against the electric field. The extrapolation yields the dissociation constant at zero electric field, the K_D. The association energy is calculated from the E_A–K_D relation.

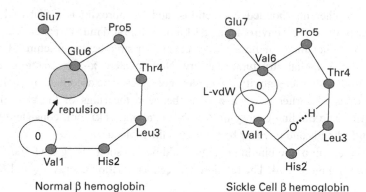

Normal β hemoglobin   Sickle Cell β hemoglobin

Figure 2.15   Models of the first seven amino acids in normal and sickle cell mutated β chain of hemoglobin. The shaded field at Glu6 in normal hemoglobin represents the negative charge on the carboxyl side group, which will repel the hydrophobic field around the side group of Val1. The valine substitution at position 6 leads to a cascade of new bond formation and structural changes, with the London attraction between the valines bringing the peptide bonds between Val1–His2 and Leu3–Thr4 sufficiently close to form a hydrogen bond. The combination of bonds has a total energy sufficiently greater than kT that the structure is maintained for a significant duration.

H-bonds and one π–π bond, each with 5 kJ/mol, is a good fit with this calculation (Dillon *et al.*, 2006).

Small changes in the bonds between molecules, even within molecules, can have profound effects. In the case of sickle cell anemia, the root cause of the condition is a genetic mutation in the β hemoglobin chain, leading to a cascade of binding events. The glutamic acid with a negatively charged side group at position 6 is replaced by a valine with a neutral side group (Figure 2.15). The hydrophobic valine side group at position 6 forms an intramolecular non-covalent association using London attraction with the hydrophobic side group of the valine in position 1. This association would have an energy well below that of the local ambient energy, and would not withstand any significant strain on its own. This association, however, forms a cyclic structure that brings the peptide bonds between residues 1–2 and 3–4 sufficiently close that they form a hydrogen bond, providing enough stability to the structure that it will have a longer duration. This cyclic structure protrudes from the β subunit. This protrusion fits into a slot on the surface of the α subunit of an adjacent, deoxygenated hemoglobin, linking two consecutive hemoglobins. This process continues, forming stacks of hemoglobin that causes the altered, sickle shape (Muriyama, 1966). The β protrusion only fits into the deoxygenated form because oxygen binding alters the shape of the α subunit in such a way that the β protrusion cannot bind. For this reason, a decrease in oxygen in the blood of a sickle cell individual triggers the sickle cell crisis. Interestingly, a different mutation of hemoglobin, Hb C Georgetown, produces effects similar to sickle cell, but in this case the mutation at the 1 position of the β chain produces an ionic bond. In normal hemoglobin, no bond would form between the valine at 1 and the glutamate at 6, and the cascade of events would not occur.

Biophysical control mechanisms must always involve transient binding: nothing will happen if there is no binding, and permanent binding removes the possibility of multiple states. Transient binding in the physiological system uses hydrogen bonds, ionic bonds, and non-covalent attractions. There are multiple methods for measuring binding, using solution chemistry, capillary electrophoresis, osmotic pressure, enzyme kinetics and molecular fluorescence. Molecular and ionic binding must always be considered against the background of water binding. The additive nature of the energy of molecule bonds produces structures of substantial duration even in a system where all the individual bonds are near or below ambient energy. Deviations from normal association energy can produce pathological conditions, like sickle cell, when the duration of binding increases and new structures appear. The relation of binding energy to ambient energy ultimately controls the duration of biological structures.

References

Busch E, Hohenester E, Timpl R, Paulsson M and Maurer P. *J Biol Chem.* **275**:25 508–15, 2000.

Delmonte C. *Advances in AFM and STM Applied to the Nucleic Acids.* King's Lynn, UK: Invention Exploitation Ltd., 1997.

Delmonte C S and Mann L R B. *Curr Sci.* **85**: 1564–70, 2003.

Dillon P F, Root-Bernstein R S and Lieder C M. *Biophys J.* **90**:1432–8, 2006.

Dillon P F, Root-Bernstein R S, Sears P R and Olson L K. *Biophys J.* **79**:370–6, 2000.

Glaser R. *Biophysics.* Berlin: Springer-Verlag, 2001.

Klug A and Schwabe J W. *FASEB J.* **9**:597–604, 1995.

Kothinti R, Blodgett A, Tabatabai N M and Petering D H. *Chem Res Toxicol.* **23**:405–12, 2010.

Krishna S S, Majumdar I and Grishin N V. *Nucleic Acids Res.* **31**:532–50, 2003.

Maret W. *J Trace Elem Med Biol.* **19**:7–12, 2005.

Markovitch O and Agmon M. *J Phys Chem A.* **111**:2253–6, 2007.

McGee M and Wagner, W D. *Arterioscler Thromb Vasc Biol.* **23**:1921–7, 2003.

Moore, W. *Physical Chemistry*, 4th edn. Englewood Cliffs, NJ: Prentice Hall, 1972.

Muriyama M. *Science.* **153**:145–9, 1966.

Rand R P, Fuller N L, Butko P, Francis G and Nicholls P. *Biochemistry.* **32**:5925–9, 1993.

Romero-Isart N and Vasák M. *J Inorg Biochem.* **88**:388–96, 2002.

Root-Bernstein R S. *Art J.* **55**:47–55, 2008.

Root-Bernstein R S and Dillon P F. *J. Theor. Biol.* **188**:447–79, 1997.

Root-Bernstein R S and Dobblestein C. *Autoimmunity.* **33**:153–69, 2001.

Sears P R and Dillon P F. *Biochemistry.* **38**:14 881–14 886, 1999.

Spyracopoulos L, Lavigne P, Crump M P, *et al. Biochemistry.* **40**:12 541–51, 2001.

Tinoco Jr I, *et al. Physical Chemistry: Principles and Applications in the Biological Sciences*, 3rd edn. Upper Saddle River, NJ: Prentice Hall, 1995.

3 Diffusion and directed transport

The movement of material within the cell and across membranes always requires a driving force. For diffusive processes, the driving force is the electrochemical gradient. The electrical component of this force requires a separation of charge: the negative and positive charges must be kept from one another until a conductive channel opens and the charged species can flow down their electrical gradient. Intact membranes, whether the cell membrane or those of organelles, are needed to provide the voltage buildup that will allow current to flow when conduction becomes possible. Within the cytoplasm or in the extracellular fluid, the charges are not kept separate, and without a voltage there will be no electrical gradient. In these cases, diffusion is entirely driven by concentration gradients. The generation of these gradients is an active process: functions linked to ATP hydrolysis are ultimately responsible for all diffusion gradients. Once the gradients are generated, however, they will produce the movement of material without further ATP input. Across the membrane, of course, there can be both concentration and electrical gradients for charged moieties. In some cases, like Na^+ across the resting cell membrane, these gradients will be in the same direction, inward in this case, with a higher Na^+ concentration on the outside and a negative charge on the inside attracting the positive sodium. In others, like K^+ across the resting cell membrane, the electrical and concentration gradients are in opposite directions, with the higher K^+ concentration on the inside driving K^+ out countered by the negative charge on the inside drawing K^+ in. If diffusion down a gradient does not require ATP directly, other intracellular transport processes do require ATP. The movement of vesicles, organelles, and other cargo by kinesin, dynein and non-polymerized forms of myosin requires the direct hydrolysis of ATP to power each step. Transport within the cell therefore comes in two forms: diffusion allows material to move in all directions, while ATP-driven processes move material in particular directions. In this chapter, both of these mechanisms, and the principles behind them, will be discussed.

3.1 Forces and flows

All physiological systems have to obey the laws of physics. A system is not exempt from the law of gravity, or Newton's laws of motion, because it happens to be living. The importance of any physical law will depend on the particulars of the system. When you trip, the law of gravity will be an important factor in the next few seconds, but the resistance of air to your fall will be negligible. Gravity will influence the distribution of blood in the lungs, but to red blood cells moving through a capillary the resistance of the capillary wall and the blood pressure will be of primary importance. To an alveolus expanding during inspiration, gravity will be of only minor significance, while surface tension of water will be a major concern. All of these physical parameters, gravity, air resistance, applied pressure, surface tension, and more, will be present at all times, but in most cases only a limited set of physical laws will dominate. None of our molecules move near the speed of light, so there will be no change in our mass as described by the special theory of relativity, nor will there be any bending of light waves leading to non-Euclidian geometries. Movement in the physiological world is Newtonian and Euclidian.

A detailed derivation of the laws governing the thermodynamic and kinetic properties of biophysical systems was presented by Katchalsky and Curran (1975). In covering the salient points of this area, the force and flow equations presented here have a detailed, formal development in that book. Biophysical systems are not static: they are subject to various forces, and respond to those forces with movement. Thus, there is a direct link between force and flow. Students are exposed to this concept in their initial exposure to electrical physics, in the relation between current I, resistance R, conductance C and voltage V:

$$I = \frac{V}{R} = CV. \tag{3.1}$$

The voltage is the force, and the current is the flow. An analogous equation applies to the flow of blood F, the blood pressure P and the resistance R:

$$F = \frac{P}{R} \tag{3.2}$$

and the relationship of cardiac output CO, total peripheral resistance, TPR, and blood pressure BP:

$$CO = \frac{BP}{TPR}. \tag{3.3}$$

Each of these relates a flow to a particular driving force. Newton defined the relation of force, mass and acceleration for an object in a vacuum with no friction,

$$f = ma \tag{3.4}$$

conditions that do not apply to physiological systems. When friction is also considered, the relation becomes

$$f = ma + hv \tag{3.5}$$

with the force f having both acceleration and velocity components. When the acceleration is damped out or the mass is so small as to make ma insignificant, the relation relates a flow v and a force f. The h factor, as seen in the equations above, is a resistance or frictional factor.

Physiological systems do not exist in isolation. There is linkage between the hydrolysis of ATP and the movement of ions against their gradient, the movement of ions down their gradient and the production of second messengers, or the contraction of muscle cells and the movement of blood. The principles of linked forces and flows were formalized by Onsager (1931). He modeled the relation between a flow J and force X as

$$J = LX \tag{3.6}$$

with L as a numerically positive phenomenological coefficient, a relationship applicable when the system is near equilibrium and the coefficient is independent of the flow. As systems move away from equilibrium, the total coefficient L_T will change as a function of the flow

$$J = L_T X = (L_o + L_J J)X \tag{3.7}$$

where L_o is the equilibrium coefficient and L_J is the flow-dependent coefficient.

Rearranging the non-linear equation to solve for J,

$$J = \frac{L_o X}{1 - L_J X}, \tag{3.8}$$

it is clear that the system is linear when X is small, and becomes non-linear as the driving force X increases. Figure 3.1 shows that energy systems are linear near equilibrium, but become increasingly non-linear the farther the system is from equilibrium.

Friction is the ratio of force/flow, X/J, and the dependence of friction on the flow rate is shown in Figure 3.2. At low flow rates, L_J has little influence, and the system is linear. As the flow rate increases, the system becomes progressively more non-linear. The degree of non-linearity is determined by the ratio of the coefficients L_J/L_o, with the friction produced increasing as the ratio increases.

For coupled systems, the relation between two forces and two flows is

$$J_1 = L_{11}X_1 + L_{12}X_2 \tag{3.9}$$

$$J_2 = L_{21}X_1 + L_{22}X_2 \tag{3.10}$$

where the flows J_1 and J_2 are driven by forces X_1 and X_2, with the relationships connected by the phenomenological coefficients L_{ik}. L_{11} and L_{22} are the direct coefficients linking the flow and force of systems 1 and 2, respectively, as the conductance linked voltage and current above. In systems that interact, force X_1 will have an influence on flow J_2, and force X_2 will have an influence on flow J_1. The linkage is described by the coupling

Figure 3.1   The filled circle state is within kT, but removed from the lowest energy state. The force returning it to the lowest energy state will be small, with a linear Onsager coefficient. The open circle state is far from equilibrium. It will have large force and a non-linear Onsager coefficient.

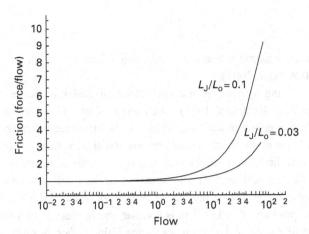

Figure 3.2 Non-linearity of force/flow system. Increased force produces increased flow away from near-equilibrium conditions, resulting in increased friction within the system.

coefficients L_{12} and L_{21}. The higher the value of the coupling coefficients, the stronger the linkage between the systems. If the systems are not linked, the coupling coefficients will be zero, and the paired equations above will reflect two entirely separate systems.

Onsager proved that in coupled systems, the value of

$$L_{12} = L_{21} \tag{3.11}$$

as long as the local entropy production σ of the two-equation system is

$$\sigma = J_1 X_1 + J_2 X_2, \tag{3.12}$$

a condition that will apply when all parts of the systems are in the same local environment. The product of a force and flow has units of energy. The equivalence of the coupling coefficients reduces the number of factors that have to be experimentally determined to measure the degree of coupling. In the two-force/flow system, only three coefficients, L_{11}, L_{22}, and $L_{12} (= L_{21})$ need to be measured. This advantage is even greater in more complex systems with more than two forces and flows, only needing to measure 6 out of 9 coefficients in a three-part system. The equation governing the number of measured coefficients L_m is

$$L_m = (N_i)^2 - \Sigma(N_i - 1) \tag{3.13}$$

where N_i is the number of force/flow systems. In practical terms, the equivalence of the coupling coefficients means that if the two systems interact, the force of A on B is equal and opposite to the force of B on A (Dillon and Root-Bernstein, 1997). An interesting extension of Onsager's work showed that in a two-equation system

$$L_{11} L_{22} \geq L_{12}^2 \tag{3.14}$$

which limits the magnitude of L_{12}. At least one of the direct coefficients must be greater than L_{12}. This relation allows calculation of the degree of coupling q between the systems:

$$q = \frac{L_{12}}{\sqrt{L_{11}L_{22}}} \tag{3.15}$$

where $1 \geq q \geq 0$, with maximum coupling occurring when $q = 1$ and no coupling occurring when $q = 0$ (Glaser, 2001).

A characteristic of living systems is the conversion of scalar processes to vector processes (Dillon and Root-Bernstein, 1997). The energy of an ATPase reaction can be converted into muscle movement or an ion gradient. A chemical reaction in solution is a scalar: it has magnitude but no direction. Muscle contraction and ion flux are both vectors, with both magnitude and direction. If all the chemical components of actomyosin ATPase are placed in a flask, the chemical reaction will occur, but the flask will not move. Coupled systems in physiology must be able to link scalar and vector systems (Dillon and Root-Bernstein, 1997). The product of two scalars is a scalar, the product of two vectors is a scalar, and the product of a scalar and a vector is a vector. Using *italics* for scalars and **bold** for vectors, a coupled system of a scalar reaction R and a vector process C is

$$J_R = L_{RR}X_R + \mathbf{L_{RC}X_C} \tag{3.16}$$

$$\mathbf{J_C} = \mathbf{L_{RC}}X_R + L_{CC}\mathbf{X_C}. \tag{3.17}$$

The direct coefficients in both equations are scalars, but the coupling of a vector system with a scalar system requires that the coupling coefficient be a vector. This requires that the coupling coefficient have directionality. In an isotropic system, such as a solution in a beaker, there is no directionality, so that scalar systems such as reactions cannot be coupled to vector systems. This is the Curie–Prigogine principle applied to biological systems: the coupling of scalar to vector systems requires that the system as a whole must be anisotropic; that is, the system is not physically equal everywhere (Dillon and Root-Bernstein, 1997). For example, if an ion pump were floating free in the cytoplasm, it could hydrolyze ATP and translocate an ion from one side of the molecule to the other, but since both sides are in easy communication, there would be no ion gradient produced (Figure 3.3). The cell membrane produces an anisotropic system where the physical

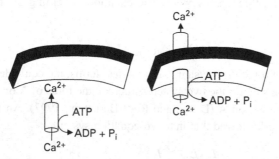

Figure 3.3　　The isotropic intracellular system on the left, linking a scalar ATPase reaction with an ion pump in the isotropic cytoplasm produces no net change in the calcium concentration inside the cell. Having the ion pump span the membrane makes the system anisotropic, linking the scalar ATPase reaction to a vectorial ion gradient.

environment can be different on the two sides of the membrane. The scalar ATPase reaction translocates ions and produces a concentration gradient across the membrane, thus coupling a scalar reaction to a vector ion gradient (Dillon and Root-Bernstein, 1997).

3.2   Fick's laws of diffusion

The diffusion of molecules is dependent on the gradient of those molecules over a distance. The measurement of flux is the amount of material moved through a cross-sectional area in a given time. As shown in Figure 3.4, this area is assumed to be of uniform consistency, so that the movement through all parts of the area are under the same conditions. These conditions, such as the viscosity of the medium or the temperature, will affect how fast the material will diffuse. The unit of flux is $mol/cm^2 \cdot s$. This is intuitively satisfying, for it tells you the number of molecules (mol) moving through a given area (cm^2) over a given time (seconds). The flux J_x through area A in the x-direction for a given concentration gradient dC/dx is

$$J_x = -DA\frac{dC}{dx} \tag{3.18}$$

where D is the diffusion coefficient, the proportionality constant determined by the conditions of the medium and the physical properties of the molecule, such as the molecular weight. The reason for the negative sign is illustrated in Figure 3.4. The concentration gradient, always measured from high to low concentration, goes from left to right in the figure; that is, it has a negative slope. Material will diffuse down the concentration gradient, so that the flux will also go from left to right. Since the flux will have a positive value, and the direction of the flux will be in the same direction as the negative slope, the negative sign is needed to reconcile these factors. This is Fick's first law of diffusion.

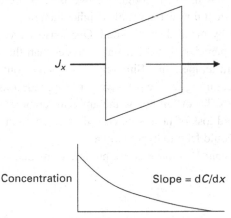

Figure 3.4   Fick's first law of diffusion. The flux of material through a plane of area A in the x direction will be proportional to the concentration gradient dC/dx.

In physiological systems it is often not sufficient to measure just the rate of material passing through a given area as related by Fick's first law. Since the material diffusing, such as oxygen, may be consumed within a given cell and also supplied by the entry of more oxygen across the membrane, the rate of change of concentration over time is also important. If the rate of consumption equals the rate of supply, then the system is in a steady state, the condition required over the long term for a physiological system to remain alive. There are conditions in which supply and consumption are not equal, and these conditions often trigger important physiological responses, such as increased respiration during exercise. If the flux into a given area $\partial J/\partial x$ is > 0, then the concentration over time $\partial C/\partial t$ will also be > 0. This process is described by Fick's second law of diffusion

$$\left(\frac{\partial C}{\partial t}\right)_x = D\left(\frac{\partial^2 C}{\partial x^2}\right)_t \qquad (3.19)$$

where D is again the diffusion coefficient. The second law says that the change in concentration over time at x is proportional to the second derivative of the concentration with respect to x at time t. Partial derivatives are used because the concentration is dependent on both time and distance. The utility of the second law lies in determining the rate at which adjacent compartments will approach equilibrium (even if they never reach equilibrium) when the molecule can diffuse between the compartments, or, when heat is used instead of concentration, how fast two adjacent volumes will approach thermal equilibrium. You can see by examining the equation for the second law that when the derivative of the concentration difference between the compartments is zero, that there will be no net change in the concentrations.

When the diffusion constant is unchanging across the system, the second law yields a linear gradient across the interface between two areas of different concentration or temperature (Figure 3.5). Changes in the diffusion coefficient, however, produce different perceptions in physiological receptors. When considering the touch receptors in the skin, touching a metal object at room temperature will draw heat away from the body at a higher rate than a wooden object will, even though the two objects are at the same temperature. The diffusion coefficient for heat in metal is higher than in wood, so heat is drawn away from the skin faster by metal than by wood. Our sensory receptors detect *changes* in temperature, and we perceive that the metal is colder than the wood, even though they are at the same room temperature. Similarly, wind across our skin draws away heat faster than still air, accounting for the perceived colder conditions associated with wind chill factor. Our body will not fall below the ambient temperature, but in a windy environment the more rapid loss of heat is perceived as colder, and does reflect more dangerous conditions that could lead to hypothermia.

The value of the diffusion constant D is made up of multiple elements:

$$D = \frac{kT}{f} \qquad (3.20)$$

where f is the frictional coefficient and kT is the molecular ambient energy. Because it is proportional to kT, an increase in the local temperature will increase the value of D and

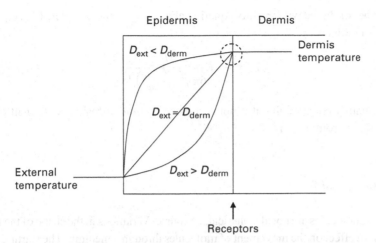

Epidermis   Dermis

$D_{ext} < D_{derm}$ Dermis temperature

$D_{ext} \neq D_{derm}$

External temperature $D_{ext} > D_{derm}$

Receptors

Figure 3.5 Effect of external diffusion coefficients on perceived temperature. With equal diffusion constants ($D_{ext} = D_{derm}$), there is a linear temperature gradient across the epidermis. With differences between the diffusion coefficients there will be a curvilinear temperature profile in the epidermis. Temperature receptors at the edge of the dermis (dashed circle) will see the greatest change in temperature, and percieve the coldest temperature, when D_{ext} is much higher than the D_{derm}, such as when touching metal.

thus increase the flux due to diffusion. The frictional coefficient f is a measure of the resistance a molecule will have as it moves through its environment. The value of f is

$$f = 6\pi\eta r \qquad (3.21)$$

where η is the viscosity of the medium and r is the radius of the molecule. Viscosity is synonymous with the "thickness" of a liquid: in the sense that as maple syrup is thicker than water, it has a higher viscosity. Both viscosity and radius increase the value of the frictional coefficient, and thus decrease the value of D. The larger the molecule, and the more viscous the medium it is moving through, the lower will be the flux. Combining the different elements in Fick's first equation we get:

$$J_x = -DA\frac{dC}{dx} = -A\frac{kT}{f}\frac{dC}{dx} = -A\frac{kT}{6\pi\eta r}\frac{dC}{dx}. \qquad (3.22)$$

In some cases, the molecular weight of the molecule is used instead of the radius and an appropriate correction factor related to the density of the molecule is used, since the relation between the molecular weight MW and the density ρ for a spherical molecule of radius r is

$$\rho = \frac{mass}{volume} = \frac{MW}{\frac{4}{3}\pi r^3} \qquad (3.23)$$

which when solved for r yields

$$r = \sqrt[3]{\frac{3MW}{4\pi\rho}}. \qquad (3.24)$$

This can be combined with the flux equation above to give the complete Fick equation for a spherical molecule:

$$J_x = -A \frac{kT}{6\pi\eta\left(\sqrt[3]{\frac{3MW}{4\pi\rho}}\right)} \frac{dC}{dx} \tag{3.25}$$

such a daunting equation that it is not surprising that the diffusion coefficient D form of the equation is routinely used.

3.3   Brownian motion

Biological molecules are not all spherical, of course. Variations in the shape of the molecule will have an effect on the movement of molecules through a medium. The starting point for assessing the influence of molecular shape is to first assume the molecule is a sphere moving through a medium and measure the frictional force F as it moves with velocity v, a relation known as the Stokes equation:

$$F = 6\pi\eta r v \tag{3.26}$$

which is clearly related to frictional coefficient f above:

$$F = fv. \tag{3.27}$$

When a frictional force is measured in this way, the calculated radius r is referred to as the Stokes radius R_S. The value of R_S is often compared with another experimentally determined factor, the radius of gyration R_G (Glaser, 2001). The radius of gyration is the mean square distance $\langle r^2 \rangle$ of the atoms in a molecule from the center of gravity of the molecule, when all of the atoms are linked together in a chain formation of n segments each of length l. The distance d between the end points of the chain is

$$d \approx \sqrt{\langle r^2 \rangle} = l\sqrt{n}. \tag{3.28}$$

The value of $\sqrt{\langle r^2 \rangle}$ has a practical meaning. If you were at the gravitational center of a molecule, the radii in half the directions would be positive and in the other half negative. If you just add up the radii in this way, the average value would be zero, which would tell you nothing. By first squaring all the radii, you make all the values positive, then the square root does give you information on how far R_G is from the center of gravity. For the chain molecules, R_G is

$$R_G = \sqrt{\frac{\langle r^2 \rangle}{6}}. \tag{3.29}$$

For molecules with compact shapes, however, the n segments will not behave as a chain, and will be related to R_G as

$$R_G \approx \sqrt[3]{n} \tag{3.30}$$

and for rod-shaped molecules the relation is

$$R_G \approx n. \tag{3.31}$$

When both R_S and R_G have been experimentally measured, the ratio of R_G / R_S provides information about the general structure of the molecule. A perfect sphere would have a value of $R_G/R_S = 1$, while higher values would indicate an increasingly elongated molecule. Surprisingly, it is possible for the ratio to have a value of less than one. This would indicate that the molecule is not only spherical, but more compact within the medium than predicted, due to internal interactions within the molecule.

Molecules diffuse in solution by Brownian motion. First described by botanist Robert Brown, Brownian motion occurs because of random collisions between molecules. In hydrophilic physiological systems, the predominant molecule, at almost 55 M, is water. Water molecules will constantly, and randomly, collide with other molecules in solution. Recalling Newtonian physics, the effect that a collision between a water molecule and another molecule will have depends on the other molecule's size: the larger the molecule, the less it will be affected by a collision with a water molecule.

The movements of molecules in a solution due to Brownian motion will, on average, have zero net movement. Since the random collisions can move the molecule in any direction, over a long time all directions will have collisions. To measure Brownian motion then, the root mean square distance is used, similar to the discussion above on the radius of gyration. During observations of objects moving by Brownian motion, the specific position of the object is noted at each time interval Δt, and the distance moved x_i from the previous position measured. After n such measurements, the mean square of the distance moved is

$$\langle x^2 \rangle = \frac{\Sigma x_i^2}{n} \tag{3.32}$$

and the mean path length of the movement is

$$\overline{x} = \sqrt{\langle x^2 \rangle}. \tag{3.33}$$

All the factors that influence diffusion will affect $\overline{x}$: size and shape of the molecule, the viscosity of the medium, and the temperature. When considering the Brownian motion of spherical molecules (the conditions used in the Stokes equation above), Einstein (1905) and Smoluchowski (1906) calculated the square of the displacement as

$$\langle x^2 \rangle = \frac{kT\Delta t}{3\pi\eta r} \tag{3.34}$$

and that the mean path length would be the square root of this:

$$\overline{x} = \sqrt{\frac{kT\Delta t}{3\pi\eta r}}. \tag{3.35}$$

This is a very useful equation, as it relates the time Δt it would take a molecule of radius r to diffuse distance $\overline{x}$. Consider a few examples. An average cell is about 20 μm across, while the axon of a neuron extending from the spinal cord to the toes is about 1 meter long. A glucose molecule has a radius of 0.5 nm, and a medium-sized protein has a radius of about 5 nm. How long would it take a glucose molecule to diffuse 20 μm, or 1 meter? How long would it take a protein?

To answer these questions, first rearrange the equation to solve for Δt:

$$\Delta t = \frac{\langle x^2 \rangle 3\pi \eta r}{kT}. \tag{3.36}$$

The viscosity of water is 0.00089 Ns/m^2. Inside cells, the viscosity is about twice that of water, so the value of η in the equation will be 0.00178 Ns/m^2. The value of k is 1.38×10^{-23} J/K. The value of T at body temperature is 310 K. The radii are converted to meters: 0.5×10^{-9} m for glucose and 5×10^{-9} m for the protein, as are the diffusion distances, 20×10^{-6} m and 1 m. The calculation for the 20 μm diffusion of glucose is

$$\Delta t = \frac{\left(20 \times 10^{-6} \text{m}\right)^2 (3\pi)(0.00178 \text{ Ns/m}^2)(0.5 \times 10^{-9}\text{m})}{(1.38 \times 10^{-23}\text{J/K})(310\,\text{K})} = 0.784\,\text{s}. \tag{3.37}$$

The protein radius is larger than the glucose radius by a factor of 10, so its diffusion time over 20 μm is 7.84 seconds. For a 1 meter diffusion, the glucose would take 1.96×10^9 seconds or 62.2 years, and the protein would take 1.96×10^{10} seconds or 622 years. The importance of these calculations depends on the physiological processes they are involved with. Consider protein synthesis, energy utilization in skeletal muscle, and the action potential. Protein synthesis takes tens of minutes, so within an average cell diffusion alone could supply both glucose and protein to support protein synthesis with plenty of time to spare. Energy utilization in skeletal muscle works on a timescale of seconds, so while glucose diffusion might be sufficient to support a large fraction of energy metabolism, requiring a diffusion time of 7.84 s for a protein would not be sufficient to control the energy production needed to support continuous contractions. Action potentials are over in 1–2 ms in neurons, and last for up to 200 ms in cardiac muscle cells. In no case could diffusion alone supply glucose or a protein to control an action potential. As for the diffusion of glucose and proteins down an axon, the decades that would be required preclude diffusion as a significant supply mechanism to the axon terminal, recalling the discussion of gravity and air resistance during a fall. There must be other systems involved in axonal transport that are much faster. These processes, using the energy of ATP and the molecular transporters kinesin and dynein, will be discussed in a later section as part of directed transport.

3.4   Physiological diffusion of ions and molecules

Diffusion plays a key role in multiple physiological processes. The movement of gases across the alveolar walls is driven by the partial pressure of the gases, partial pressure being the equivalent of concentration for a gas. The movement of oxygen into a cell will be accelerated when myoglobin, capable of binding oxygen, is present. The release of calcium from the sarcoplasmic reticulum of skeletal muscle also has interesting diffusional characteristics, as the release site and the calcium pump re-uptake sites are at different locations in the sarcoplasmic reticulum (SR).

The movement of transcription factors along the chromosome is faster than would be predicted by diffusion, but no direct transfer of ATP energy appears to be involved. This apparent contradiction is due to the structural confinement of the transcription factor, limiting three-dimensional diffusion, but allowing linear diffusion. Each of these processes will be discussed below.

The partial pressures of nitrogen, oxygen and carbon dioxide in the atmosphere are 600 mmHg, 160 mmHg and 0.3 mmHg, respectively. Oxygen and carbon dioxide are in constant exchange with the body through the lungs, while nitrogen is in equilibrium between the body and the atmosphere. Nitrogen does not react with any substance in the body, and thus nitrogen plays no role in normal physiological function. There is only one pathological circumstance in which dissolved nitrogen becomes a factor. In physical chemistry, perhaps the must useful equation is the ideal gas equation:

$$PV = nRT \qquad (3.38)$$

describing how the number of gas molecules n will have a particular pressure P and volume V. Useful in understanding why a cold basketball doesn't bounce well, it also plays a role, in its Boyle's law form, PV = constant, in describing how the lungs inflate and deflate. The nitrogen dissolved in the blood must also obey the ideal gas law. When a scuba diver descends to depth, the body is under greatly increased pressure, and the number of nitrogen molecules dissolved in the blood will increase. If the diver surfaces too quickly, the drop in pressure in the fixed volume of the body fluid means the number of dissolved nitrogen molecules will rapidly decrease and create bubbles in the extracellular fluid, similar to the way carbon dioxide bubbles come out of solution when a carbonated beverage is opened. Decompression sickness, the bends, occurs when nitrogen bubbles form in the body, most commonly causing pain in the large joints, and less commonly in neural tissues and in the lungs. Recompression drives the bubbles back into solution, and slow decompression thereafter allows the excess nitrogen to re-equilibrate with the atmosphere.

The movement of oxygen into the body is driven by its partial pressure gradient. The alveoli of the lungs, the sites of gas exchange, have an oxygen partial pressure of about 100 mmHg. Having a lower partial pressure than the atmosphere insures a constant flux of oxygen into the alveoli from the larger bronchioles. The blood in the capillaries coming to the alveoli has an oxygen partial pressure of 40 mmHg when it arrives. Thus, there is a significant oxygen gradient across the thin (1 μm) alveolar epithelium, the thin (1 μm) capillary endothelium, and the narrow (1 μm or less) interstitial space between the epithelium and the endothelium (Figure 3.6). It is this interstitial space, and the plasma between the endothelium and the RBCs, which constitute the major resistance to oxygen diffusion (Maina and West, 2005). Under normal conditions, the gases in the capillaries come very close to equilibrating with the alveolar gases as the blood passes. Oxygen in the arterial blood leaving the lungs via the pulmonary vein has a partial pressure of oxygen of about 95 mmHg (Ganong, 2005) compared with the 100 mmHg in the alveolar air space. Since most oxygen, 98.5%, in the blood is bound to hemoglobin and does not contribute to the partial pressure, this represents a very large flux of oxygen in the few seconds the blood is within the capillary.

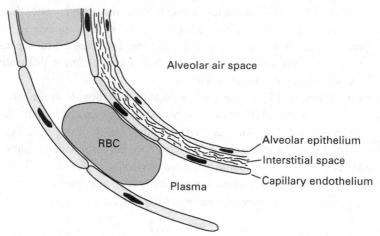

Figure 3.6   Alveolar–capillary oxygen exchange site. The alveolar and capillary cells have higher oxygen diffusion coefficients than the hydrophilic interstitial space, which contains the elastic and reticular fibers of the basal lamina. The red blood cells (RBCs) are wider than the capillaries, and are deformed and pressed against the capillary endothelium, enhancing oxygen transfer across the closely opposed membranes. Pathological increases in fluid at the alveoli, in the interstitial space or lining the inside of the alveolar air space would decrease oxygen transport into the capillary, increasing the alveolar–arterial ($A - a$) oxygen difference.

The resistance to oxygen diffusion by the hydrophilic fluid compartments in series with the cellular components is due to the partition coefficient or constant of oxygen. The partition constant is the ratio of concentrations of a particular molecule between a hydrophobic domain and a hydrophilic domain. In measurements of the oxygen ratio between the lipid bilayer oxygen and water oxygen, the ratio at 37 °C was 3, with oxygen far more soluble in the bilayer than in water. The low solubility of oxygen in water domains is most responsible for the resistance to oxygen diffusion. In healthy people, despite the barrier presented by the fluid layers that oxygen must diffuse through, the difference ($A - a$) between the alveolar oxygen partial pressure (A) and the arterial oxygen partial pressure (a) is only about 5 mmHg. In a wide range of pathological conditions, congestive heart failure, sarcoidosis, and some forms of pneumonia, the value of $A - a$ is greater than 30 mmHg. This increased difference arises primarily from decreased diffusion between the alveoli and the capillaries (Cunha *et al.*, 2007), as fluid buildup in the interstitial space decreases the diffusion constant D of Fick's equation for oxygen flux. In severe cases, the diffusion of oxygen is so compromised that death results.

In a normal tidal volume of 500 ml, the anatomic dead space of the mouth, pharynx, trachea, bronchi and bronchioles constitutes about 150 ml, leaving 350 ml of alveolar air space used during each breath. As noted above, this is sufficient to supply the body's oxygen needs. Only a fraction of the lung alveoli are used during normal breathing, and the vital capacity, the total available volume of the respiratory system, can increase up to 4500 ml during exercise. During very strenuous exercise, the flux of oxygen into the blood can increase by more than tenfold. The body has substantial reserves of lung tissue available to increase the area A component of Fick's equation when the need arises.

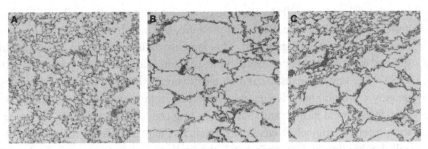

Figure 3.7 Images of (A) healthy lung alveoli, (B) homogeneous distribution of airspace in emphysema and (C) heterogeneous distribution of airspace in emphysema. (Reproduced from Parameswaran *et al.*, 2006, with permission.)

The lungs have a large number of resident white blood cells (WBCs). Their presence is designed to attack pathogens that may be breathed into the lungs. Among the weapons WBCs have are a range of proteolytic enzymes, such as elastase, that can attack pathogens. Healthy lung tissue release a substance, α-antitrypsin, that blocks the activity of proteolytic enzymes released by immune tissue cells. The tar components of tobacco smoke and coal dust initiate a cascade that decreases the activity of α-antitrypsin, allowing the proteolytic enzymes to attack healthy lung tissue, leading to emphysema and "black lung" disease. Because of the large reserve of pulmonary tissue, it takes decades in many cases before the destruction of alveolar tissue, resulting in fewer, larger alveoli and a great reduction in surface area, leads to decreased oxygen delivery under normal breathing circumstances (Figure 3.7).

In terms of Fick's equation, the flux of oxygen is compromised due to a decrease in area A. This decrease does not of course decrease the body's need for oxygen, and to counter the decreased area there must be compensation by a different factor of the equation. Oxygen is about 20% of the atmospheric air. Breathing pure oxygen will increase the flux of oxygen by increasing dC/dx. This increased gradient will help oxygen delivery, but the progressive loss of lung tissue inevitably leads to death. For healthy individuals, in whom the normal flux of oxygen virtually fills their hemoglobin, breathing pure oxygen will have no significant effect on oxygen delivery. While athletes breathing pure oxygen will not find any increased oxygen delivery effect, the psychological benefit they derive from pure oxygen breathing may produce improved performance.

The diffusion of oxygen into cells is driven by the oxygen gradient across the cell membrane. The external partial pressure or concentration of oxygen in the interstitial fluid is in equilibrium with the oxygen dissolved in the blood. The partial pressure of oxygen in the arterial blood of an individual is constant, even under most pathological conditions in which the partial pressure is lower than in healthy individuals. Variations in the flux of oxygen in an individual are therefore driven by intracellular oxygen concentration changes. In tissues with high metabolic activity, such as actively contracting skeletal muscle, the steady-state decrease in the intracellular oxygen concentration will increase the oxygen gradient and increase flux into those cells that need it most. Further, it has been shown that the presence of myoglobin inside cells increases the flux of oxygen into the cells (Meyer, 2004). Binding of oxygen to myoglobin in the cell reduces the free concentration of oxygen, its partial pressure, and thereby increases the flux of additional oxygen into the cell (Figure 3.8). This increase in flux induced by molecular binding is a general property, also applicable to the effect of calcium binding on calcium flux (Feher, 1983). The key elements for assisting diffusion by binding are that the binding element, usually a protein, must have a concentration on the same order of magnitude or greater than the element in flux, and that the binding constant must be near the concentration of the protein, a subject discussed in detail in Chapter 2. While it is apparent that intracellular binding will contribute to the storage and buffering of the bound element, whether oxygen, calcium or something else, the influence on diffusion can also be of considerable physiological significance.

The physical parameters of muscle contraction compose a major section of Chapter 5. There is, however, one area in which the diffusion of calcium ion plays an important role, and so is included here. The T-tubules in mammalian skeletal muscle invaginate from the cell surface at the A-band/I-band junction, corresponding to the ends of the thick filaments (Figure 3.9).

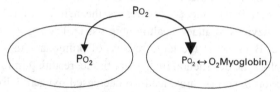

Figure 3.8   Influence of myoglobin on oxygen diffusion. By binding to myoglobin, the intracellular P_{O_2} decreases, increasing the oxygen gradient across the cell membrane and the rate of oxygen flux, making the flux on the right side greater (thicker arrow) than on the left.

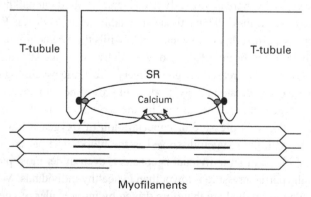

Figure 3.9   Calcium release and diffusion in skeletal muscle. Action potentials traveling down the T-tubules activate dihydropyridine receptor calcium channels (black ovals), causing calcium entry that activates ryanodine receptor calcium channels in the sarcoplasmic reticulum (SR; gray ovals). Calcium leaving the SR activates the myofilaments and diffuses to the Ca-ATPase pumps in the middle of the SR (hatched oval).

Dihydropyridine receptor calcium channels in the T-tubules are linked to ryanodine receptor calcium channels in the sarcoplasmic reticulum (SR). Calcium entering from the T-tubules triggers calcium release from the SR. Calcium then binds to troponin on the thin filaments to initiate contraction. Relaxation occurs when the action potential has passed and calcium ATPase pumps return the calcium to the cisternae of the SR. The calcium pumps are in the middle of the SR, as far from the T-tubules and the calcium release sites as possible. Since the thick filaments are 1.5 μm long, the maximum distance from the release site to the calcium pumps would be 0.75 μm. Using the calculation for diffusion time from the previous section, calcium with an ionic radius of 0.1×10^{-9} m would take 0.22 ms to diffuse that distance. This would be the fastest diffusion, as hydration of the calcium ion will increase its effective radius and decrease its diffusion rate. A muscle action potential passes in 2–3 ms, so that soon after the calcium is released from one part of the SR, it will diffuse to the far part of the SR and begin the process of resequestration. As the calcium diffuses from its source to its sink in the SR, it must pass through the region of overlap of the thick and thin filaments, and thus be available to bind to troponin and initiate contraction. The diffusion process is sufficiently fast, however, that for a single action potential producing a twitch there is not enough time for all the troponin sites to bind calcium and allow all the overlapping myosin heads to contact actin and generate force. A twitch lasts for about 20 ms, until the calcium has returned to the SR by diffusion to the pump sites and Ca ATPase activity translocates the calcium across the SR membrane. When a train of action potentials is sufficiently close in time, the calcium release process overwhelms the re-uptake process, and maximum force, tetanus, is produced. The calcium concentration in the SR is in the millimolar range, while in the cytosol its highest concentration is about 10 μM. There is never equilibrium between the calcium in the SR and in the cytosol.

A key process in many physiological systems is the requirement that two molecules come into contact. Two molecules free in the cytoplasm will both diffuse through this environment. What is the probability that two molecules will come into contact? Estimation of the collision frequency of two molecules was made by Smoluchowski (1917). He derived the calculation for the second order rate constant k_D

$$k_D = 4\pi N_o (D_1 + D_2)(r_1 + r_2) \tag{3.39}$$

where N_o is Avogadro's number, D_1 and D_2 are the diffusion coefficients for the two molecules, and r_1 and r_2 are the radii of the two molecules. The interaction of the two molecules will be

$$M_1 + M_2 \Rightarrow M_1 \cdot M_2 \tag{3.40}$$

and the rate of $M_1 \cdot M_2$ formation is

$$\frac{dM_1 \cdot M_2}{dt} = k_D [M_1][M_2] \tag{3.41}$$

where the concentrations of the molecules are within the brackets.

One of the interesting developments in molecular contact involves the exchange of substrates between enzymes. Research continues to find more and more protein clusters

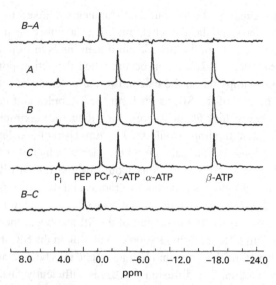

Figure 3.10 Saturation transfer of phosphate between pyruvate kinase and creatine kinase. Saturation of the phosphate of PCr is transferred forming PEP and saturation of PEP is transferred forming PCr via direct ATP exchange between the enzymes without dissociation of ATP into the solution. (Reproduced from Dillon and Clark, 1990, with permission from Elsevier.)

in which substrate channeling occurs; that is, the product of one enzyme reaction is passed directly to the next enzyme in the metabolic pathway without diffusing into the cytoplasm. An example of this type of exchange is shown in Figure 3.10 (Dillon and Clark, 1990). This figure shows NMR saturation transfer for the exchange of phosphate between creatine (Cr) and pyruvate (Pyr) by the enzymes creatine kinase (CK) and pyruvate kinase (PK). As we saw in Chapter 1, in saturation transfer the magnetization on the phosphate of one molecule is electronically saturated or canceled out, and as that phosphate is enzymatically transferred to another molecule, the saturation follows and the phosphate peak in the second molecule is decreased proportionally to the kinetic rate of the enzyme. For CK, saturation of the phosphate of PCr should decrease the size of the γ-ATP phosphate as the phosphate is added to ADP:

$$P^*Cr + ADP \xrightarrow{CK} Cr + ATP^* \tag{3.42}$$

where the asterisk indicates the saturation exchange. For PK, saturation of the phosphoenolpyruvate (PEP) phosphate should also decrease the size of the γ-ATP phosphate as phosphate is added to ADP:

$$P^*EP + ADP \xrightarrow{PK} Pyr + ATP^*. \tag{3.43}$$

When both enzymes and substrates are present, as in Figure 3.10, the control spectrum B shows all of the unsaturated peaks of the substrates. Spectrum A shows saturation of the PCr peak, which is electronically canceled. In the difference spectrum $B - A$, where the peak height shows the amount of peak decrease, the PCr peak is decreased as expected by the saturation, but the γ-ATP peak has not changed, and the PEP peak has decreased. Similarly, when the PEP peak is saturated in spectrum C, the PEP peak is canceled. In the difference spectrum $B - C$, the PEP is decreased, but again the γ-ATP has not decreased, and the PCr peak has decreased. These results occur because the ATP produced by either enzyme is transferred directly to the other enzyme without being released into solution, as there is no direct phosphate exchange between creatine and pyruvate without an ATP intermediate. Enzymes with this type of exchange were termed diazymes (Dillon and Clark, 1990).

The exchange of substrate by PK and CK in vivo can only occur if the enzymes contact one another, a process that can be estimated using the Smoluchowski equation to calculate k_D and the measured cellular concentrations of PK and CK. For a calculated k_D of 5.4×10^9/M·s and skeletal muscle PK and CK concentrations of 129 μM and 512 μM, respectively, the PK–CK collision rates for ATP-bound enzymes are about 50 times more frequent than the fastest rates of ATP dissociation for either enzyme, and the absolute collision rate for all forms of the enzymes is almost 900 times higher than the highest rates of ATP dissociation for both enzymes combined (Dillon and Clark, 1990). These calculations assume that both enzyme concentrations represent the free concentration: enzyme binding, such as CK binding to thick filaments, will of course decrease the diffusion of the enzymes. Still, as these enzymes are not a covalent part of any polymerized structure, the calculations indicate that the exchange of ATP between PK and CK is highly possible, making this exchange important in the buffering of ATP in skeletal muscle cells.

Binding of a diffusing molecule to sites other than a specific site of action also plays a role in genetics. Transcription factors must find a specific site of the approximately 10^7 available sites on double-stranded DNA (von Hippel, 2007). As expected, the binding of the transcription factor to its specific site will have a much higher association constant than to non-specific sites, usually by about a factor of 10^8, effectively countering the greater number of non-specific sites. If a mutation occurs lowering the affinity of the transcription factor for its specific site, the non-specific sites will then effectively compete en masse with the mutated site, lowering its activation. This could have interesting evolutionary effects. In order to activate protein production in the mutated gene, there would have to be a higher transcription factor concentration to insure its binding to the promotor site. This would increase the possibility of the transcription binding to and altering the activation, up or down, of some other gene previously unrelated to that transcription factor. Even if this is very, very rare, over millions of years the possibility of altering survival through multiple gene activation by one transcription factor becomes real.

In addition to classical diffusion and intersegment exchange of transcription factors, sliding of the factor along the double-stranded DNA has additional diffusional characteristics. Transcription factors will slide along DNA at a rate faster than predicted by three-dimensional diffusion, yet without the input of external energy from ATP. The explanation for this is that the diffusion of the factor is not equal in all directions; that is, D in Fick's equation is not the same everywhere. The factor has a preferred diffusion along the DNA, essentially one-dimensional diffusion, with decreased diffusion laterally, away from the DNA (von Hippel, 2007). This tunneling effect, regardless of whether there is an actual physical tunnel alongside the DNA, reduces the diffusional volume covered by the transcription factor and accounts for its accelerated diffusion.

The collision of a transcription factor with its promotor site, in its most complete form, would have to account for the different diffusion coefficients of the transcription factor in the different physical locations in the cell, yielding a composite diffusion coefficient. The Smoluchowski equation for k_D of the transcription factor contacting the chromosome would be reduced to

$$k_D = 4\pi N_o D_{tf} r_c \qquad (3.44)$$

since the diffusion coefficient of the chromosome would approach zero, and the radius of the transcription factor would be insignificant compared with the radius of the chromosome.

Many transcription factors, such as steroid hormones and thyroid hormone, are very hydrophobic, and will be bound to protein. Their diffusion through a watery environment will be very low. As we will see in the next section, directed transport using molecular motors can carry molecules between the periphery of a cell and the perinuclear area. Despite the phenomenal growth in our knowledge of genetics, there is not a list of the dissociation constants of transcription factors for particular promotor sites, nor a comprehensive measurement of transcription factor concentrations. There is also not a comprehensive list of individual transcription factor transport mechanisms, either by diffusion or directed transport, or, as is surely the case, a combination of the two for some factors. The dearth of biophysical knowledge of these areas awaits the interest of scientists. There is one thing that can be predicted with some confidence, however. When both dissociation constants (derived from their time constant equivalents) and their transcription factor concentrations are known, they will have similar values. If the K_D values were much higher than the factor concentrations, there would be minimal binding and the gene would never get turned on. If the K_D values were much lower than the factor concentrations, the factors would never come off the promotor site, and the gene would be continuously activated. For a well-controlled genome, where the gene is activated only when necessary, the K_D values and the transcription factor concentrations must be similar.

3.5   Molecular motors

Directed transport inside cells involves the conversion of the energy in ATP to drive molecules along filamentous highways. Kinesin and dynein both move along microtubules, kinesin in the positive direction away from the microtubule organizing center (MTOC) near the nucleus, and dynein in the negative direction toward the MTOC. Both kinesin and dynein will reversibly bind and transport cargo, from proteins to entire organelles. Some isoforms of myosin are not arranged in polymers such as the thick filaments of muscle, but exist as free molecules transporting cargo along actin filaments. The common characteristic of all these transporters, kinesin, dynein and free myosin, is their two-headed structure, with movement occurring through alternate

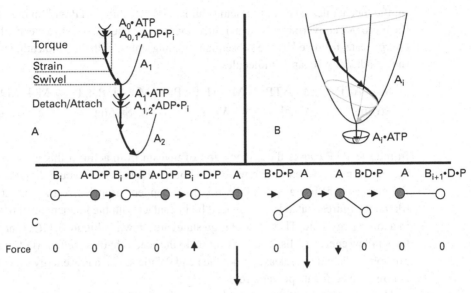

Figure 3.11 General principles of dynein stepping. (A) Energetic sequence of two steps by the same head (Numata *et al.*, 2008). The energy of the B head is not shown, but would be the same sequence as A, offset with the Detach/Attach state of the A head occurring as the B head swivels. ATP binding to A_i produces a low force binding state. Conversion of ATP to bound ADP•P_i retains the low force binding. Loss of products P_i and ADP produces a high force state of A_i, with initial torque on the binding of A_i to the filament. The torque is transformed into strain as the head rotates pulling on the other head B_i, with B in the B•ADP•P_i state initially resisting. As A rotates, B•ADP•P_i reaches the limit of its binding site, and with a lower binding energy than A, B•ADP•P_i detaches and follows the swivel of A. A reaches its lowest energy position, and binds ATP, producing the low force state with the B head now reattaching to the filament at position B_{i+1}. A•ATP then converts to A•ADP•P_i which will sequentially detach, swivel and re-attach (Detach/Attach) as the B head goes through its power stroke. (B) The power stroke must go through a swivel stage in order to advance the step. The energy path must go through progressively lower energy points until it reaches an energy minimum in a rotated position relative to the original detachment angle. The following step of A_i to A_i•ATP does not require rotation. (C) Sequential rotation of the A head relative to the B head. After B has rotated to the lowest energy position, it will have no vectorial force, nor will the weakly bound A•ADP•P_i (A•D•P) have any force component. B will bind ATP and convert it to B•ADP•P_i (B•D•P), but will still have no net force, nor will the A head in the A•ADP•P_i (A•D•P) state. The loss of ADP and P_i resulting in the A state produces a net rotational force on the A head, resisted by the weakly bound B•ADP•P_i state. The greater the difference in the forces A and B•ADP•P_i the faster the B-filament bond will be broken. As A rotates, its force will progressively fall as it approaches its lowest energy configuration. Unattached B•ADP•P_i will have no net force. When A reaches its lowest energy position, the rotation will stop, and B•ADP•Pi will then re-attach at B_{i+1}, and the cycle will repeat as the B head goes through its power stroke.

binding and dissociation of the heads, essentially "stepping" along the filament. In this section, the general principles of these molecular motors will be addressed using dynein as the model.

The different aspects of molecular motors are presented in Figure 3.11. In many cases, the molecular details mirror those of the well-studied interactions of myosin with actin in muscle. The driving energy of directed motion is ATP. In the model presented here based on the biochemical states of dynein (Numata *et al.*, 2008), ATP will bind to the head of the motor M, converting it from a tightly bound, high energy state to a weakly bound, low energy state, followed by the subsequent binding states, with the · indicating binding and the + indicating separate molecules:

$$M + ATP \rightarrow M \cdot ATP \rightarrow M \cdot ADP \cdot P_i \rightarrow M \cdot ADP + P_i \rightarrow M + ADP + P_i.$$

Strong   Weak   Weak   Strong   Strong

(3.45)

In muscle, ATP causes the dissociation of myosin from actin. If this were to occur in a molecular motor when only one head is bound, the molecule would dissociate from the filament and diffuse away. While this may happen, progressive movement along the filament requires that at least one head be in contact with the filament at all times, even if in a low energy state. These low energy states are shown in Figure 3.11(A) and 3.11(B) as the shallow energy wells. The ATP bound to the head is hydrolyzed to ADP and inorganic phosphate, P_i, but not released from the head. While still in a low energy state, the system is now primed for its power stroke.

The power stroke of the molecular motor involves dissociation of the products, ADP and P_i, leaving the head in a high energy state. The dissociation produces an internal torque on the head (Figure 3.11A). Note that the energy change is vertical, without any horizontal component indicating physical movement. Since the head is no longer at its lowest free energy position, there will be a force on the head. The force is the first derivative of the energy: that is, the slope of the energy curve. At all points except the lowest energy position, there will be a force on the molecule. The head must rotate. Without rotation, the motor would just "run in place," consuming ATP but not going anywhere. Always moving toward its lowest energy position, the head will start to rotate. The movement, or strain, will decrease the force on the high-energy head, but also strain the other head of the molecule, which is in the low energy state. The dashed vertical line in Figure 3.11(A) shows that as the force in the high-energy head (A) falls, the movement would cause an energy increase in a bound, low-energy ADP•P_i head (B). Although the A head is shown in the figure, the attached B head in the ADP•P_i state would have a similar energy profile. The countering forces on the A and B heads are shown in Figure 3.11(C).

The movement of the high energy A head will also move the low energy B head. When pulled sufficiently far, the B head will detach, as long as the energy in the A head is greater than the energy in the B head at that position. When free of binding to the filament, the head will have no force on it at all, rapidly assuming its lowest energy position. The high energy A head, no longer restricted by the B head, will now assume its lowest energy position. The position must have rotated the head through some angle. If the angle is 180°, then the molecular motor will proceed directly along the filament. If the angle is less than this, the forward motion will be the cosine of the angle multiplied by the step size. The rotational motion is shown in Figure 3.11(B). The angled planes within the three-dimensional energy profile indicate that at the head–filament junction, there will be a preferred path to the lowest energy position, the path of least resistance.

This path will result in the rotation of the bound head, with the force on the head continually decreasing as it rotates (Figure 3.11B,C) As the head reaches its lowest energy position, there will no longer be any internal force on the head, and the rotation will cease. The binding of another molecule of ATP will convert the head into the low energy, weakly bound ADP•P$_i$ state, and the process will be repeated with the other head rotating.

For kinesin, some of the details are different, but the principles will be the same. The trailing, low-energy head has P$_i$ dissociated prior to detachment and rotation to the front position in the B•ADP state. The release of ADP converts the head into the high force state (Block, 2007). The key to force generation on the myosin head is the dissociation of P$_i$ (Cooke and Pate, 1985), but the dissociation of both products is so close in time that for practical purposes they can be considered to dissociate together. There are other proposed mechanisms for molecular motor movement, and further research may show other changes in the details of the process, but the general principles of maintaining at least one head attached, having the high energy head break the attachment of the low energy head, and rotation of the high energy head, must be present in any system.

Not all motors will move with the same speed. The factors controlling the speed of movement are the size and direction of each step, the ATPase rate of the motor, and the load on the motor. The size and direction of the step are determined by the structure of each motor. The legs connecting the heads to the common cargo area have different lengths, and therefore each motor has different steps. The closer each rotation is to 180°, the straighter the complex of motor and cargo will move along the filament. Every motor will have a maximum ATPase rate, but this maximum has only theoretical value. If the motor is moving at its maximum rate, it would have to have the lightest possible load: that is, it is carrying no cargo. The addition of cargo will decrease the rate of movement, just as muscles move more slowly the greater the load they carry. The other load on the motor is internal. At the time of the power stroke, the rotating head will be resisted by the bound, low energy head. The force required to break the low energy bond may be different for each type of motor, even each isoform of each motor. The greater this internal load, then statistically the high energy head will take longer to break the low energy bond and the net effect will be slower movement of the complex. Regardless of these restrictions, the movements induced by molecular motors produce movements far faster than those driven by diffusion.

Our understanding of molecular motors has been greatly advanced by the use of optical tweezers, also called laser traps, to study the behavior of individual motor molecules. These systems use an electric field generated by a laser beam to attract and hold in place an electric dipole. This dipole, usually a bead, is attracted to the narrowest point, or waist, of the focused laser beam (Figure 3.12) where the electric field is highest. This method is used in many fields, using a range of lasers, but those used to study biological materials employ lasers that minimally excite water molecules, to minimize heating of the sample. The bead will be attached to one of the elements of the motor system, either the filament as in Figure 3.12 (Rüegg *et al.*, 2002), or to the motor molecule, such as kinesin (Visscher *et al.*, 1999). The force on the bead is a function of

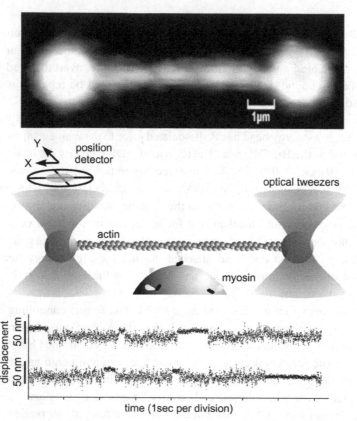

Figure 3.12   Laser trap measurement of actin–myosin interaction. The upper image is of a fluorescently labeled actin filament connecting two fluorescently labeled beads held by laser beams. This arrangement is shown diagramatically in the middle part of the figure, along with bead mounted myosin. Power strokes by the myosin head will place strain on the actin filament and the laser-trapped beads, shown in the lower traces as the low noise section, in contrast to the higher noise region in which the laser-trapped beads are subject to Brownian motion. (Reproduced from Rüegg *et al.*, 2002, with permission.)

the electric field, which will vary with the bead's distance from the waist of the focused beam. The force on the bead is

$$F = \frac{1}{2}\alpha\nabla E^2 \qquad (3.46)$$

where E is the electric field, α is proportionality constant particular to each physical system, and ∇ is the nabla operator, also called the del operator. This operator can be viewed as the derivative of a function in multidimensional space, such as the x, y, and z dimensions in Euclidian space:

$$\nabla = i\frac{\partial}{\partial x} + j\frac{\partial}{\partial y} + k\frac{\partial}{\partial z}. \qquad (3.47)$$

In one dimension, the ∇ reduces to the derivative of calculus. For a system in which the function, in this case the square of the electric field, is changing value differently in multiple directions, the ∇ is the appropriate descriptor of the gradient, such as the gradient of the square of the electric field (grad E^2):

$$\text{grad } E^2 = \frac{\partial E^2}{\partial x}\mathbf{i} + \frac{\partial E^2}{\partial y}\mathbf{j} + \frac{\partial E^2}{\partial z}k = \nabla E^2. \tag{3.48}$$

This is just a more complex case of the force being the slope of the energy function that was shown in Figure 3.11 for the dynein motor. By holding the bead at a set distance from the laser beam using a computer-controlled system, the electric field will be constant and the force on the bead will be constant. This force is then the force the molecular motor will be working against.

In Figure 3.12, two separate laser beams holding fluorescent beads are connected by a fluorescently labeled actin filament. Myosin motors move along actin filaments. When single myosin molecules are brought into contact with the actin filament, the myosin can go through an ATP-dependent power stroke. The lower part of the figure shows the time course of movements of the actin filament, detected by position changes in the direction parallel to the actin filament direction. The time course shows two phases, one with high noise and one with low noise. The high noise regions are due to Brownian motion of the beads within the laser traps, moving due to random thermal motion. The low noise regions are due to the myosin power stroke displacing the actin filament, reducing the range of movement. Note that the myosin-generated displacements are not uniform in duration: the dissociation of myosin heads from actin is a stochastic process having a probability of occurring over any given time.

The value of this type of system is shown in Figure 3.13. Myosin has many different isoforms, each using the energy of ATP to produce a power stroke by the release of the products ADP and P_i. In Figure 3.13 the behavior of three different myosins using the laser trap system in Figure 3.12 is shown. The upper two experiments show two Class 1 myosins, rat liver Myr-1a (upper) and brush border myosin I (BBM1; middle). By superimposing the displacements from multiple runs, the myosins show a biphasic behavior, indicative of two separate biochemical steps. From a knowledge of the biochemistry, the first phase is thought to be due to the release of P_i, and the second movement phase due to the release of ADP. This is in sharp contrast to the single phase displacement generated by skeletal myosin (lower part of Figure 3.13), which is associated with the dissociation of phosphate. Without these single molecule techniques, the details of these molecular motors could never be determined.

In other laser trap experiments, a bead attached to kinesin was used to determine the step size, load dependence and ATP dependence of this motor (Visscher, 1999). Microtubules were mounted on a surface, and the kinesin moved across the microfilament. The bead attached to the kinesin moved along with it, and was kept at a constant distance from a laser beam that moved along with the kinesin. This motility assay then measured the speed of kinesin movement along the microtubule. The bead was held at different distances from the laser trap, varying the force on the bead and therefore the load on the kinesin. These experiments found that the step size of kinesin was 8 nm, a distance consistent with the size of the kinesin molecule. The speed of

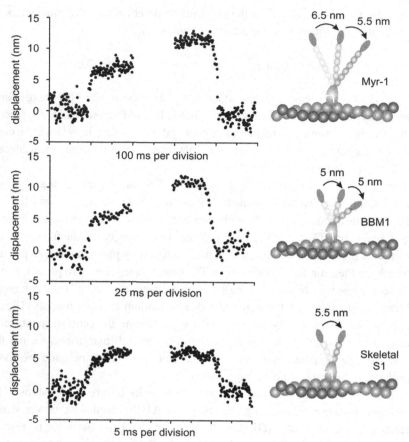

Figure 3.13   Measurement of actin filament displacement by different isoforms of myosin using a laser trap. The upper traces are from two myosin 1 isoforms, rat liver Myr-1a and brush border BBMI, while the lower traces are from S1 heads of skeletal muscle myosin. The upper traces show two steps, while the skeletal myosin shows a single step. (Reproduced from Rüegg *et al.*, 2002, with permission)

movement was inversely related to the force on the bead: the greater the load the slower the movement, just as in muscle. The velocity increased with increasing ATP concentrations, plateauing above 300 μM for all the loads measured. Since the ATP concentration in all cells is higher than 300 μM, the movement of kinesin inside cells will be highly load dependent but minimally ATP dependent. Further experiments like these will continue to expand our knowledge of the kinetic and thermodynamic details of molecular motors.

Single molecule techniques have also been applied to molecular motors inside cells. Fluorescently labeled kinesin heavy chain, the part of the molecule with the ATPase activity, was studied inside the cytoplasm of COS cells and under control in vitro conditions outside cells (Cai *et al.*, 2007). Figure 3.14 demonstrates the behavior of kinesin under both circumstances. The fluorescent label allowed detection of the

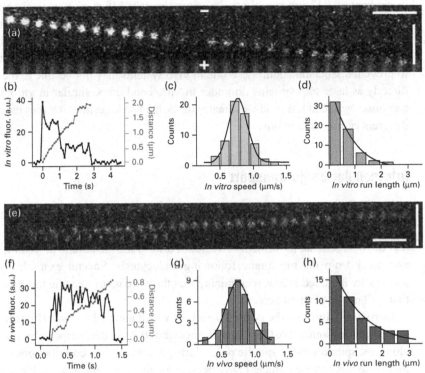

Figure 3.14 Motile properties of single kinesin heavy chains molecules in vitro and in vivo. (a–d) Motile properties of kinesin in vitro. (a) Taxol-stabilized microtubules, fluorescently labeled at their plus ends, were incubated with fluorescently labeled kinesin. The superimposed kymograph shows one representative fluorescence spot moving progressively along a microtubule. (b) Graph of displacement and fluorescence intensity over time for the same fluorescence spot shown in panel (a). (c) Gaussian fitting of the kinesin speed histogram shows the motors move at 0.77 ± 0.14 μm/s ($N = 54$). (d) Single exponential decay fitting of the run length histogram shows the same motors move processively for 0.83 ± 0.29 μm/run. (e–h) Motile properties of kinesin in vivo. (e) Kymographs show kinesin movement. (f) Graph of displacement and fluorescence intensity over time of the same fluorescent spot shown in panel (e). (g) Gaussian fitting of the speed histogram shows that kinesin motors move at 0.78 ± 0.11 μm/s ($N = 54$). (h) Single exponential decay fitting of the run length histogram shows the same motors move processively for 1.17 ± 0.38 μm/run. (Reproduced from Cai et al., 2007, with permission from Elsevier.)

movement of the kinesin along a fluorescently labeled microtubule in the + direction. Under in vitro conditions shown in the upper part of the figure, the kinesin motor moved at an average speed of 0.77 μm/s, and stayed on the filament for an average of 0.83 μm/run before dissociating from the microtubule. Under in vivo conditions, the kinesin motor had an average speed of 0.78 μm/s and stayed on the microtubule for an average of 1.17 μm/run.

The data indicates that intracellular movements of kinesin are neither increased nor hindered by molecular interactions. The measurements made inside cells closely reflect kinesin's behavior under in vitro conditions. With a step size of 8 nm, an intracellular movement of 1.17 μm would require approximately 146 steps before dissociating, and the average speed 0.78 μm/s run would take 1.5 s. This indicates that the energy wells holding kinesin under its lowest force conditions are sufficient to resist the dissociating effects of random thermal energy for relatively long periods of time, but are not so deep as to prevent dissociation entirely. While in vivo systems may not be able to measure force directly as laser trap systems do under in vitro conditions, similar in vitro and in vivo responses would at least allow estimates of the load on kinesin as it varies its speed under different cellular conditions.

3.6   Intracellular cargo transport

The molecular motors are involved in myriad intracellular movements, including the separation of chromosomes during mitosis and meiosis, the positioning of organelles, and the bidirectional movement of vesicles, toward the membrane preceding exocytosis and away from the membrane following endocytosis. Several examples of different aspects are included below, recognizing that there are many specific transport processes that will not be covered here.

During cell division, the general details of which are well known (Brunet and Vernos, 2001), chromosomes must line up at the center of the cell during metaphase and progress along the spindles to the spindle poles during anaphase. The contributions of molecular motors to this complex process have been ascertained in some aspects, but are the object of continuing research in others. Microtubules connect the spindle poles to the kineto-chores of the chromosomes, with the negative end of the microtubule at the pole and the positive end of the microtubule at the chromosome end. Mitosis first has the paired chromosomes lining up together at the positive end of the microtubules. The kineto-chores are structures at the centrioles of the chromosomes. Kinetochores are complexes of dozens of proteins, some of which are molecular motors, as well as DNA-containing regions that contribute to the binding of the kinetochore to the chromosome. Kinetochores are bidirectional, their reversal of polarity being essential to sending each paired chromosome to the opposite spindle poles.

Kinesin moves in the + direction along microtubules, and there is evidence that kinesin-like proteins propel the chromosomes in the + direction, leading to the alignment of the chromosomes at metaphase (Brunet and Vernos, 2001). The contribution of microtubular growth and degradation to chromosome alignment, called chromosome congression, may also play a role. Spindle microtubules are highly unstable, and will go through alternating phases of assembly and disassembly after the chromosomes have gathered during metaphase (Stumpff *et al.*, 2007). The spindle pole distance is maintained at a constant distance due to the bridging of opposite polarity microtubules by kinesin and dynein, exerting balanced, opposing forces on the microtubules (Skowronek *et al.*, 2007). At anaphase, tension in the spindle network leads to the activation of dynein, which contributes to the separation of the bidirectional kinetochores and actively moves the chromosomes toward the spindle poles (Brunet and Vernos, 2001). Kinesin motors play a key role in regulating microtubule length by acting as a microtubule depolymerase (Mayr *et al.*, 2007). Kinesin depolymerization of microtubules occurs at a faster rate in long microtubules than in short microtubules. The longer microtubules are able to capture and translocate more kinesin molecules to their positive end where the depolymerization takes place. This leads to a balance of assembly and disassembly of kinetochore-bound microtubules at a particular microtubule length, a process that maintains all the chromosome-attached microtubules in a spindle at similar lengths in a steady state. Thus, the movement of chromosomes to the metaphase state appears to have elements of both kinesin activity and microtubule dynamics. As chromosomes move toward the spindle during anaphase, kinesin depolymerizes trailing microtubules, ensuring unidirectional movement of the chromosomes.

Because of the polarization of the microtubule, kinesin will carry cargo away from the MTOC near the nucleus and toward the + end of the microtubule: that is, toward the cell membrane. Among the cargo that kinesin carries are exocytotic vesicles produced by the Golgi apparatus. Vesicles containing neurotransmitters can be carried near the membrane along microtubules, then transferred to free myosin and carried along actin filaments for merger with the cell membrane (Bi *et al.*, 1997; Huang *et al.*, 1999). Dynein will carry endocytotic vesicles away from the cell membrane toward the nucleus (Suikkanen *et al.*, 2002). While in general the scheme in Figure 3.15 is thought to occur in most cells, with kinesin and dynein moving cargo radially in the positive and negative directions respectively and myosin carrying it laterally along the membrane, there are exceptions, such as zymogen granules carried to the cell membrane by dynein (Kraemer *et al.*, 1999).

The calculations for the extended time that molecules would take to traverse axonal distances were known long before the discovery of kinesin and dynein, so their discovery was not a surprise. Nor was it surprising that two separate molecules were involved. Given the polarity of the microtubule, a molecular motor would be expected to move in only one direction, a prerequisite for net movement along the microtubule. If there were

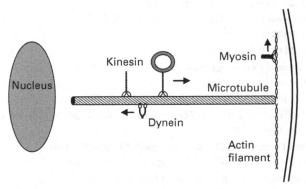

Figure 3.15   Intracellular transport. In general, kinesin will carry cargo, such as the vesicle attached to the right-hand kinesin, toward the cell membrane. Free myosin can receive cargo from kinesin and carry it along actin filaments, which can run parallel to the cell membrane. Dynein carries cargo away from the cell membrane toward the nucleus. Both kinesin and dynein can carry each other as cargo.

only one molecular motor, all the motors would soon build up at one end of the micro-tubule and be unable to return, a very inefficient process. A second molecular motor, moving in the opposite direction, could carry the first motor back to the other end of the microtubule, making the first motor available to carry more cargo. Kinesin and dynein can each carry the other, ensuring the availability of molecular motors to move cargo around the cell. This is currently an area of intense research, and additional details of these systems should become known in subsequent years.

Movement within cells requires one of two energy sources, either the concentration gradients driving diffusion or ATP driving directed transport. The physical conditions of diffusing substances and the environment they diffuse through determines the rate of diffusion, the laws of which apply to ions, molecules and thermal gradients. The time course of different biophysical processes determines the degree to which diffusion can contribute to those processes. Systems that require long distances cannot be supported by diffusion alone, and use molecular motors to carry cargo along polymerized filaments. The use of single molecule techniques continues to advance our knowledge of intracellular-directed transport. There is currently limited knowledge of the biophysics of genetic transcription factors and promotors, an area that awaits experimental exploitation.

References

Bi G Q, Morris R L, Liao G, et al. J Cell Biol. **138**:999–1008, 1997.
Block S M. Biophys J. **92**: 2986–95, 2007.
Brunet S and Vernos I. EMBO Rep. **2**:669–73. 2001.
Cai D, Verhey K J and Meyhöfer E. Biophys J. **92**:4137–44, 2007.
Cooke R and Pate E. Biophys J. **48**:789–98, 1985.
Cunha B A, Eisenstein L E, Dillard T and Krol V. Heart Lung. **36**:72–8, 2007.
Dillon P F and Clark J F. J. Theor. Biol. **143**:275–84, 1990.
Dillon P F and Root-Bernstein R S. J. Theor. Biol. **188**:481–93, 1997.
Einstein A. Ann Phys. **17**:549–60, 1905.
Feher J J. Am J Physiol. **244**:C303–7, 1983.
Ganong W. Review of Medical Physiology. New York: Lange, 2005.
Glaser R. Biophysics. Berlin: Springer-Verlag, 2001.
Huang J D, Brady S T, Richards B W, et al. Nature. **397**:267–70, 1999.
Katchalsky A and Curran P F. Nonequilibrium Thermodynamics in Biophysics, 4th edn. Cambridge, MA: Harvard University Press, 1975.
Kraemer J, Schmitz F and Drenckhahn D. Eur J Cell Biol. **78**:265–77, 1999.
Maina J N and West J B. Physiol Rev. **85**:811–44, 2005.
Mayr M I, Hümmer S, Bormann J, et al. Curr Biol. **17**:488–98, 2007.
Meyer, R A. Am J Physiol Regul Integr Comp Physiol. **287**:R1304–5, 2004.
Numata N, Kon T, Shima T, et al. Biochem Soc Trans. **36**:131–5, 2008.
Onsager L. Phys. Rev. **37**:405 and **38**:2265, 1931.
Parameswaran H, Majumdar A, Ito S, Alencar A M and Suki B. J Appl Physiol. **100**:186–93, 2006.

Rüegg C, Veigel C, Molloy J E, *et al. News Physiol Sci.* **17**:213–18, 2002.

Skowronek K J, Kocik E and Kasprzak A A. *Eur J Cell Biol.* **86**:559–68, 2007.

Smoluchowski M. *Z. Physik Chem.* **92**:129, 1917.

Smoluchowski M. *Ann Phys.* **21**:756–80, 1906.

Stumpff J, Cooper J, Domnitz S, *et al. Methods Mol Biol.* **392**:37–49, 2007.

Suikkanen S, Sääjärvi K, Hirsimäki J, *et al. J Virol.* **76**:4401–11, 2002.

Visscher K, Schnitzer M J and Block S M. *Nature.* **400**:184–9, 1999.

von Hippel P H. *Ann Rev Biophys Biomol Struct.* **36**:79–106, 2007.

4 Energy production

Nothing happens without the expenditure of energy. Humans use the energy in ATP to drive all of the reactions in the body, either directly by using the energy of the phosphate bond between the second and third phosphates of ATP, or by siphoning energy from chemical and electrical gradients produced by ATP hydrolysis. Before ATP can be utilized it has to be produced. ATP is made anaerobically through glycolysis, and aerobically using oxidative phosphorylation in mitochondria. Once made, its concentration can be buffered in many cells through the creatine kinase reaction, with phosphocreatine contributing its phosphate to ADP to maintain the ATP concentration during periods of high energy expenditure. Energy expenditure is not solely dependent on the supply of ATP, however. The removal of the products of energy use, especially inorganic phosphate, are as important to function as the supply of ATP. The different components of ATP production and buffering, and inorganic phosphate removal, directly influence human athletic performance. The different phases that phosphocreatine, glycolysis, and oxidative phosphorylation control are graphically demonstrated in the rate of running in world track records as a function of the log of the distance run.

4.1 Energetics of human performance

The most interesting races on the track take place when someone goes out fast, takes an early lead, and then is run down by the field. Will the leader hang on, or will someone catch up with a strong finishing kick? The factors involved in human performance include ability, training and motivation. For most people, psychological limitations play a major role in reducing performance. Training and experience produce the confidence needed to reach the fastest time possible. World class runners are "world class" precisely because they have overcome the psychological limitations on performance, and their racing is only limited by physiological factors. Physiological factors include some elements that can be controlled, like rest and diet, and others that cannot, like leg length. The variable elements ultimately involve energy. How much energy is available to drive the myosin ATPase reactions that move muscle? At what rate are the energy supplies used? Can the energy supplies be increased to prolong top flight performance? The starting point for answers to these questions is to look at the relation of the rate of running of world track records as a function of the race distances in Figure 4.1.

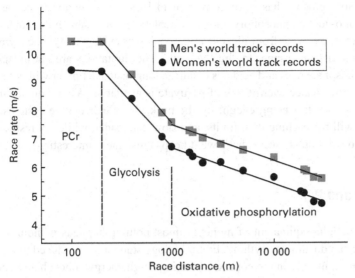

Figure 4.1 Rate of running vs. log(distance) for men's and women's world track records. The lines through the data are the linearized least squares best fits. The three phases correspond to the use of different energy supplies within skeletal muscle.

For both men and women, Figure 4.1 shows three phases. For races up to 200 meters, which take up to 20 seconds, there is a flat line, with no fall in the running rate as the distance increases. The first break point occurs at about 20 seconds, when the running rate starts to fall as a function of distance. There is a second break as the race distance approaches 1000 meters, which takes just over 2 minutes. Thereafter, there is a slow decline in running rate as a function of distance. Men have higher rates at all distances compared with women, primarily based on longer average stride lengths and higher filament content in their muscles, both a consequence of testosterone. Interestingly, the slopes of rate vs. log(distance) are virtually identical for the two genders, indicating that the factors involved in the decline in rate are independent of testosterone-controlled factors. The primary controllers of the phase transitions are the energy sources available to drive muscle contraction: phosphocreatine (PCr), glycolysis, and oxidative phosphorylation, each of which has different control elements. PCr buffers the concentration of ATP, and glycolysis and oxidative phosphorylation make ATP.

Before examining the three energy sources in detail, what does Figure 4.1 *not* tell us? It may appear for a race just longer than 20 seconds that the runner would expend all of their phosphocreatine, then use a little bit of glycolysis to finish the race. This strategy would not maximize performance. If you have ever run a 400 meter race and started too fast, you feel like you have a piano on your back for the last 100 meters. You have not used all of your PCr in a race that goes for more than 20 seconds, but you have used much of the available PCr, and the products of that system are now limiting your performance. The key to overall performance, epitomized by people who have world records, is to maximize energy use without reaching the limitations of any of the energy supply systems. For PCr, this is the buildup of inorganic phosphate. For glycolysis, limitation occurs when available glucose has been converted to lactate and acidosis decreases neural function. For oxidation phosphorylation, the depletion in available mitochondrial-usable energy sources, pyruvate from glucose and, to a lesser extent, acetyl coenzyme A from fats, causes a distance runner to "hit the wall." The limitations are not independent: if inorganic phosphate rises and acidosis occurs, a long distance runner will not reach their maximum performance even if lots of pyruvate is available. When you see that leader who went out too fast being caught by the pack, those exhaustible systems, PCr and glycolysis, will be reaching their limits, and slow that leader, while not restricting those runners who went out slower. That's what keeps those races interesting.

4.2 ATP, ADP and P_i

ATP is the cellular equivalent of money: almost nothing happens without it, and with enough you can do almost anything. Every biology student is introduced to ATP early in their education, usually in the context of having "high energy" phosphate bonds. While this is consistent with the unique nature of ATP as the currency of the cell, it is unfortunate that it implies a special nature to the phosphate bonds of ATP. The anhydride bonds of ATP are indicated by the wavy lines in Figure 4.2. When hydrolyzed to ADP and inorganic phosphate, P_i, there is a release of -31 kJ/mol under standard conditions, the $\Delta G°$ of ATP (Renger, 1982). This energy is between those of glycolytic intermediates PEP ($\Delta G° = -61.8$ kJ/mol) and glycerol-3-phosphate ($\Delta G° = -9.2$ kJ/mol). It is this intermediate standard energy makes ATP so useful: it is easily able to either accept or donate its terminal phosphate, making it an ideal transition molecule. Perhaps the phosphate bonds of ATP should be termed "highly transferable" bonds, but this is not likely to occur.

ATP is very stable in salt solutions having near neutral pH. Since these are the conditions in living organisms, the terminal phosphate of ATP does not spontaneously dissociate into ADP and P_i. In pure water, ATP rapidly splits into its products. The four negative charges on ATP repel one another, and the products ADP and P_i are both stable.

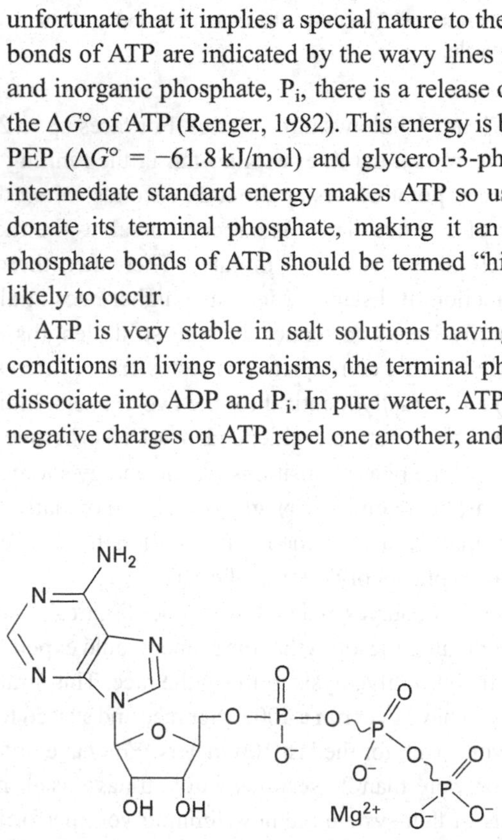

Figure 4.2 The structure of ATP. Inside cells, the divalent cation Mg^{2+} binds to ATP. The Mg^{2+} causes a chemical shift in the NMR resonances of the three phosphates, with the β peak having the largest chemical shift, indicating that Mg^{2+} binds closest to the β phosphate.

Once dissociated, the products would rapidly be hydrated by water, forming hydration shells similar to those around ions. The high water concentration would drive this reaction forward without the need for an enzyme. The usefulness of the divalent cation Mg^{2+} in stabilizing ATP is now apparent. By binding to the negative charges on the oxygens between the phosphorus atoms, the repulsions of the negative charges on the oxygens are canceled. The high charge density of Mg^{2+} means the ATP oxygens will prefer to bind with Mg^{2+} rather than the hydrogen dipoles of nearby water molecules. Thus, the driving forces for dissociation are blocked, and ATP remains stable for long periods of time.

As noted above, the standard energy of ATP, ΔG°, is −31 kJ/mol. We had previously discussed a slightly different term in the measurement of the association energy E_A,

$$E_A = -2.303RT(\log(K_D))$$ (4.1)

which yields a positive value for the association energy. The equivalent equation for the dissociation free energy E_D can also be written:

$$E_D = 2.303RT\log(K_D) = RT\ln(K_D),$$ (4.2)

the free energy for the separation of two molecules. An additional factor is necessary when molecules do not just bind, but are also involved in a chemical reaction. This factor is the ΔG°, the standard free energy of a reaction, a condition only applicable when the products and reactants are present in 1.0 M concentrations. This normalization allows comparisons between different reactions. Under cellular conditions, of course, the concentrations are never 1.0 M for the products and reactants, so the actual free energy of a reaction, ΔG, is

$$\Delta G = \Delta G^\circ + RT\ln\left(\frac{\Pi(p^i)}{\Pi(r^j)}\right)$$ (4.3)

where $\Pi(p^i)$ is the multiplication (Π) of all the concentrations of the molecular products of the reaction with each product concentration raised to the power equal to the number of molecules of that product involved in the reaction (p^i). $\Pi(r^j)$ is the equivalent multiplication of all the molecular reactants r^j. For the systems of association/dissociation, there is no chemical reaction between the molecules, so $\Delta G^\circ = 0$, and that term is omitted from the energy equations. For a reaction such as the hydrolysis of ATP, ΔG° represents the starting point for the reaction energy. Reactions with a negative value of ΔG are exergonic and will spontaneously progress to their products, while those with a positive value of ΔG are endergonic and will not spontaneously go forward. While knowing the ΔG° for a reaction does provide the starting point, the concentrations of the reactants and products will determine how much change there will be from ΔG°, and determine the ΔG. In any physiological system, it is the value of ΔG that will determine the degree to which a reaction will occur.

For the hydrolysis of ATP, the concentrations of ATP, ADP and P_i will determine ΔG:

$$\Delta G = \Delta G^\circ + RT\ln\left(\frac{\Pi(p^i)}{\Pi(r^j)}\right) = -31\,\text{kJ/mol} + RT\ln\left(\frac{[\text{ADP}][\text{P}_i]}{[\text{ATP}]}\right).$$ (4.4)

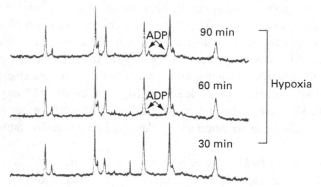

Figure 4.3   31P-NMR spectra of hypoxic pig carotid arteries. During hypoxia, the β-ADP appears to the right of the γ-ADP resonance, both being the terminal phosphate. The weaker binding of Mg^{2+} to ADP relative to ATP allows the β-ADP peak to be seen. The α-ADP is hidden by the α-ATP peak. This spectrum has the ATP, ADP, and PCr peaks. The position of the intracellular P_i peak gives the pH of the cells. By chemically measuring the creatine concentration, all the components of the creatine kinase reaction are known. The intracellular CK reaction was found to be at equilibrium. (Reproduced from Fisher and Dillon, 1988, with permission.)

The concentrations of ATP, ADP and P_i will vary from cell to cell. 31P-NMR spectroscopy gives us the relative free concentrations of ATP and P_i in many cells, with chemical measurements of ATP used to give the standards for the NMR spectra. ADP is harder to determine. It is present in the low μM range in most cells, a concentration difficult to detect by NMR, and in any case the α- and β-ADP peaks may be hidden by the α- and γ-ATP resonances, respectively, unless the free Mg^{2+} is sufficiently low. Despite these difficulties, free ADP has been seen in 31P-NMR spectra of smooth muscle (Figure 4.3).

Calculation of the free ADP is usually done by assuming that the creatine kinase reaction

$$PCr + ADP \overset{CK}{\longleftrightarrow} Cr + ATP \qquad (4.5)$$

is at equilibrium. Measuring the components of the CK reaction in smooth muscles where free ADP has been detected using 31P-NMR, the creatine kinase reaction was shown to be at equilibrium, so this assumption is reasonable (Fisher and Dillon, 1988). Chemical measurements of ADP do not reflect the free concentration, as ADP binds to actin filaments (Hozumi, 1988; Greene and Eisenberg, 1980). During chemical extraction, this bound ADP is released in amounts that far exceed the free ADP concentration, making chemical measurements of free ADP unreliable.

There is a wide range of reported ATP, ADP and P_i values. The highest ΔG values will occur when the concentration of ATP is high and ADP and P_i are low. In general, white glycolytic skeletal fibers will have the highest ATP concentration of the striated muscles, with red oxidative and cardiac muscle having lower concentrations. Smooth muscle has significantly less ATP. There will also be differences in ADP and P_i. The range of ΔG can be estimated by using the highest ATP concentrations with the lowest ADP and P_i concentrations to calculate the highest ΔG, and the lowest ATP and the highest ADP

Table 4.1 Free energy range in striated and smooth muscle

Energy	Striated muscle		Smooth muscle	
	Highest	Lowest	Highest	Lowest
ATP	10 mM	4 mM	2 mM	0.7 mM
ADP	20 μM	100 μM	20 μM	70 μM
Pi	0.5 mM	2 mM	0.5 mM	1.5 mM
ΔG	−48.8 kJ/mol	−38.7 kJ/mol	−44.7 kJ/mol	−35.9 kJ/mol

and P$_i$ for the lowest ΔG. This will bracket the ΔG range, and all muscles should fall within this range. Table 4.1 shows that despite the chemical differences, there is a relatively narrow range of free energy of ATP hydrolysis, and considerable overlap of ΔG between striated and smooth muscles. This implies that the mechanical characteristics are not initially determined by the free energy of ATP, but by the characteristics of the myosin ATPases in the different muscles, and the ability to regenerate ATP using glycolysis and oxidative phosphorylation.

The ΔG of ATP hydrolysis represents the immediate chemical energy available to drive reactions in the cell. There is an alternative method of representing the energy state of the cell, the energy charge, EC,

$$EC = \frac{[ATP] + \frac{1}{2}[ADP]}{[ATP] + [ADP] + [AMP]}. \tag{4.6}$$

This equation has been associated with the metabolic state of the cell. The equation is dominated by ATP under normal conditions when ADP and AMP are lower by 1–2 orders of magnitude. Under metabolically stressed conditions, a decrease in ATP will correspond with an increase in ADP and AMP, lowering the energy charge. This will stimulate the metabolic machinery of the cell to return the cell to its normal energy charge. A disadvantage of this measurement in studying a specific ATP-dependent process is the absence of the P$_i$ term. Since ΔG depends in part on P$_i$, its absence from the energy charge equation eliminates using free energy as an analytical tool. The energy charge is most useful in assessing the overall energy state of a tissue, rather than the energetics of a particular reaction.

Using the ΔG value allows us to consider one of the most important concepts in energetics: both the reactants and products are determinants of a reaction's activity. In some cases, energy studies focus solely on the concentration of ATP, or PCr in those tissues in which PCr buffers ATP. Imagine a car with a full gas tank. The car will run, as long as the exhaust gases are removed from the engine. Block the exhaust pipe, and the car stalls out. It has plenty of gasoline to power the car, but without removal of the waste products, the system stops. The same is true of biological systems: product removal is necessary to maintain enzymatic functions. In muscles, where it has been studied extensively, inorganic phosphate has been shown to significantly influence muscle contraction. The products of the myosin ATPase reaction are ADP and P$_i$. As we saw in Figure 3.13, skeletal muscle myosin has only a single step, corresponding to the nearly

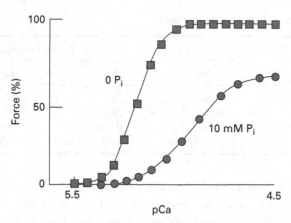

Figure 4.4   Phosphate sensitivity of skinned frog skeletal muscle fibers. High P_i decreases the rate of P_i dissociation from the actin–myosin complex, inhibiting the force development step in the crossbridge cycle. High P_i also decreases the calcium sensitivity of force development. (Redrawn from Brozovich *et al.*, 1988)

simultaneous release of both ADP and P_i. The force-generating step has been biochemically shown to be the release of P_i (Cooke and Pate, 1985), while in non-muscle myosins both ADP release and P_i release produce mechanical steps. Increasing concentrations of P_i cause a progressive decrease in tension in both skeletal muscle (Cooke and Pate, 1985) and smooth muscle (Dillon, 2000). Increased P_i also shifts the skeletal calcium activation curve to the right, decreasing the calcium sensitivity (Figure 4.4). ATP-dependent reactions therefore are not solely dependent on the ATP concentration. The relative concentrations of ATP, ADP and P_i determine the net energy available to drive these reactions.

4.3   Phosphocreatine

Phosphocreatine (PCr) is present in high concentrations in many cells. In skeletal muscles, its concentration is only exceeded by water and K^+. PCr buffers the concentration of ATP when cells have high energy demands. Creatine is present in many foods and is manufactured in the liver, but the absence of creatine kinase in the liver prevents its phosphorylation there. Creatine is exported from the liver and taken up by many other tissues: all forms of muscle, neural tissue, the kidneys and many others (Walliman and Hemmer, 1994). As noted above, the enzyme creatine kinase (CK) catalyzes the reaction transferring a phosphate from PCR to ADP, thereby maintaining the ATP concentration during times of high ATPase activity. The concurrent reactions under these conditions are

$$\text{PCr} + \text{ADP} \xrightarrow{\text{CK}} \text{Cr} + \text{ATP} \tag{4.7}$$

$$\text{ATP} \xrightarrow{\text{ATPase}} \text{ADP} + P_i \tag{4.8}$$

yielding the net reaction

$$PCr \rightarrow Cr + P_i. \tag{4.9}$$

The net reaction shows why the initial phase of the human race performance (Figure 4.1) only lasts for 20 seconds. The maximum rate of the myosin ATPase reaction uses ATP at such a rate that the PCr within skeletal muscles is consumed in about 20 seconds. This produces a concentration of P_i which is in excess of 10 mM. As was seen in Figure 4.4, this concentration of P_i significantly decreases the force generation of skeletal muscles, and therefore limits human performance. It also shows why races of greater duration need alternative energy sources. The ATP generated by glycolysis and oxidative phosphorylation replenishes both the PCr stores and ATP stores, if they have been depleted. These ATP-generating processes are always occurring, and will be accelerated as soon as the high energy activity starts, not when PCr is exhausted. If glycolysis only started after 20 seconds of activity, the high P_i concentration would drop the record rates precipitously, not gradually. The key to outstanding performance is keeping the P_i concentration low. Thus, there will be a competition between the energy-consuming reaction and the energy-producing reactions aimed at avoiding phosphate inhibition of the myosin ATPase.

Other tissues also have PCr to buffer the concentration of ATP, of course. In the case of cardiac muscle, limited oxygen delivery due to atherosclerosis will limit oxidative phosphorylation, and at times of high energy demand may lead to net decrease in PCr, a buildup of P_i, and decreased cardiac muscle performance, resulting in myocardial infarction, a heart attack. Tissues other than striated muscle do not consume energy at the same high rates. PCr is still present in many of these tissues, and will buffer changes in ATP. In smooth muscle, for example, both the PCR and ATP concentrations are 5–10 times lower than they are in striated muscle. Yet, the myosin ATPase rate, and the shortening velocity, is 10–100 times slower. The lower concentrations of available phosphate stores are offset by the lesser energy utilization. While smooth muscle force is also inhibited by increased P_i (Dillon, 2000), this inhibition would only occur under pathological conditions.

The enzyme creatine kinase is present in two forms in cells: the cytosolic dimer and the octameric mitochondrial form. The cytosolic CK exists in three forms, MM in skeletal muscle, MB in cardiac muscle, and BB in brain, smooth muscle and other non-muscle tissues. The MB form is sometimes used to measure tissue damage after a heart attack, as its presence in the blood only occurs when cardiac muscle membranes have been ruptured and the enzymes leak out. (Measurements of troponin in plasma are also used to measure cardiac muscle damage.) The cytosolic CK forms may be bound to either glycolytic enzymes such as pyruvate kinase (Sears and Dillon, 1999), or to energy-consuming structures, such as the filaments of muscle (Kraft *et al.*, 2000). The mitochondrial forms of CK are associated with the inner mitochondrial membrane (Walliman and Hemmer, 1994). The differential locations of CK have led to the formation of the creatine shuttle model (Figure 4.5).

The CK shuttle occurs during the oxidative state of energy supply. It relies on the anisotropic distribution of the forms of CK that produce PCr, primarily in the mitochondria during oxidative stress, and those CK molecules that catalyze the buffering of ATP

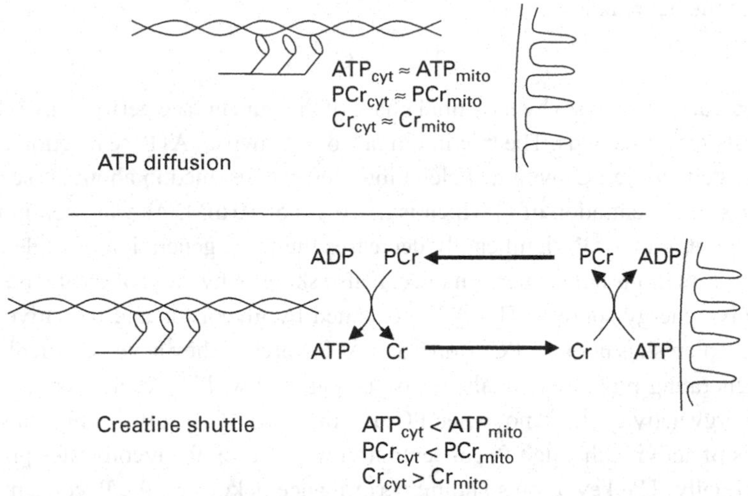

Figure 4.5   Diffusion equilibrium and creatine shuttle models. In the ATP diffusion model, the rate of ATP diffusion between the mitochondria and the cytoplasmic ATPase sites is sufficient to maintain nearly equal concentrations of ATP at both sites. In the creatine shuttle model, the faster rates of PCr and Cr diffusion carry the phosphate exchange between the mitochondria and the ATPase sites.

by contributing the PCr phosphate to ADP near the sites of ATPase activity. PCr and Cr are smaller than ATP and ADP, and thus will diffuse faster than the nucleotides. The shuttle concept has ATP that is generated by the mitochondria contribute a phosphate to Cr, generating PCr, which then diffuses to ATP-consuming areas, such as muscle filaments. There, the PCr transfers its phosphate to ADP, that is, generating ATP for an energy-consuming reaction, and the Cr diffuses back to the mitochondria to be reloaded with phosphate. For the shuttle to be effective, the diffusion distance and the rate of ATP consumption would have to create a window in which PCr and Cr diffusion, but not ADP and ATP diffusion, would be fast enough to support the ATP-consuming activity. If both were fast enough (i.e., either the ATPase rate was low or the diffusion distance was small), a creatine specific shuttle would not be necessary. Also, if either the distance or the ATPase rate were too great, then neither creatine- nor adenosine-based diffusion would be sufficient to maintain the ATPase reaction. This would lead to a decline in energy resources, but the rate of energetic decline could be influenced in part by diffusion rates. The best candidates for a creatine shuttle would be large cells with localized mitochondria and potentially high ATPase rates, such as striated muscle cells that have mitochondria in the perinuclear region. The importance of the PCr/Cr shuttle, as opposed to ATP/ADP diffusion, has been a point of scientific contention, and appears to have been considered settled by both those who regard the creatine shuttle as important (Aliev and Saks, 1993) and those who do not (Meyer *et al.*, 1984).

   PCr levels can be modified by dietary creatine supplementation. Adding creatine to the diet increases the amount of PCr in cells. There are many studies showing that increases in PCr produce an increase in muscle performance, particularly in short-duration, high-intensity activities (Bembem and Lamont, 2005). In exercise bursts with intermittent recovery times, increases in free creatine following CK activity to maintain ATP caused an increase in the rate of PCr resynthesis during the rest periods, as it would in an enzyme at equilibrium. The more rapid PCr resynthesis then was better able to support further high-intensity muscle contraction (Yquel *et al.*, 2002). There does appear to be a window of activity in which creatine supplementation will increase physical performance. Pursuit bicycle racing, in which one rider in the front fights wind resistance during a high pedaling spell, creates a slipstream for the following teammates pedaling with less wind resistance. After 10–15 seconds, the first rider peels off and moves to the end of the line, recovering in the slipstream while a teammate does the hard work up front. This sequence has the riders dependent on PCr availability, and supplementation makes more energy available initially. The downside of PCr supplementation occurs during prolonged activity. More PCr would mean an increase in P_i during PCr utilization, causing the decrease in myosin ATP activity that was seen in Figure 4.4. Increased PCr may prolong the high-intensity duration by a few seconds, but prolonged high-intensity activity will produce a more profound fatigue. The increase in PCr would only remain as long as the supplementation continued, since as soon as there was no additional creatine available the cells would return to their normal PCr levels.

The other aspect of creatine supplementation is the degree to which creatine supplementation makes muscles bigger. There are two aspects to this. First, as the muscles acquire creatine and it is phosphorylated to PCr, there will be an increase in the concentration of non-diffusable (across the membrane) molecules in the cell, making it hyperosmotic and hypertonic. To counter this, water will diffuse across the cell membrane until the osmolarity of the cell matches the normal osmolarity of the interstitial fluid, 300 mOsm. (Diarrhea is often a consequence of creatine supplementation and disruption of the normal osmolarity of the body can have severe consequences, as are present in the hyperosmolarity produced in diabetes mellitus.) Muscle cells will be hydraulically inflated, like a water balloon. There is no increase in filament number, no increase in crossbridge number. If the overhydration causes changes in the muscle filament architecture, then the performance increase caused by having more PCr could be offset by a decrease in actomyosin ATPase efficiency. The second aspect of creatine-dependent muscle size would occur during muscle hypertrophy. While hypertrophy will be covered in detail in the next chapter, its creatine dependence will be covered here. Hypertrophy depends in part on high-intensity isometric or near isometric contractions that cause microdamage to muscle filaments. The repair of these filaments results in more filaments than were originally present. The greater force output in creatine-supplemented muscles could cause more microdamage during a training bout, and thus more filament replacement. This would allow muscles to develop more force. Since maintenance of muscle size requires continuous training, the return to a normal training regimen would cause the muscles to return to their normal trained size. Further, the amount of pain associated with this type of training is proportional to the amount of microdamage, so there will be more pain in store for the creatine-supplemented athlete during these high-intensity training regimens.

While creatine supplementation is legal, it is discouraged by many school authorities. Coaches are often prohibited from recommending creatine supplementation to their athletes. As we have seen, there are only limited circumstances in which it would help a performance, and there are short-term consequences in terms of muscle pain, hyper-osmolarity, and unknown consequences from long-term supplementation. Ethical coaches and parents appropriately discourage young people from using chemicals to enhance their athletic performance. Embracing chemically enhanced performance will lead some athletes toward other chemicals beyond creatine, including illegal ones such as anabolic steroids, erythropoietin and human growth hormone. There are certainly elite athletes that use creatine supplements, but as there is no way of detecting exogenous creatine, since it is present in many foods, this is not an area that is likely to be regulated anytime soon.

4.4   Glycolysis

The details of glycolysis are covered in most general biology classes. Anaerobic energy production by the glycolytic pathway produces energy without the use of oxygen:

$$Glu + 2ADP + 2P_i \rightarrow 2Pyr + 2ATP. \tag{4.10}$$

The process is inefficient, only producing 2 net ATP molecules per molecule of glucose (Glu), in contrast to the 34 additional ATP that can be produced by oxidative phosphor-ylation. More advanced studies discuss the regulation of glycolysis by the initial phos-phorylation reactions. Phosphorylation of glucose by hexokinase traps the glucose within the cell, since the phosphorylated form cannot be carried by glucose transporters. The flux through the glycolytic path is regulated by phosphofructokinase (PFK), an enzyme that has a large, negative ΔG, and therefore will be strongly favored to go in the forward direction. Some discussions of glycolysis dwell on the central role of $NAD^+/NADH + H^+$. The general scheme of glycolysis, in more complex form,

$$Glu + 2ATP + 2ADP + 2P_i + 2NAD^+ \rightarrow$$
$$2Pyr + 4ATP + 2H_2O + 2NADH + 2H^+ \tag{4.11}$$

is a one way reaction. The $NADH + H^+$ must be recycled back to NAD^+ in order to sustain the reaction. In the absence of oxygen, the lactate dehydrogenase (LDH) reaction

$$pyruvate + NADH + H^+ \xrightarrow{\text{LDH}} lactate + NAD^+ \tag{4.12}$$

maintains the availability of NAD^+ in order to continue ATP production by glycolysis. Both pyruvate and lactate are present in their dissociated form inside cells. The protonated forms, pyruvic acid and lactic acid, while often referred to in academic and lay texts, are present in very small concentrations, as the pK_a for the carboxyl hydrogen dissociation is about 4.0. This dissociation accounts for the fall in pH inside cells when glycolytic flux is high, falling more than 0.5 pH units in skeletal muscle. Since PCr will routinely be hydrolyzed under these conditions, the increase in P_i, while having a detrimental effect on muscle contraction, will help buffer the changes in pH within the cell.

The ΔG values for the glycolytic enzymes (Garrett and Grisham, 2005) are shown in Figure 4.6. The $\Delta G°$ values are quite different from the ΔG values for these reactions, as the concentrations are not at 1.0 M inside cells. Only three of the reactions, hexokinase, PFK and pyruvate kinase, have ΔG values far removed from the RT thermal equilibrium value of 2.58 kJ/mol. The exergonic reactions will spontaneously move in the forward direction. All of the other reactions have ΔG values that are near equilibrium. For reactions at equilibrium, the Le Chatelier principle applies (Moore, 1972):

For a system at equilibrium, a change in the one of the variables of the system will be countered by a change in another variable, returning the system to equilibrium.

The application of the Le Chatelier principle means that as the substrate or product of one of the equilibrium reactions is altered, the activity in the other enzymes will return the first reaction to its equilibrium condition. This principle has been applied to many scientific systems, and to social sciences, such as economics, as well. Perhaps its most common application to physiological systems is in the carbonic anhydrase (CA) reaction

$$CO_2 + H_2O \underset{CA}{\longleftrightarrow} H_2CO_3 \leftrightarrow HCO_3^- + H^+ \tag{4.13}$$

controlling the loading and unloading of carbon dioxide by the circulatory system. In glycolysis, the hexokinase and PFK reactions will drive the glycolytic flux forward, increasing first the reactants in the equilibrium reactions, then the products, as the enzymes obey the Le Chatelier principle. The exergonic PK reaction at the end ensures net flux from the pathway. While the low concentration of the glycolytic intermediates means both that the $\Delta G°$ values will not be directly applicable to cells and that exact ΔG values will be subject to a significant amount of experimental uncertainty, the structural aspects of glycolytic enzyme clusters make energetic analysis even more difficult.

Earlier we discussed how the protein concentration in cells is far higher than concentrations that can go into a laboratory solution. The formation of protein complexes was

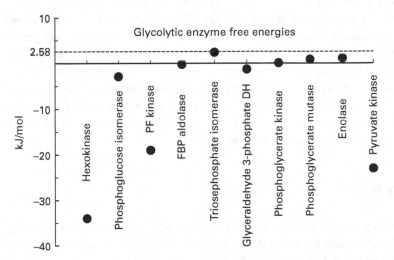

Figure 4.6   Free energy ΔG values for the glycolytic enzymes. Only hexokinase, PFK and PK are removed from equilibrium. Each has a large negative ΔG driving the reaction forward. These reactions ensure forward flux through the equilibrium reactions.

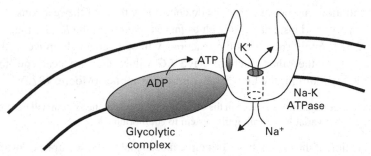

Figure 4.7   Direct linkage of glycolysis to the Na–K ATPase. The ATP product of membrane-bound glycolysis directly transfers ATP to the ATP-binding site (side gray oval) of the Na–K ATPase.

not just structural, but also functional: the diazymatic coupling of pyruvate kinase to creatine kinase was just one example of the channeling of metabolites between sequential enzymes. Glycolytic enzymes were among the first to exhibit protein complex behavior. Glycolytic enzymes bound to the cell membrane were found to directly supply the ATP for the Na–K ATPase in erythrocytes (Solomon, 1973) and smooth muscle (Paul *et al.*, 1979). In Figure 4.7, the ATP product of membrane-bound glycolysis directly contributes its ATP product to the sodium pump without releasing the ATP into the cytoplasm.

These early reports of glycolytic coupling were followed by theoretical proposals of a glycolytic supercluster, with all of the glycolytic enzyme bound together and tethered to the cell membrane (Kurganov, 1986). Experimental evidence of glycolytic clusters has found that the sequential enzymes aldolase, glyceraldehydephosphate dehydrogenase and triosephosphate isomerase form a complex (Beeckmans *et al.*, 1999). An intriguing finding showed a 1:2 ratio between the hexose and triose enzymes (Maughan *et al.*, 2005), a structure that would accommodate the greater flux through the triose enzymes following the aldolase reaction. These findings present difficulties in the energetic analysis of glycolytic flux. What is the effective concentration of a metabolite that is being passed from one enzyme to another? From an experimental perspective, the metabolite concentration exposed to an isolated enzyme could be altered until the flux through the enzyme matched that of the channeled flux, but that would not be very useful information. More important questions would ask what controls complex formation? To what extent are bound metabolites exchanged with the cytosol? And what other functional processes, like the Na–K ATPase, are linked to glycolytic complexes? The answers to these questions await further research.

4.5   Mitochondria

Steady-state human performance requires aerobic ATP production. The high rates of energy utilization that occur during the first two minutes of activity, energy costs borne by PCr and glycolysis, result in phosphate buildup, acidosis and depletion of glucose that limits muscular contraction. The use of oxygen by the mitochondria efficiently draws energy from carbohydrates and fats, resulting in the maintenance of ATP concentrations.

The energy for oxidative phosphorylation comes from pyruvate, the product of glycolysis. The glycolytic complexes, with the exception of those in the membrane supplying the ATP for the active transport of ions, are located throughout the cytoplasm. Pyruvate has to travel to the mitochondria for its oxidation. Pyruvate is a small molecule, and due to its dissociated carboxyl group has a net negative charge. Both its oxygen side group with a weak dipole and its dissociated carboxyl oxygen will attract and form hydrogen bonds with the positive hydrogen dipoles of water. Pyruvate's methyl group will not form any strong bonds, so that as a molecule pyruvate will not form the long duration bonds needed for directed transport. If the energy of directed transport exceeded the ATP production from pyruvate, there would be no net ATP production anyway. Pyruvate only reaches the mitochondria by diffusion. Just as with the creatine shuttle above, any limitation of energy production caused by pyruvate's diffusion time will be a combination of the rate of pyruvate production, its average diffusion distance to the mitochondria, the viscosity of the cytoplasm, and the rate of pyruvate metabolism at the mitochondria. Since few cells have energy usage that exceeds their energy production, only in striated muscle cells is the diffusion of pyruvate likely to be critical. It is possible that in some special cases, such as neurons with very long axons, there could be circumstances where diffusion is a limiting factor.

The other transport element important at the mitochondria is the NADH produced by glycolysis. Although this only represents about one-sixth of the total aerobic energy production, the two modes of transfer of the NADH-reducing equivalents across the inner mitochondrial membrane can effect the efficiency of ATP production. The malate–aspartate shuttle begins with the malate dehydrogenase (MDH) reaction:

$$\text{Cytoplasm:} \ \text{oxaloacetate} + \text{NADH} \xrightarrow{\text{MDH}} \text{malate} + \text{NAD}^+ \qquad (4.14)$$

followed by

$$\text{malate}_{\text{cytoplasm}} \rightarrow \text{malate}_{\text{mitochondria}} \qquad (4.15)$$

$$\text{Mitochondria:} \ \text{malate} + \text{NAD}^+ \xrightarrow{\text{MDH}} \text{oxaloacetate} + \text{NADH}. \qquad (4.16)$$

Mitochondrial malate dehydrogenase converts the malate back to OAA and NADH. Thus, there is no loss of reducing equivalents during this transfer. In contrast, the 3-P-glycerol DH reaction

$$\text{Dihydroxyacetone phosphate} + \text{NADH}_{\text{cytoplasm}} \rightarrow \text{3-P-glycerol} + \text{NAD}^+ \quad (4.17)$$

and

$$\text{3-P-glycerol} + \text{FAD} \rightarrow \text{DHAP} + \text{FADH}_{2 \ \text{mitochondria}} \qquad (4.18)$$

results in the loss of reducing equivalents, as now FADH_2 will enter the electron transport chain rather than NADH, with the loss of one ATP. Since both transfer processes can occur, there will be a reduction in the efficiency of ATP production to the degree that the 3-P-glycerol process occurs rather than the malate process.

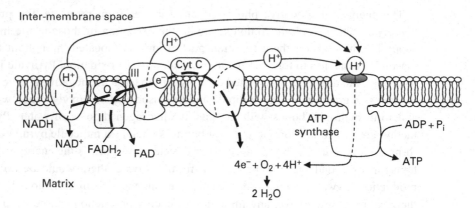

Inter-membrane space

Matrix

Figure 4.8 ATP production at the inner mitochondrial membrane. The hydrogen donated by NADH to Complex I, the NADH coenzyme Q reductase, is split into the proton and electron. The electron is passed to the other cytochromes: Q (ubiquinone), Complex III (cytochrome bc$_1$ complex), Cyt C (cytochrome C), and Complex IV (cytochrome C oxidase), before combining with protons and molecular oxygen to form water. FADH$_2$ enters the chain at Complex II (succinate dehydrogenase). Protons are transported across the membrane by complexes I, III, and IV. The protons re-enter the matrix using ATP synthase, sometimes called Complex V, which uses the energy of the re-entering protons to phosphorylate ADP and make ATP.

Virtually every biology and biochemistry book states that there are three ATP molecules produced by the ETS for each NADH oxidized, and two ATP produced for each FADH$_2$ oxidized. ATP production occurs because of the energy gradient produced by the transfer of protons from the matrix of the mitochondria to the intermembrane space by the cytochromes (Figure 4.8). Initial understanding of this process was based on Mitchell's chemiosmotic model (Prebble, 2000), the details of which have proven to be essentially correct more than 50 years later.

There are approximately ten protons moved across the membrane with the passage of an electron from Complex I to Complex IV, and six protons moved as an electron passes from Complex II to Complex IV. The pH of the intermitochondrial space is about 0.75 pH units negative to the matrix pH, so the protons are moved against their gradients, an endergonic process (Voet *et al.*, 2002). The energy required to transport a proton, taking into account both the chemical and electrical gradients, can be calculated as

$$\Delta G = 2.303 RT(pH_{matrix} - pH_{IMS}) + ZF\Delta\Psi \tag{4.19}$$

where Z is the charge on the proton, F is Faraday's constant, and $\Delta\Psi$ is the membrane potential across the inner mitochondrial membrane, which is 168 mV, with the matrix negative. The free energy for the translocation of one proton out of the matrix is

$$\Delta G = (2.303)(2.58)(0.75) + (+1)(96840\,C/mol)(0.168\,V) = 4.6 + 16.2$$
$$= 20.8\,kJ/mol. \tag{4.20}$$

The energy of the proton gradients is dissipated, a negative ΔG exergonic reaction, when they pass back into the matrix through ATP synthase. Since 20.8 kJ/mol is not sufficient to produce an ATP molecule with a terminal phosphate energy in the range of -40 kJ/mol

(Table 4.1), the energy of multiple protons is harnessed for this process. The 10 protons moved across the inner mitochondrial membrane are energetic enough to make 3 ATP molecules, and the 6 protons moved by $FADH_2$ enough to produce 2 ATP molecules, processes which would still have energetic efficiencies of greater than 50%, tremendously efficient processes. The theoretical maximum for ATP production in the best possible case is 38 ATP/glucose. This value has been called into question, with values of 2.5 ATP/NADH and 1.5 ATP/$FADH_2$ calculated (Hinkle, 2005). Research in this area is extremely difficult, as any process that perturbs the area near the inner membrane will significantly disrupt the components involved, a minor variant of the Heisenberg uncertainty principle. Given the combination of differential NADH transport, potential leakage of the proton gradient, the possibility of superoxide production dissipating a portion of the proton gradient, and the use of protons for purposes other than ATP synthase, many people conclude that 32 ATP/glucose is a reasonable estimation. Local conditions of course may (or may not) alter this number. In any case, regardless of what the actual net ATP production is at the mitochondria, it far exceeds that of glycolytic ATP production, and makes most of the processes we associate with life possible.

The energy driving biophysical processes requires both the supply of substrate and the removal of waste products. The energetic limits on human performance are determined by the rate that ATP can be generated, sequentially using the creatine kinase reaction, glycolysis, and oxidative phosphorylation. Conditions that increase the concentration of inorganic phosphate produce decreased myosin ATPase activity and decreased calcium sensitivity in muscle, resulting in reduced mechanical performance. The recycling of NAD is critical in maintaining both glycolysis and oxidative phosphorylation. Glycolysis is inefficient at producing ATP, but does provide much of the energy for ion transport through direct transport of newly made ATP from glycolytic enzymes to ion ATPases. Survival requires mitochondrial ATP production, using the energy of proton gradients across the inner mitochondrial membrane to drive ATP synthesis. The accounting of ATP production by oxidative phosphorylation is an area of active debate, but under any circumstances mitochondrial energy production far exceeds glycolytic energy production.

References

Aliev M K and Saks V A. *Biochim Biophys Acta*. **1143**:291–300, 1993.
Beeckmans S, Van Driessche E and Kanarek L. *J Cell Biochem*. **43**:297–306, 1999.
Bembem M G and Lamont H S. *Sports Med*. **35**:107–25, 2005.
Brozovich F V, Yates L D and Gordon A M. *J Gen Physiol*. **91**(3):399–420, 1988.
Cooke R and Pate E. *Biophys J*. **48**:789–98, 1985.
Dillon P F. *J Vasc Res*. **37**:532–9, 2000.
Fisher M J and Dillon P F. *NMR Biomed*. **1**:121–6, 1988. www.interscience.wiley.com
Garrett R and Grisham C M. *Biochemistry*, 3rd edn. Belmont, CA: Thomson Brooks/Cole, 2005.
Greene L E and Eisenberg E. *Proc Natl Acad Sci U S A*. **77**:2616–20, 1980.

Hinkle P C. *Biochim Biophys Acta*. **1706**:1–11, 2005.

Hozumi T. *J Biochem*. **104**:285–8, 1988.

Kraft T, Hornemann T, Stolz M, Nier V and Wallimann T. *J Muscle Res Cell Motil*. **21**:691–703, 2000.

Kurganov B I. *J Theor Biol*. **119**:445, 1986.

Maughan D W, Henkin J A and Vigoreaux J O. *Mol Cell Proteomics*. **4**:1541–9, 2005.

Meyer R A, Sweeney H L and Kushmerick M J. *Am J Physiol*. **246**:C365–77, 1984.

Moore, W. *Physical Chemistry*, 4th edn. Englewood Cliffs, NJ: Prentice Hall, 1972.

Paul R J, Bauer M and Pease W. *Science*. **206**:1414–16, 1979.

Prebble J. *Nature*. **404**:330, 2000.

Renger G. In *Biophysics*, ed. Hoppe W, Lohmann W, Markl H and Ziegler H, 2nd edn. New York: Springer-Verlag, pp. 347–71, 1982.

Sears P R and Dillon P F. *Biochem*. **38**:14 881–6, 1999.

Solomon A K. In *Membrane Transport Processes*, ed. Hoffman J. F. New York: Raven, p. 31, 1973.

Voet D, Voet J G and Pratt C W. *Fundamentals of Biochemistry*, Upgrade Edition. New York: Wiley, p. 513, 2002.

Walliman T and Hemmer W. *Mol Cell Biochem*. **133–4**:193–220, 1994.

Yquel R J, Arsac L M, Thiaudiere E, Canionic P and Manier G. *J Sports Sci*. **20**:427–37, 2002.

5 Force and movement

The ability of cells to develop force and to shorten are commonly associated with skeletal muscles. These processes are not limited to skeletal muscles, however, but extend to cardiac muscles that pump blood, the smooth muscles that both maintain forces for extended durations and empty the contents of cavities, and in the migration of individual cells such as leukocytes around the body. All of these processes rely on the interaction of actin and myosin, the primary components of thin and thick filaments, respectively. The structure of the filaments converts the scalar myosin ATPase into vectorial force and shortening. Individual differences produce the diverse mechanical activities in the body. With experimental observations going back hundreds of years, the technical details known about contraction rival those of any other biophysical system, details that are covered below.

5.1 The skeletal length–tension relation

Now, the greater the extenſion the more is the tone and the vigor of the action of a muſcle increaſed – the leſs, the weaker will be its powers … But this tonic power, as far as it depends upon extenſion, is confined within certain limits; for, so far is a great and continued extenſion of muſcular fibres from rendering their contraction eaſier and ſtronger, that often it weakens and deſtroys it – and thus muſcles commonly loſe their power and are rendered incapable of either ſuddenly or with facility recovering their former ſtrength. (John Pugh, *Treatise on Muscular Action*, 1794)

It was only with the discovery of the overlapping filaments of striated muscle in 1954 (Huxley and Niedergerke, 1954; Huxley and Hanson, 1954) that the molecular basis for muscle contraction was formed. This restriction did not keep previous scientists, certainly going back to Pugh, from observing the behavior of muscle. His description of the length–tension curve, although not by that name, is clear to us today. Extend a muscle, and its force increases, but stretch it too far, and it weakens: accurate, and even perhaps implying the damage that occurs if muscles are overstretched. We do have the advantage of having considerably more information than Pugh had.

The striation patterns in skeletal and cardiac muscle consist of alternating dark and light bands, the A and I bands. The bands form because of the overlapping thin and thick filaments. The thin filaments, consisting of the proteins actin, troponin and tropomyosin, do not significantly block the transmission of visible light. The thick filaments, consisting primarily of the protein myosin, do block light. There can be regions of just thin

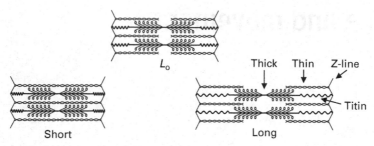

Figure 5.1   Filament overlap in striated muscle. The muscle sarcomere, running from Z-line to Z-line, is the unit of contraction in striated muscle. At the optimum length for force development, L_o, the maximum number of crossbridges, extensions of the molecule myosin in the thick filament, can contact actin molecules on the thin filaments. At shorter lengths, overlap of the thin filaments and compression of the thick filament against the Z-line, the thin filament anchor structure, reduces force generation. At long lengths the reduction in thick–thin overlap results in decreased force. The very large protein titin tethers the thick filament to the Z-line, keeping the thick filament centered in the sarcomere.

filaments, just thick filaments, or areas where the thick and thin filaments overlap. Wherever there are thick filaments, light will not pass, producing the dark A band. The light I band occurs wherever thick filaments are not present. As a muscle shortens during contraction, the amount of filament overlap increases (Figure 5.1). Since the filaments have fixed lengths, the width of the A band is constant. But as a muscle shortens, the overlap increase reduces the areas where no thick filaments are present, so that the I band width decreases.

Figure 5.1 shows the arrangement of filaments at short and long lengths, and at L_o, the optimum length for force development. At this length, the maximum number of cross-bridges, extensions of the protein myosin, span the space between the filaments and bind to actin molecules on the thin filament. The release of the products of ATP hydrolysis, ADP and, as we saw in the previous chapter, inorganic phosphate, P_i, results in force development and, if the load on the muscle is less than the force the muscle can develop, contraction to a shorter length. At lengths longer than L_o, the amount of filament overlap will decrease. There will be fewer actin–myosin contacts, and the force will fall. At shorter lengths, there are three processes that decrease force. First, the thin filaments from opposing Z-lines progressively overlap, blocking the efficient binding of crossbridge heads to actin. Second, the thick filaments are compressed against the Z-line structures, creating an internal resistance to filament sliding and a decrease in force generation. Third, there is a decrease in calcium release from the sarcoplasmic reticulum, reducing the number of available actin sites and therefore reducing the force. Calcium binds to troponin in the thin filament, causing the filamentous protein tropomyosin, which runs along the double helix of actin monomers forming the backbone of the thin filament, to shift into the groove of the actin double helix. This movement exposes the myosin binding site on actin and allows the myosin ATPase activity to generate force and shortening.

Structurally, the physical transmission of force requires two reversals of polarity within the sarcomere, the repeating contractile unit. Without these reversals, the interaction of filaments would cause the filament to pull in only one direction. The system must pull in both directions in order to generate external force and to pull the two ends of a muscle cell closer together, shortening the cell. Thin filaments are anchored to the Z-line. The thin filaments extending from the two Z-lines at the end of a sarcomere have opposite polarity: myosin binding to each filament will move toward that filament's Z-line. The other reversal of polarity occurs at the M-line, the center of the thick filament. The crossbridges extending left and right from the M-line will have opposite orientations, so that each crossbridge will draw a thin filament toward the M-line. This arrangement results in increased filament overlap when the muscle is activated, and allows the muscle fiber, the muscle cell, to shorten.

The quantitation of the skeletal muscle length tension curve is shown in Figure 5.2. The quantitation of muscle force with sarcomere length and structure (Gordon *et al.*, 1966) did not take titin into account, as this structural protein had not yet been discovered. The model system derived from experiments has a thick filament length of 1.65 μm and thin filament lengths of 1.0 μm. With thin filaments extending from the Z-lines on each side of a sarcomere, the overlap would reach zero at 3.65 μm, corresponding to the tension reaching zero when there was no overlap. The force reaches zero on the short end at 1.27 μm, or 63.5% of the L_o length of 2.0 μm. This zero force length occurs much closer to L_o in skeletal muscle than it does in smooth muscle.

Force generation requires that the force on the two sides of the thick filament be equal, otherwise the thick filament would just move to one end of the sarcomere and stay there. This centering of the thick filament is not a problem on the short end of the length–tension curve. If one side of the thick filament got closer to the Z-line than the other end, it would

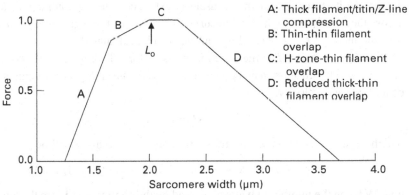

Figure 5.2   Length–tension relation in skeletal muscle. The quantitation of the filament overlap changes that alter tension generation have a thick filament length of 1.65 μm and thin filament length of 1.0 μm extending from each Z-line. The muscle will be at L_o when the sarcomere length is 2.0 μm. As the muscle shortens through A, there is compression of the thick-filament/titin/Z-line structure resulting in reduced force. As the muscle shortens through B, opposing thin filaments have increased overlap, interfering with crossbridge binding to thin filaments. As the muscle is stretched through C, the ends of the thin filaments are separated through the H-zone of thick filament myosin reversal, so there is no change in crossbridge attachment or force. As the muscle is stretched through D, there is reduced overlap between thick and thin filaments, resulting in decreased force. Force reaches zero when there is no overlap between thick and thin filaments.

develop less force, and the other end would be able to pull it back to the center. This is not the case on the long end. Here, if one end of the thick filament got closer to the Z-line, it would reduce the actin–myosin contacts on the other end, and the thick filament would rapidly move to the Z-line. This does not occur because of the protein titin (Figure 5.1). Titin is a very large protein, with a molecular weight of almost 3M. It is anchored to the Z-line and the M-line, and keeps the thick filament centered in the sarcomere after contraction or stretch (Fukuda *et al.*, 2008) due to its large elastic domains. Titin bears most of the passive tension in striated muscle in the length range near L_o.

5.2   Muscle contractions after windup

Common experience shows that as athletes become more adept and well trained, they consistently develop more power, whether it is throwing a baseball, kicking a soccer ball, or swinging a tennis racket. An important element in this process is the windup. The windup has several effects. First, by activating stretch receptors, the alpha motor neurons of a stretched muscle are activated, increasing the number of working motor units, the combination of an alpha motor neuron and the muscle fibers it innervates. Second, the stretching of connective tissue prior to shortening will assist in the shortening process. Third, by stretching a muscle before its contraction, the muscle will use its length–tension curve more effectively. Skeletal muscles at rest sit near L_o, the length at which they develop their highest force. This occurs because their passive tension curve reaches zero force near L_o, as well as their structural arrangement through the attachment to bones. As a muscle contracts from L_o, it will slide down the length–tension curve (Figure 5.2), decreasing its force output. If a muscle is extended prior to contraction, it will move across the top of the length–tension curve as it shortens. For a given length of shortening, the amount of power generated by muscles can be calculated.

A muscle's tension T is measured in units of force/length. As it shortens, its length changes, ΔL. It will shorten at a given velocity V. A muscle's power output is the product of its force and velocity:

$$P = F \times V \tag{5.1}$$

which for the calculation of power as a function of shortening will be

$$P = T \times \Delta L \times V = (N/m) \times m \times (m/s) = J/s = W \tag{5.2}$$

with W being the number of watts generated by the contraction. From the length–tension curve for skeletal muscle (Figure 5.2), the $T \times \Delta L$ value is the area under the curve. This area can be calculated from the values in Figure 5.2. Using shortening velocity V as a constant, the relative power as a function of pre-stretching is shown in Figure 5.3. There is a substantial increase in power when a muscle is stretched before shortening. This contributes heavily to the increased performance when someone winds up before throwing or kicking. Note that all the curves reach a maximum, after which a longer extension starts to decrease the power output. This longer extension will also require a more vigorous contraction of the paired muscle in doing the stretching, as the stiffness of the

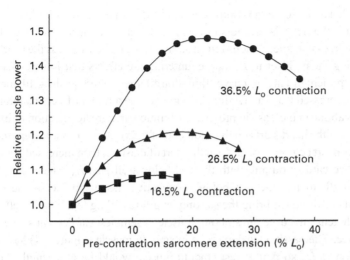

Figure 5.3 Effect of pre-stretching on muscle power output. Calculation of the power output for contractions of given lengths, 36.5, 26.5 and 16.5% L_o. Extending the sarcomere length past L_o prior to contraction uses the top of the length–tension curve. The greater the length of the contraction, the more effective pre-contraction extension, or windup, is at increasing muscle power output.

passive connective tissue will be increasing exponentially. For contractions that progress to very short lengths, the difference in power will be even greater, as the decrease in Ca^{2+} release from the SR that occurs at short lengths may decrease the maximum velocity due to tropomyosin head trapping. Proper use of the skeletal muscle length–tension curve is just one tool in the athlete's arsenal.

The length–tension relation in cardiac muscle has slightly different characteristics than the skeletal muscle curve, primarily due to the mechanical effects of connective tissue. When a skeletal muscle is stretched past L_o, there is an increasing amount of tension. The initial portions of this tension near L_o is borne by titin, as was noted above. Connective tissue, particularly the protein polymers collagen and elastin, then bear progressively more tension as the tissue is stretched. These proteins will be analyzed in more detail in the next chapter. Here, they contribute to the resting length of the muscle. In skeletal muscle, collagen and elastin are not significantly engaged until the muscle is past L_o. If there were tension on connective tissue at L_o, the rest length of the tissue would be shorter than L_o, minimizing the length until there was no force on the connective tissue when the muscle is not contracting. Using L_o as the starting point allowed the power analysis following a windup in Figure 5.3.

5.3 Cardiac and smooth muscle length–tension relations

In cardiac muscle, the tension on titin in skinned fibers, with no external connective tissue, reaches zero when fiber is near L_o, just as it does in skeletal muscle (Linke *et al.*, 1994). One of the most well known aspects of cardiac performance is Starling's law, which states that an increase in venous return produces an increase in cardiac output. This occurs because the heart is on the short end of the cardiac muscle cell length–tension curve at the start of diastole. Venous return stretches the muscle cells as the heart fills with blood, bringing them closer to L_o, and countering the effects that lower tension on the short end of the length tension curve: thin–thin filament overlap, thick filament–Z-line contact, and decreased Ca^{2+} activation. The greater the amount of venous return, within limits, brings about an increase in pressure generation and greater ejection of blood. This can only occur if the heart starts on the short end of the volume–pressure (length–tension) relation. If the heart cells were near L_o at the start of diastole, then increased venous return would decrease cardiac output, with the cell being well past L_o. Since titin brings the sarcomere length to L_o, and not below, connective tissue outside of the sarcomere (collagen and elastin) must bring the starting length for filling to a length well below L_o.

The length–tension curve in smooth muscle resembles cardiac muscle more than skeletal muscle. The passive tension curve in smooth muscle (Figure 5.4) has significant passive tension at L_o, so that at rest smooth muscle would be at a length shorter than L_o. Figure 5.4 also shows that smooth muscle will reach zero active tension at about $0.35\ L_o$, in sharp contrast to the zero active tension length in skeletal muscle at more than $0.60\ L_o$ (Figure 5.2). Smooth muscle does not have the organized sarcomeres of striated muscle. The thin filaments are attached to dense bodies, analogs of the Z-line. While there are some large proteins in the titin family in smooth muscle (Keller *et al.*, 2000),

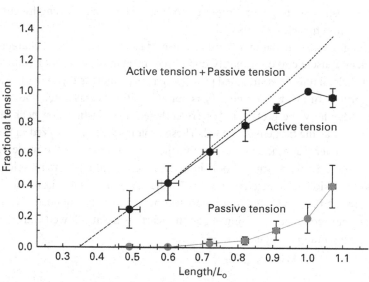

Figure 5.4   Smooth muscle length–tension relation in rabbit bladder strips. Smooth muscle does not have troponin, and does not have the sarcomere structure of striated muscle. The thin filaments are bound to dense bodies in the cytoplasm and on the membrane. The thick filament crossbridges will bind to actin and generate force. The smooth muscle length–tension curve is broader than the skeletal muscle length–tension curve. The passive tension in smooth muscle is borne by elastin and collagen.

mammalian smooth muscle thick filaments do not appear to have the same tethered arrangement as those in striated muscle. This may allow the filaments to rearrange themselves as the tissue shortens, and play a role in the greater breadth of shortening that occurs in smooth muscles.

Despite the lack of register of the filaments in smooth muscle, the force-generating capacity of arterial smooth muscle is almost twice that of skeletal muscle (Dillon and Murphy, 1982a). There is no evidence that individual smooth muscle crossbridges generate more force than individual skeletal muscle crossbridges. The force generated by muscle filaments is in part determined by the length of the filaments. The longer the thick filament, the more crossbridges will pull on the same thin filaments, resulting in a higher force. While the smooth muscle thick filaments are slightly longer than in striated muscle, they are not long enough to account entirely for the much higher force/cross-sectional area in smooth muscle. The other element controlling maximum force is the duty cycle of the myosin ATPase. The duty cycle is the fraction of time in the crossbridge cycle that the myosin head is actually generating force. For example, if the total cycle time of a crossbridge is 10 ms, and it is generating force for 3 ms before detaching, then this system has a duty cycle of 30%. While there are variable estimates of the duty cycle in skeletal muscle, 10–50%, tonic smooth muscle contractions appear to have a very high duty cycle (Dillon and Murphy, 1982b), a finding consistent with a decrease in shortening velocity during prolonged contractions (Section 5.7 below). The difference in skeletal and smooth muscle duty cycles appears to account for the much higher force/cross-sectional area of smooth muscle.

5.4   The Hill formalism of the crossbridge cycle

There have been a number of models of muscle force transduction. The model pioneered by T. L. Hill (1974) has proven to be particularly robust. In a more complex form, this formalism was used for the model of dynein cycling in the previous chapter. In its simplest form, the crossbridge cycle has four stages, shown in Figure 5.5. A is actin and M is myosin; the · indicates bound molecules and the + indicates separate molecules. The loss of the products ADP and P_i corresponds to the power stroke. After the power stroke, new ATP binds to the myosin in A·M complex, causing the separation of the

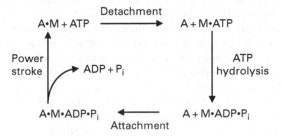

Figure 5.5   The crossbridge cycle

proteins. The bound ATP splits into its products ADP and P_i, which remain bound to the myosin. There is only a small change in the free energy in this step. The product bound myosin ($M \cdot ADP \cdot P_i$) collides with the actin at a high frequency, but only rarely binds and then releases the products. The release of the products, especially P_i (as discussed in the previous chapter), results in a drop in free energy and the generation of force. This is the rate-limiting step in the cycle. If the load on the muscle is less than the force the attached myosins can generate, the filaments will slide by one another until there is no force on the crossbridge. Another ATP will then bind and the cycle repeats.

The Hill formalism incorporates these different elements (Figure 5.6). The starting point of the model is recognition that at its lowest free energy, there will be no force on a crossbridge. Stretching or shortening a muscle rapidly produces an approximately linear relation between the change in length and the change in force. Since integrating a linear function yields a parabola, the energy of a bound crossbridge is a U-shaped function of filament position. When actin and myosin are not attached, there is no change in force as the filament slide by one another, yielding a flat energy profile for the $A + M \cdot ADP \cdot P_i$ state. The attachment form $A \cdot M \cdot ADP \cdot P_i$, before the release of product, sits at the bottom of its energy well (State 1 in Figure 5.6). The release of products results in a twist of the myosin head while still attached to actin (State 2). At this point, the crossbridge is not at its lowest energy, and will have generated a force. States 1 and 2 are at the same position, indicating that when force is generated, there is no sliding of the thick and thin filaments relative to one another. Entropy drives the system from State 2 to State 3, still in the

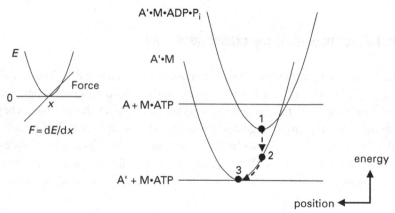

Figure 5.6   Hill curves: force generation and filament sliding. (Left) An actin–myosin complex will have an energy minimum at a particular position x. The force on the complex will be the length derivative of the energy, $F = dE/dx$. At the energy minimum, the force will be zero. (Right) Myosin (M) separates from actin A when ATP attaches. The myosin-attached ATP, $M \cdot ATP$, splits into its products, $M \cdot ADP \cdot P_i$, which remain attached to myosin. This complex can attach to a different actin, A', forming $A' \cdot M \cdot ADP \cdot P_i$. which sits at its free energy minimum at 1. When ADP and P_i dissociate, the complex $A' \cdot M$ is formed. This complex is not at it lowest free energy when it is formed, creating a force at 2. The complex will move to its lowest energy by sliding the filaments, until it reaches 3, the free energy minimum where there is no force on the complex. ATP will bind to the myosin and cause the actin to dissociate at its new position. The cycle then repeats. (Redrawn from Hill, 1974, with permission from Elsevier.)

attached A·M form, causing the filaments to slide by one another until the force on the crossbridge reaches zero. Thus, force generation and muscle shortening are separate processes. During isometric contractions, the muscle will not shorten after force generation, staying at position 2 until either the force resisting the contraction is decreased or muscle activation ends.

With the exception of insect flight muscle, the repeat distances of actin monomers on the thin filament and myosin monomers on the thick filament are not in register, so that within a muscle there will be an average force produced by the crossbridges. Each individual crossbridge will have a force either higher or lower than this average, representing the distribution of crossbridges. An isometric contraction will have a specific position on the Hill diagram, but that position will not be at the highest energy, because the average force cannot be at the highest force. As a muscle is shortening under an isotonic load, the average force and position will be displaced toward the zero force position by a fraction of the isometric force position based on the linear force/position relation.

There are elements in muscle contraction that are not present in the Hill model. When the shortening muscle has its crossbridges reach the lowest point on the energy curve, experimental evidence in striated muscle indicates that the crossbridge detaches. If it remained attached, it would slide up the opposite side of the energy curve and produce a resistive force on the shortening crossbridges, an internal load that would decrease the velocity. Crossbridge detachment at the lowest point is not a feature of the Hill model. Also, an attached crossbridge is not infinitely stiff: it has some elasticity. This would produce variations in attachment force and the free energy curve, elements incorporated in more complex forms of the Hill formalism, as would the different curves representing the slight temporal difference in the release of ADP and P_i. There are other models of crossbridge function: any model is as useful as the insight it provides. The generality of the Hill model, as evidenced by its application to dynein in Chapter 3, indicates its utility.

5.5   Muscle shortening, lengthening and power

Everyday experience tells us that we can throw light objects faster than we can throw heavy objects. There is an inverse relation between the load we place on our muscles and the velocity the muscles can contract. By varying the load on a muscle, the shortening velocity will also be changed. From this data, the maximum shortening velocity can be determined by extrapolation to zero load (see Figure 5.8a). An alternative method is to rapidly shorten an activated muscle until the tissue gets slack and then measure the time taken to redevelop force while under zero load. This method, shown in the lower part of Figure 5.7, does not give information on shortening velocities at intermediate loads.

A muscle contraction under a constant load is an isotonic contraction. If the load the muscle is trying to pull is heavier than the force the muscle can generate, the contraction will be isometric: there will be no change in muscle length. The upper part of Figure 5.7 shows how an isotonic force–velocity curve is generated. A muscle is activated isometrically, generating its maximum force F_o. The muscle is then released using a lever system

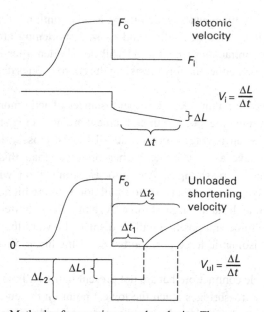

Figure 5.7 Methods of measuring muscle velocity. The upper panel shows isotonic velocity: the muscle is activated to its isometric force F_o, then released to a lower isotonic force F_i; there is a rapid recoil of the series elasticity followed by steady state change in length ΔL over time Δt, which yields the isotonic velocity V_i. Different isotonic velocities are plotted and the maximum velocity V_o determined by extrapolation, as shown in Figure 5.11. The lower panel shows the unloaded shortening velocity method: the activated muscle is shortened by different lengths sufficient to make the force go to zero and the tissue go slack. As the muscle contracts and takes up the slack it will eventually redevelop force: the muscle is contracting at V_o during the unloaded phase. To account for internal connective tissue stretching, the different length and time data points are plotted and the maximum velocity calculated from the slope of this data set, as shown in Figure 5.10.

to a load less than F_o, the isotonic force F_i. The muscle will rapidly recoil, with its passive tension element quickly coming to a new length determined by the new force F_i. Since the muscle can generate a force greater than F_i, the muscle will then shorten at a velocity $V_i = \Delta L/\Delta t$. By varying the isotonic load, a series of force/velocity points are produced. When plotted, these points yield the inverse relation between load (force) and velocity that we are familiar with (Figure 5.8a). Because muscles can never have zero load (they must shorten their own weight), the value of a muscle's maximum velocity V_o must be calculated.

Observing the curvilinear nature of the force–velocity curve, A. V. Hill (no relation to T. L. Hill above) fit isotonic points to an offset hyperbola, using the formula

$$(F + a)(V + b) = (F_o + a)b \qquad (5.3)$$

where F and V are the isotonic force and velocity, respectively, F_o is the isometric force, and a and b are constants (Hill, 1938). In the days before the availability of computers for non-linear line fits, Hill used a linearization of the offset hyperbola to calculate V_o. At V_o, the force F will be zero, so the FV equation becomes

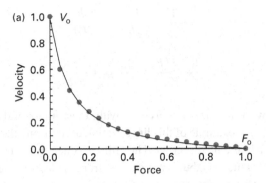

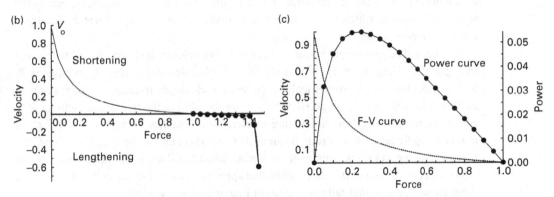

Figure 5.8   Skeletal muscle isotonic force–velocity and power curves. (a) A muscle is isometrically contracted to its maximum force F_o, then rapidly released to a lower isotonic force F_i. The muscle has a rapid shortening due to the recoil of its series elastic elements, then a steady shortening ΔL over a time Δt. Its isotonic shortening velocity V_i is $\Delta L/\Delta t$. By varying F_i, a curve is constructed and the maximum velocity V_o is calculated from the Hill velocity equation. (b) When an activated muscle is lengthened by the action of its paired muscle, it will extend very slowly until the stretching force approaches $1.5 \times F_o$, when it will yield and rapidly lengthen. (c) Muscle power is the product of its force and velocity. At F_o ($V=0$) and V_o ($F=0$), the power is zero, but at every other F_i and V_i, there is a positive power output. The power curve peaks at about $0.25\ F_o$.

$$aV_o + ab = F_ob + ab \tag{5.4}$$

$$aV_o = F_ob \tag{5.5}$$

$$V_o = \frac{F_ob}{a}. \tag{5.6}$$

Then, rearranging the original *FV* equation,

$$FV + Fb + aV = F_ob \tag{5.7}$$

$$\frac{FV}{F_ob} + \frac{F}{F_o} + \frac{aV}{F_ob} = 1 \tag{5.8}$$

$$\frac{FV}{F_o b} + \frac{aV}{F_o b} = 1 - \frac{F}{Fo} \qquad (5.9)$$

$$\frac{1 - F/F_o}{V} = \frac{1}{b} \cdot \frac{F}{F_o} + \frac{a}{F_o b}, \qquad (5.10)$$

the equation is now in the form $y = mx + b$, with, most importantly, the y-intercept of $a/F_o b$ equal to $1/V_0$. An example of this linearization used to calculate V_0 is shown in the inset of Figure 5.11 in the next section. The Hill FV equation is phenomenological, not thermodynamic: it fits the data, but is not derived from the laws of thermodynamics. The data it provides is useful in comparing FV behavior from different muscles, especially the values of V_0 and a, which reflects how much curvature the FV relation has, a factor that reflects the power output of the muscle.

An alternative method of measuring the maximum velocity is shown in the lower panel of Figure 5.7. The principle behind using the unloaded shortening velocity to measure V_0 is to activate the muscle isometrically, then rapidly release the muscle by a length change sufficient to make the tissue go slack. Under this condition, the tissue will contract with zero load, regathering itself and getting shorter, until it reaches a length at which it starts to redevelop force. By measuring the time Δt taken to take up the change in length ΔL that made the tissue get slack (this method is also called the slack test), an unloaded shortening velocity $V_{ul} = \Delta L/\Delta t$ is determined that approximates V_0. For different values of ΔL, there should be a Δt that falls on a straight line, whose slope is V_0.

Anatomically, almost all skeletal muscles exist in pairs, the contraction of one muscle opening a joint and its paired muscle closing a joint, as with the triceps and biceps opening and closing the elbow joint. Muscles only actively contract: they are lengthened by the action of the paired muscle stretching them out. If both muscles of a pair are activated at the same time, the muscle with the greater force will stretch out the other muscle in what is termed an eccentric contraction. When muscles are damaged during athletics, much of the time that damage occurs during eccentric contractions. Muscles can resist stretch with more force than they can generate. This resistance is at the heart of many athletic contests. If two athletes are pushing each other, one will probably be able to generate more force than the other. If muscles behaved like weights on a balance, the stronger force, like the heavier weight, would always win. Every contest would be boring. But because muscles can resist more force than they can generate, if two athletes have similar but not necessarily the same strength, elements like stamina, motivation and technique also play a role in the outcome. The limits of muscle resistance are shown in Figure 5.8(b).

When an activated muscle is stretched, the stretching force must be greater than F_o. The velocity of the eccentric stretch will be very low if the stretching force is less than $1.5 \times F_o$. Above this, the muscle will rapidly yield, and the stretched velocity will be very high. Using the Hill model from Figure 5.6, the eccentric contraction will extend the A·M attachment from State 2 up the energy curve, resisting with higher force, until the limit of the A·M attachment is reached, the upper end point of the energy parabola. Beyond this point, the crossbridge would break and the muscle will yield.

Eccentric contractions are thought to be linked to muscle hypertrophy. Stretching an activated muscle has been shown to cause microdamage in muscle. These focal lesions show that in some sarcomeres the filaments have been pulled apart so that there is no overlap of thick and thin filaments on one side of a sarcomere, and in others there is no overlap on either side. These damaged filaments, which would be dysfunctional, are interspersed among many intact filaments (Brown and Hill, 1991). This type of damage occurs when sarcomeres are stretched beyond L_o. At lengths past L_o, the longer the sarcomere, the weaker the force it generates. If some sarcomeres are weaker than others, they will be further extended by the stronger sarcomeres, causing the weak sarcomeres to rapidly extend to the point of pulling the filaments apart and eliminating the overlap – the popping hypothesis of muscle damage (Morgan and Proske, 2006). While this micro-damage does not kill muscle fibers, it does lead to hypertrophy.

Hypertrophy is a consequence of lifting heavy weights. Looking at the Hill model, when the muscles are producing forces near F_o, they will be farthest from their zero energy position and, from the length–tension curve, should be at a length near L_o. At lower forces, the average crossbridge position will be nearer to the minimum energy position. This positional difference means that when muscles exerting high forces are stretched, they are more likely to pass L_o and reach lengths where popping can occur. This damage has to be repaired. Since muscles do get stronger after this kind of exercise, there must be more filaments present. The enzymatic removal of the damaged filaments must in some way trigger the genetic production of filament-forming enzymes (Stewart and Rittweger, 2006). While the links between the microdamage and the induction of the enzymes have not yet been determined, the physical results tell us that they must exist. The microdamage repair is also consistent with the pain associated with lifting heavy weights. Partially digested proteins are strong activators of pain receptors, and the temporal profile of muscle soreness parallels the time course of hypertrophy. The process takes several days to complete, and as all well-trained athletes know: working the same muscle groups hard on consecutive days significantly increases soreness without increasing the rate of muscle hypertrophy.

The isotonic FV curve also gives us information on the muscle power output, its energy/time. The formula for power is

$$P = F \times V. \tag{5.11}$$

At F_o, the velocity is zero, so the power is zero. At V_o, the force is zero, so the power is zero. At every other F_i and V_i, the product of the force and velocity will be a positive number, as shown in Figure 5.8(c). Since the two ends of the power curve must be zero, and every other value is positive, the power curve must have a maximum. In skeletal muscle, this occurs at about 0.25 F_o. This figure has practical value. Suppose you have to move a pile of bricks. You can lift 100 pounds. How much should you carry on each trip? While there are other factors such as balance to consider, as a first approximation, if you carry 25% of your maximum, or 25 pounds at a time, you will maximize your energy output, getting the most work done with the least effort. Carry more, you slow down a lot. Carry less, and you have to make more trips. The practical side of muscle mechanics.

5.6   Calcium dependence of muscle velocity

Calcium is necessary to activate all muscles. In striated muscle, calcium binds to troponin on the thin filament. Calcium-bound troponin causes tropomyosin to shift into the groove of the actin filament, exposing the actin–myosin binding sites on the thin filament. This removal of inhibition allows the myosin head to bind to actin, generating force and shortening. The level of activation of striated muscle is determined by the calcium concentration: more calcium, more crossbridges, more force. Unlike force, the maximum velocity V_o should be independent of the calcium concentration. If all the crossbridges are active, and there is no load, the muscle will contract at V_o. If there were only half as many active crossbridges, but still no load, the muscle should still contract at V_o. This is analogous to a horse pulling a cart with no mass: adding extra horses will not make the cart go faster if there is no load to pull. Experiments measuring the velocity of skeletal muscle at different levels of calcium activation should have found, theoretically, that there was no calcium dependence of muscle velocity, as in the left half of Figure 5.9. This is what Podolsky and Teichholz found (1970). In contrast, Julian (1971) found that there was a calcium dependence of velocity: at low calcium, there was a lower maximum velocity than the V_o at high calcium, as on the right side of Figure 5.9. This difference produced quite a controversy at the time, for it appeared that both could not be right.

It actually took decades for this matter to be resolved (Swartz and Moss, 2001). Using the unloaded shortening velocity method, a skinned skeletal muscle fiber was activated using different concentrations of calcium (Figure 5.10). The velocity measured up to 40 ms was independent of calcium, but for longer times was calcium dependent. Experiments show that the binding of a non-tension-producing molecule to the thin filaments eliminates the calcium dependence. It appears that when a muscle is partially activated, the initial crossbridge cycles move at V_o, but that the movement of tropomyosin, which in partially activated muscle will be moving back and forth between the "on"

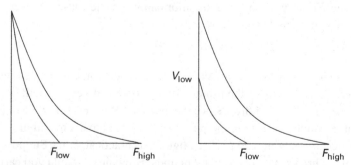

Figure 5.9   Calcium independent and calcium dependent skeletal muscle force–velocity curves. Force development is calcium dependent, with high calcium producing high force and low calcium producing low force. (Left) In theory, at zero load the maximum velocity V_o should be independent of the isometric force, whether at high or low calcium. (Right) Calcium-dependent maximum velocities at zero load imply a second mechanism separate from the calcium dependence of crossbridge formation.

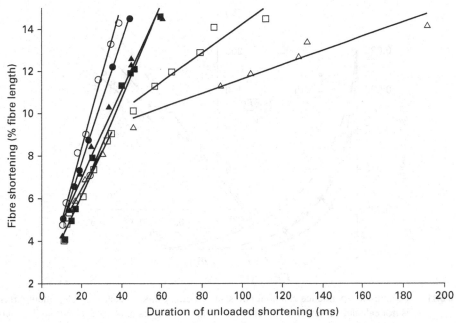

Figure 5.10 Calcium dependence of skeletal velocity during unloaded shortening. Using the unloaded velocity method, skinned single skeletal muscle fibers (open symbols) had an initial calcium-independent velocity phase, followed by a calcium-dependent velocity phase. Treatment with N-ethyl-maleimide subfragment-1 (closed symbols), which binds tightly and randomly to the thin filaments and does not develop or bear tension, eliminated the calcium-dependent phase. The slowing of velocity after 40 ms may be related to some tropomyosin moving to the "off" position and creating an internal load on the muscle. (Redrawn from Swartz and Moss, 2001, with permission.)

and "off" states, creates an internal load on the muscle, perhaps briefly trapping the crossbridge heads, and so causes the decrease in velocity. Tropomyosin takes about 40 ms to move toward the off position, so that in the short duration shortenings remained calcium independent.

There are several aspects about tropomyosin that can be inferred from experiments. In Chapter 2 we saw that calcium binds to troponin with an energy of 8 kJ/mol. Since calcium binding to troponin is the driving force for tropomyosin movement, this places an upper limit of 8 kJ/mol on the energy needed to move tropomyosin. Tropomyosin is a filamentous protein that spans six to seven actin monomers on a thin filament. Since that entire molecule appears to move into the groove, the binding energy of tropomyosin to actin at one calcium-controlled site cannot exceed 8 kJ/mol, or less than the energy of two hydrogen bonds. An internal load due to tropomyosin binding would be relatively small, but if there are stearic limitations on the movement of the myosin head when tropomyosin is partially restricting myosin's movement, there could be a greater internal load and a further decrease in velocity. Depending on when velocity is measured, both possibilities in Figure 5.9 could be correct.

The possibility of a tropomyosin influence on shortening velocity is supported in part by results from smooth muscle (Figure 5.11). Smooth muscle has tropomyosin, but does

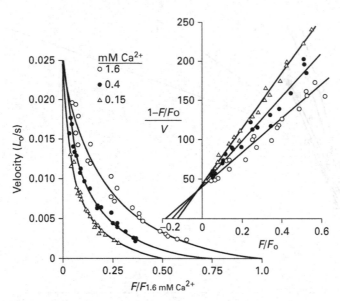

Figure 5.11 Calcium independence of smooth muscle velocity. Unlike skeletal muscle, smooth muscle velocity is not calcium dependent. Smooth muscle has tropomyosin, but it does not inhibit myosin ATPase activity. The absence of troponin in smooth muscle means that tropomyosin does not change position relative to the actin groove, and cannot trap the myosin heads. (Redrawn from Dillon and Murphy, 1982b, with permission.)

not have troponin. The tropomyosin in smooth muscle cannot be blocking the actin–myosin binding site, since there is no mechanism for moving it out of the way. Smooth muscle is activated by the calcium-dependent phosphorylation of the myosin regulatory light chain, also called LC20. Light chain phosphorylation activates the myosin ATPase, which then goes through the same crossbridge cycle as striated muscle. Smooth muscle force is sensitive to extracellular calcium, as shown on the force axis of the force–velocity curves in Figure 5.11. V_o in smooth muscle is not calcium dependent: even at a calcium concentration that only produces 50% of the maximum force, the value of V_o is unchanged (Dillon and Murphy, 1982b). Since the tropomyosin presumably does not change its position in smooth muscle, it does not place an internal load on the system. This does not prove that tropomyosin is responsible for the calcium dependence of skeletal muscle velocity (there could be other mechanisms), but neither does it refute that idea.

The controversy sparked by the different results of experiments on the calcium dependence of skeletal velocity is an object lesson for both young and old scientists. While some scientists may fake their data, in most cases differences in results will arise from small, and at the time unappreciated, differences in techniques. Small differences in the time at which velocity was measured, or conditions that altered the rate of tropomyosin movement, could have produced the different results. The gold in this type of controversy is in finding what made the difference. If a scientist is aware of apparently divergent results, looking for those small differences may produce the wedge that opens a solution. Rather than saying a pox on both your houses, assuming that both sides are honest creates an opportunity for the one clever enough to resolve it.

5.7   Smooth muscle latch

The biochemical activation of smooth muscle was shown to require the phosphorylation of the 20 kDa regulatory light chain of myosin (Gorecka *et al.*, 1976; Chacko *et al.*, 1977). The MLC-P form activates the myosin ATPase. In contrast to the removal of inhibition by the tropomyosin shift in striated muscle, light chain phosphorylation produces an enzymatic activation. The discovery of the ATPase activation led to measurement of MLC-P in smooth muscle tissues. It was discovered, as shown in Figure 5.12, that although MLC-P rose during the initial development of force in smooth muscle the level of phosphorylation fell over time despite the maintenance of force. Measurements of the load-bearing capacity in smooth muscle, also shown in Figure 5.12, were made to determine whether the force borne by each crossbridge changed over time, but the load-bearing capacity paralleled the force development (not shown in the figure), indicating that the light chain dephosphorylated crossbridges essentially bear the same load as the light chain phosphorylated crossbridges. It was subsequently found that the muscle-shortening velocity rose and fell with the light chain phosphorylation (Dillon *et al.*, 1981). The separation of the force and velocity characteristics indicated that a second mechanism, independent of the initial force development, must be occurring in smooth muscle. This second mechanism was termed latch in vertebrates, for its mechanical similarity to the catch muscles in molluscs that can maintain force with very low energy use, although molluscs use a mechanism involving a protein, paramyosin, that is not present in mammalian smooth muscles.

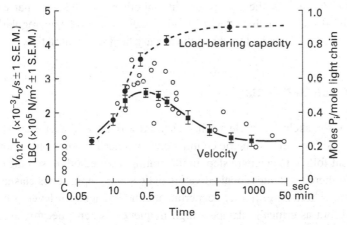

Figure 5.12   Time-dependent velocity of smooth muscle. Phosphorylation of the myosin 20 kDa light chain (open circles) activates the myosin ATPase, resulting in force generation and shortening in smooth muscle. In this data from swine carotid artery strips, the load-bearing capacity, proportional to the isometric force, increases with light chain phosphorylation, but is maintained as the phosphorylation level falls. Muscle shortening velocity tracks with the rise and fall of the phosphorylation. These experiments are the basis for the latch model. (Redrawn from Dillon *et al.*, 1981, with permission.)

The initial hypothesis for the mechanism of latch only involved the dephosphorylation process. Since the force was maintained by muscles in latch, the attached crossbridges were assumed to be dephosphorylated in the attached position. If these dephosphorylated crossbridges had a decreased rate of detachment, they could bear force with a high economy. Economy is the energy required to maintain force without the muscle having to shorten and do work. Efficiency is the energy required to do work. Smooth muscles have a very high economy, being able to maintain force for long periods of time with very low energy use. This economy would be useful in large arteries, such as the carotid artery in which latch was originally found, where the resistance to distending blood pressure could be maintained with minimal energy expenditure.

The second mechanism at work in the simplest model had force being maintained by two separate populations of crossbridges: phosphorylated and dephosphorylated. If the dephosphorylated crossbridges did not detach from actin at the lowest point in the Hill model, but detached at a position past the lowest point, they would create an internal resistance to shortening, an internal load that would decrease the shortening velocity. It is now known that light chain phosphorylation causes the myosin head to unfold in order to develop force (Sweeney, 1998). The dephosphorylation of attached crossbridges could return those crossbridges toward the folded position, altering their detachment rate and causing latch. While this is still a reasonable mechanism for latch, there have been many, many other models of how latch might work, including both thin and thick filament models as well as those invoking other cytoskeletal elements. While the mechanism remains a topic of research, many of the fundamental elements of the process, including the dephosphorylation of myosin, the maintenance of force, the fall in velocity, the decrease in energy use, and the fall in intracellular calcium leading to a decrease in myosin ATPase activity, have been firmly established. The evolutionary development of a process that minimizes the energy use in smooth muscle tissues that either resist distension for decades, such as large blood vessels, or those that regularly fill and empty, like the gastrointestinal tract and the urinary bladder, is attractive.

5.8   Muscle tension transients

As the methods in skeletal muscle mechanics improved, experiments showed anomalies during the transition from isometric contractions to either isotonic contractions or force redevelopment following a length step. In illustrating the method of isotonic contraction in Figure 5.7, there is an intentional curvature between the rapid series elastic recoil and the steady-state shortening. In early experiments, the bouncing of lever arms obscured this transient, but as critically damped, high frequency systems became available, this length transient became apparent in experiments where the force was changed and the length change followed. Experiments delineating the transient used the reverse experiment, where the isometrically contracting muscle was subjected to a length step, followed by a tension transient (Huxley and Simmons, 1971). This experiment is shown in the inset of Figure 5.13, where the isometric tension T_o is reduced by a very rapid length step shortening, followed by tension redevelopment. The lowest tension after the step

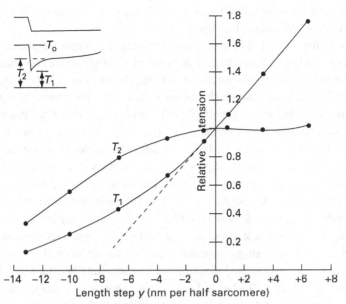

Figure 5.13 Tension transients during rapid release. As techniques for measuring muscle velocity and tension transients after release improved, a post-release non-uniformity was noted. When isometric tension T_o is very rapidly released, there is an initial rapid fall in tension to T_1, followed by a rapid tension recovery to T_2, before the muscle shows a steady state tension redevelopment. These tension transients are thought to reflect the slackening, but not detachment, of attached crossbridges. The extrapolation of the T_1 value is the length change needed to detach all crossbridges during an infinitely fast release, and it is therefore the maximum extent an attached crossbridge can be strained before detaching from actin. (Redrawn from Huxley and Simmons, 1971, with permission from Macmillan Publishers.)

was the T_1 tension, followed by the tension transient, the rapid increase in tension to a brief step at T_2, thereafter followed by the steady redevelopment of a higher tension over a much longer time frame.

The tension transient was interpreted as the behavior of attached crossbridges that were momentarily relieved of tension on the actin monomer, but that did not detach. These still attached crossbridges then rapidly redeveloped force, as they did not have to pass through an ATPase cycle to reassert tension. The longer the length step, the more crossbridges would become detached, and the lower the redeveloped force. Theoretically, if the release was of sufficient size and could be as fast as the myosin ATPase rate, all the crossbridges would break, the T_1 tension would be zero, and there would be no T_2 phase. The size of the length step that would produce zero T_1 tension, extrapolated from the dashed line in Figure 5.13, would be the farthest extent any attached crossbridge could be moved without detaching, essentially the range of reach of a crossbridge. These measurements also define the elasticity within attached crossbridge. This elasticity would reside in both the actin and myosin proteins, although most of the elasticity is presumed to lie within the myosin head and neck region. The transients in Figure 5.13 were the first experiments explicitly studying this phenomenon. Subsequent experiments with progressively higher frequency systems have pushed the experimental data down the dashed line, confirming the original Huxley–Simmons findings.

In discussing force development and muscle shortening, a single equation describing these processes has not been included. This is not an oversight: there is no such equation. This is not for lack of quality scientists in this field. The work of A. V. Hill, A. F. Huxley, and T. L. Hill has been included in this chapter. A. V. Hill (for muscle heat production) and A. F. Huxley (for the action potential) both received the Nobel Prize before spending decades researching muscle mechanics. And there has probably not been a finer theoretician in all of biophysics than T. L. Hill. The difficulty comes from the dual nature of muscle mechanics. Once a crossbridge is attached, its behavior can be analyzed thermodynamically based on its position and force. But detachment separates the actin and myosin proteins. The flat A + M·ATP line of the Hill model means there is no energetic connection between the thick and thin filaments. In mathematics, this represents a discontinuity. The attachment process is a probability, not a continuous function. Mathematical models with single function equations do not have discontinuities. Both the probabilities, and the attached transients and steady-state behavior, can be studied, but mathematically they will always be separate processes. A model can consist of deterministic behavior while bound and probabilistic behavior while detached, but there will never be a single deterministic equation defining all aspects of muscle force activation, force development, and shortening.

5.9   The Law of Laplace for hollow organs

The Law of Laplace has multiple applications in physiological systems. It relates the pressure gradient ΔP, the wall tension T and the radius r of hollow structures. For walls with significant thickness relative to the radius (Figure 5.14), such as large arteries, the bladder or the heart, the relation is

$$T = \Delta P[r_o/(r_o - r_i)] \tag{5.12}$$

where r_o and r_i are the outer and inner radii, respectively, and ΔP is the transmural pressure gradient ($P_i - P_o$). The outer part of a hollow organ wall bears more of the

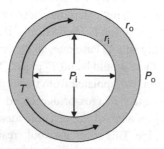

Figure 5.14   Law of Laplace in thick-walled cylinders. The internal pressure P_i and the external pressure create a pressure difference ΔP. The wall tension is equal to $T = \Delta P [r_o/(r_o - r_i)]$. The tension is greater on the outside part of the wall than on the inside.

tension than the inner part. Consider the radius r_m that is the midpoint between r_o and r_i. The pressure gradient will be uniformly distributed across the wall. The pressure gradient from r_i to r_m will be $\Delta P_{1/2}$, as will the pressure gradient from r_m to r_o. The wall tension on the inner and outer parts of the wall will be T_i and T_o, respectively, with

$$T_i = \Delta P_{1/2} \frac{r_m}{(r_m - r_i)} \tag{5.13}$$

$$T_o = \Delta P_{1/2} \frac{r_o}{(r_o - r_m)}. \tag{5.14}$$

Since $\Delta P_{1/2}$ is an element of both equations, the equations can be combined:

$$T_o \frac{r_m}{(r_m - r_i)} = T_i \frac{r_o}{(r_o - r_m)}. \tag{5.15}$$

The two divisions of the wall will be equal:

$$(r_m - r_i) = (r_o - r_m) \tag{5.16}$$

so that

$$T_o r_m = T_i r_o. \tag{5.17}$$

Since $r_m < r_o$, then $T_o > T_i$. The wall tension is borne by both muscle and connective tissue. The elements on the outermost part of a hollow organ will bear most of the pressure, and the muscle there will have to use more energy than the inner parts of the organ. We will see this principle again in the next chapter in discussing bones and aneurysms.

For thin-walled structures, the wall thickness is insignificant relative to the radius. For spheres, there are two radii of curvature perpendicular to one another, with both radii equal. Here the Law of Laplace that applies is

$$P = T\left(\frac{1}{r_1} + \frac{1}{r_2}\right) = \frac{2T}{r}, \tag{5.18}$$

a situation that approximates the conditions in an alveolus, recognizing that the entry point of the terminal bronchiole means that the alveolus is not a perfect sphere. For thin-walled cylinders such as capillaries with a cylinder radius of r, the perpendicular radius (i.e., the radius perpendicular to the circular cross-section of the cylinder) is infinite, so the Law of Laplace is reduced to

$$P = \frac{T}{r}. \tag{5.19}$$

The utility of the Law of Laplace comes into play because all the elements in it can change. There are changes in the pressure gradient in the lungs, heart, blood vessels, bladder and uterus. The wall tension varies as a function of muscle contraction and distension, engaging the connective tissue and altering the length–tension position of the muscle in the wall. Contraction of the muscle will alter both the radius and the tension. Understanding these relations applies to both normal and pathological states.

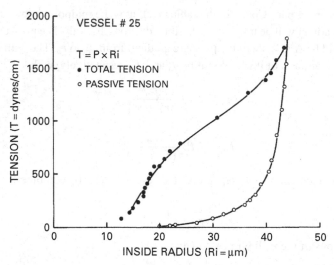

Figure 5.15 Radius–tension relation in an arteriole. The peak difference between the passive tension and the total tension occurs at 36 μm, corresponding to L_o. The radii at which the passive tension and the active tension reach zero occur at fractions of L_o similar to those of smooth muscle strips (Figure 5.4). The circular arrangement of smooth muscle in blood vessels does not fundamentally alter the length–tension relationship of the filaments. (Reprinted from Gore and Davis, 1984, Figure 4, with permission from Springer Science+Business Media.)

Figure 5.15 shows the radius–tension relation for an arteriole (Gore and Davis, 1984). For an arteriole with a diameter of 30–100 μm, its smooth muscle cells will have a significant curvature. The filament lengths are less than 2 μm, and will generate tension perpendicular to the distending blood pressure. As a whole, however, the internal structures of the cells will have to angle the successive contractile units to accommodate the curvature. Despite these requirements, the radius–tension relation is very similar to the length–tension relation for bladder smooth muscle strips in Figure 5.4, taken from the wall of the bladder where the cells would have very little curvature over the length of the cells. In both cases, there is significant passive tension at L_o and beyond, limiting the ability of any blood pressure to stretch the cells far beyond L_o. The short end of the curves approaches 35% of L_o in both cases, so that on both the large organ level of the bladder and the small scale of the arteriole smooth muscle cells can shorten to a fractionally shorter length than striated muscle.

The Law of Laplace also has important implications for the heart. While not a perfect sphere, the change in radius as it fills will have an effect on the heart's mechanics. The key variable is the change in pressure ΔP needed to drive the blood through the circulatory system. Rearranging Laplace's equation for thick-walled organs to solve for the change in pressure, we get

$$\Delta P = T((r_o - r_i)/r_o). \qquad (5.20)$$

For an increase in r_o, the value of $(r_o - r_i)$ will decrease as the wall thickness gets thinner. Thus, with increasing radius, the ratio of $(r_o - r_i)/r_o$ will always decrease. That means that in order to maintain the same pressure gradient (needed to drive blood) as the radius increases, the wall tension T must increase. Physiologically, the muscle cells must generate more force in order to increase the wall tension. This will increase the oxygen consumption of the heart, and if the energetic needs of the heart exceed its ability to generate ATP, a heart attack will result. This is what occurs in dilated cardiomyopathy. In a healthy heart, Starling's law has an increase in venous return leading to an increase in cardiac output. This appears to be at odds with the decrease in blood pressure for a constant wall tension. Starling's law works because the other factors that occur with increased venous return (better overlap of filaments and more calcium), more than compensate for the increase in radius. In congestive heart failure, the enlargement of the heart puts it closer to the top of the pressure–volume relation, and any further increase in radius will only minimally increase the Starling effect, while increasing the Laplace effect. This will diminish cardiac output, resulting in increased venous pressure. Left-side heart failure compromises lung function by increasing the water between the alveoli and the capillaries, as we saw when discussing diffusion. Right-side failure leads to blood pooling in the legs. Both increase the risk of clot formation and the possibility of thromboembolism.

5.10   Non-muscle motility

Motility in non-muscle cells is driven by a form of myosin II. This myosin is activated using the same light chain mechanism as smooth muscle, the phosphorylation of the 20 kDa light chain (Kolodney and Elson, 1993). Migrating cells come in many types, including leukocytes, growth cones in neurons, and metastatic cancer cells. Cells can move along two-dimensional surfaces such as leukocytes migrating along the inner surface of capillaries, or three-dimensionally through extracellular connective tissue (Friedl and Bröcker, 2000). The presence of chemotaxins can activate what becomes the forward edge of the cell, for example drawing leukocytes to an area of infection.

Cell migration has several common factors (Figure 5.16), regardless of the cell type (Kirfel *et al.*, 2004; Chan and Odde, 2008). The protrusion of the leading edge is

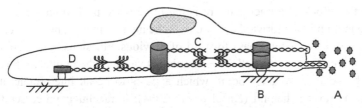

Figure 5.16   Non-muscle motility. Migrating cells use actin–myosin interactions for movement. Leukocytes are directed by chemotaxins (A), causing in response the extensions of actin filaments against the membrane-forming lamellipodia, while at the same time anchoring the cell at multiple points against an external surface (B), if available. Activation of non-muscle myosin II pulls the center of the cell toward the anchored front end (C). Prior anchor sites at the rear of the cell are broken, possibly leaving integrin molecules behind on the external surface (D). Since the cell moves forward, the weakest attachment sites must be at the rear of the cell.

associated with actin polymerization, extending the membrane without having the actin filament penetrate the membrane, forming lamellipodia (a broad front extension) or filopodia (rod-like extensions). These actin filaments must be anchored, or they would not produce the vectorial force needed to deform the membrane. There is no myosin in the extension region, so it cannot be pulling the actin filaments forward. The myosin pulls on actin filaments on the "other" side of the extending actin filaments: that is, actin filaments with an opposite polarity to those extending the membrane. Without an anchor cite, the centrally located myosin would pull the actin filaments inward, defeating the purpose of the extension. This means that there must be an anchor site between the myosin and the extending actin, forming an extracellular adhesion. There is an extensive literature on the nature of the adhesion proteins, with integrins being the most outward protein in the complex, adhering to some extracellular surface.

The nature of the adhering surface plays a role in the rate of cell migration. Very stiff surfaces, such as glass, provide significant traction to the migrating cell, while softer surfaces do not provide the same traction, the cells slip, and the forward motion is slower. Researchers in the field have described it as the difference between running on pavement and running on sand. The external proteins, anchored through the membrane to the actin-binding structures, go through a slip-and-stick mechanism. Internal activation of myosin would draw the extracellular complex along the surface until it strongly adheres, after which the myosin activity would pull the central part of the cell forward, toward the extending edge.

As the cell moves forward, the adhesion sites at what is now the back of the cell must be broken. Without disadhesion, the cell will be anchored to its present site. The cell is able to break the adhesion sites at the back of the cell, leaving part of the original adhesion complex behind, and creating a molecular trail marking its path (Kirfel et al., 2004). The internal myosin contractions must be sufficiently strong to disrupt the protein complex when pulling it forward from the back of the cell. This is in contrast to the forward motion at the front of the cell, where the myosin does not rupture the complex. While there may be protease activity that weakens the complex at the back, it cannot disrupt the actin filaments prior to disadhesion, or the myosin force could not be transmitted. The simplest model has anisotropic binding of the protein complex to the extracellular surface, a conclusion consistent with the slip-and-stick behavior. The adhesion site works much like a starting block of a track race, resisting force when pushed back against, but not restricting motion in the forward direction. As the internal myosin pulls the cell forward, the "center of force" would pass forward of previous adhesions, and would now pull them toward the front of the cell. The adhesion is weaker when pulled in this direction, and will break at its weakest point, which appears to be an interprotein attachment, leaving some protein behind (Kirfel et al., 2004). Intracellular processes then make globular actin, G-actin, subsequently available to form new filamentous actin, F-actin, at the front end of the cell to continue the process.

An interesting but as yet unstudied aspect of non-muscle motility are the temporal changes in cell activation. The phosphorylation of the myosin light chains, and the changes in velocity, in fibroblasts parallel those in smooth muscle (Kodolney and Elson, 1993). While these changes in phosphorylation and velocity have generated a great deal of work on potential latch mechanisms in smooth muscle, there is no parallel work in non-muscle motility. Are the mechanisms that slow smooth muscle contraction also operating in non-muscle cells? Do the mechanisms that control migrating cell positions also create internal loads in smooth muscle? And can experiments in non-muscle cells help resolve the many possible latch control mechanisms? These questions remain unanswered for the present.

Muscles use interactions between actin and myosin, organized into overlapping thin and thick filaments, to convert the scalar myosin ATPase into a vectorial force. The physical arrangement of the filaments produces length-dependent force generation, with decreased force at both short and long lengths. The length dependence is different in skeletal muscles compared with cardiac and smooth muscles, primarily due to the arrangement of connective tissue. The linear force generation in skeletal muscle is optimized near its resting length, but stretching prior to contraction increases the power output. Hollow organs with muscles in their walls, like the heart and bladder, work on the short end of their length–tension curves, resulting in increased force as they fill. Models of the muscle crossbridge cycle show dependence on the dissociation of the ATPase product P_i, released along with ADP. All muscles are activated by calcium-dependent mechanisms, but these are different in striated and smooth muscles. Smooth muscles have a very high economy, maintaining force for long durations with very low energy use. For hollow organ tissues, the Law of Laplace can be applied, showing that their wall tension and energy utilization is primarily borne by the outer part of the tissue. Non-muscle tissues migrate using the same biochemical mechanism as smooth muscle, moving forward by transient filament formation. Muscle is one of the most well under-stood physiological tissues, but continues to be an area of significant research.

References

Brown L M and Hill L. *J Muscle Res Cell Motil.* **12**:171–82, 1991.
Chacko S, Conti M A and Adelstein R S. *Proc Natl Acad Sci U S A.* **74**:129, 1977.
Chan C E and Odde D J. *Science.* **322**:1687–91, 2008.
Dillon P F and Murphy R A. *Circ Res.* **50**:799–804, 1982a.
Dillon P F and Murphy R A. *Am J Physiol.* **242**:C102–8, 1982b.
Dillon P F, Aksoy M O, Driska S P and Murphy R A. *Science.* **211**:495–7, 1981.
Friedl P and Bröcker E B. *Dev Immunol.* **7**:249–66, 2000.
Fukuda N, Granzier H L, Ishiwata S and Kurihara S. *J Physiol Sci.* **58**:151–9, 2008.
Gordon A M, Huxley A F and Julian F J. *J Physiol.* **184**:170–92, 1966.
Gore R W and Davis M J. *Ann Biomed Eng.* **12**:511–20, 1984.
Gorecka A, Aksoy M O and Hartshorne D J. *Biochem Biophys Res Comm.* **71**:325, 1976.
Hill A V. *Proc R Soc London B.* **126**:136–95, 1938.
Hill T L. *Prog Biophys Mol Biol.* **28**: 267–340, 1974
Huxley A F and Niedergerke R. *Nature.* **173**:971–3, 1954.
Huxley A F and Simmons R M. *Nature.* **233**:533–8, 1971.
Huxley H and Hanson J. *Nature.* **173**:973–6, 1954.
Julian F J. *J Physiol.* **218**:117–45, 1971.

Keller T C 3rd, Eilertsen K, Higginbotham M, *et al*. *Adv Exp Med Biol*. **481**:265–77, 2000.

Kirfel G, Rigort A, Borm B and Herzog V. *Eur J Cell Biol*. **83**:717–24, 2004.

Kolodney M S and Elson E L. *J Biol Chem*. **268**:23 850–5, 1993.

Linke W A, Popov V I and Pollack G H. *Biophys J*. **67**:782–92, 1994.

Morgan D L and Proske U. *J Physiol*. **574**:627–8, 2006.

Podolsky R J and Teichholz L E. *J Physiol*. **211**:19–35, 1970.

Pugh J. *Treatise on Muscular Action*. Charlottesville, VA: University of Virginia Medical Library Rare Book Room, QP 301.P8, 1794.

Stewart C E and Rittweger J. *J Musculoskelet Neuronal Interact*. **6**:73–86, 2006.

Swartz D R and Moss R L. *J. Physiol*. **533**:357–65, 2001.

Sweeney H L. *Am J Respir Crit Care Med*. **158**:S95–9, 1998.

6   Load bearing

The contractions of skeletal and cardiac muscle cause force generation and tissue move-ment. These forces and movements are borne by other tissues: bone, teeth and connective tissue. These load-bearing tissues have common elements, as well as particular character-istics in normal and pathological conditions. They do not generate forces themselves – at least, not above the cellular level during growth and response to outside forces. Load-bearing tissues will respond to changes in their length, producing stress responses. The general properties of load bearing, as well as specific topics covering bone, teeth, blood vessels, and the tendons, ligaments and cartilage of joints are included below.

6.1   Stress and strain

The most elemental of the load-bearing responses is Hooke's law, which relates stress and strain. While these terms are often used interchangeably in everyday language, they have specific definitions in biophysics. The stress σ is

$$\sigma = \frac{F}{A} \qquad (6.1)$$

where F is the force on the tissue and A is its cross-sectional area perpendicular to the applied force. Stress is therefore defined in units of pressure. Strain ε is a fractional length change of a tissue:

$$\varepsilon = \frac{\Delta l}{l} \qquad (6.2)$$

where l is the initial length of the tissue and Δl is the change in length. Since strain is the ratio of two lengths, it is dimensionless. Hooke's law is demonstrated in the upper left of Figure 6.1. Starting at the origin, the stress is zero and the strain is zero prior to any applied stretch. Hooke's law says that there will be a linear stress response to a change in length, and that after the stretch is stopped and the tissue released, the stress will return to zero along the same stress/strain line. The slope of the stress/strain line is called the modulus of elasticity, or Young's modulus. This is the most commonly used term in measuring a tissue's elasticity. As we will see, not all biological tissues obey Hooke's law, but linear approximations of non-linear behavior are still used to estimate Young's modulus in these tissues.

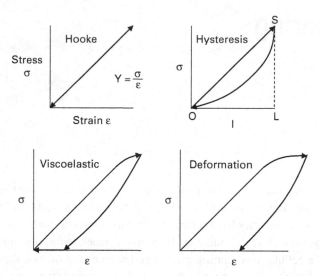

Figure 6.1   Stress–strain relations in tissues. (Upper left) According to Hooke's law, there is a linear relation between stress σ and strain ε. Young's modulus, also called the modulus of elasticity, is the ratio of stress/strain. (Upper right) In some tissues, extension of the tissue is linear, but the return to the rest length runs along a curved path, creating a hysteresis loop. The area of the loop corresponds to energy lost during the stretch and return. (Lower left) Beyond the Hooke limit, if a tissue has viscoelasticity, the initial stretch will return to zero stress at a greater length than the initial length. The tissue will slowly return to the original length. (Lower right) Beyond the viscoelastic limit, for extended strains, the tissue may deform, altering its structure so that it never returns to its original length.

In a number of biological tissues, the line of stretch does not coincide with the line of return. This produces a hysteresis loop, in which the stress is lower during the return phase than during the stretch. For this behavior, shown in the upper right of Figure 6.1, the usual convention of the strain term ε is replaced with an actual length measurement L. In a tissue with a hysteresis loop, there are two defined areas: the stretch region, defined by the triangle OSL, and the arc triangle of the return, defined by the arc OS, and the sides OL and SL. When using the actual length change L, the area of the stretch triangle is

$$\text{Area} = \frac{bh}{2} = \frac{l\sigma}{2} = \frac{L}{2} \cdot \frac{F}{A} = \frac{\text{Energy}}{\text{Cross} - \text{Section}}. \tag{6.3}$$

During the return, the arc triangle will always have a smaller area than the stretch triangle. Since the system returns to the same physical position O after the loop, the system will have done no net work. The area of the loop is the difference between the two areas:

$$\text{Loop} = \overset{\Delta}{\text{OSL}} - \overset{\cap}{\text{OSL}} = \text{Heat}. \tag{6.4}$$

This region is the heat lost, or entropy change, of moving the tissue through the hysteresis loop. As we will see, this process reflects the activity of the tissue.

Many biological tissues exhibit viscoelastic behavior, shown in the lower left of Figure 6.1. In these tissues, a rapid stretch is followed by a rapid fall in stress to zero, but the tissue reaches zero stress at a longer length than the original tissue length. Over time, the tissue will then slowly return to its original length. When a tissue obeyed Hooke's law, it acted as a pure spring, quickly returning to its original length. Viscoelastic tissues behave as though they also have a damping element, analogous to a shock absorber, which limits the rate of return without actually preventing the return of the tissue to its original length. Isolated tissues behave as though this damping element is in series with the spring, others as if the damping is parallel to the spring, and many as if there are both series and parallel elements relative to the spring. In tissues with rapid length changes, these damping elements can have a major influence on tissue behavior.

Within intact systems, load-bearing tissues are often in series with the force-generating tissues. Skeletal muscles, for example, are in series with the tendons that connect the muscles to bones. The force exerted by the muscle is equal to the force in the tendon when that muscle is moving a bone. It is possible for muscles to exert so much force and shortening that the load-bearing elements in series with them will be stretched beyond their viscoelastic limit. In these cases, the stretched tissues may yield, and upon release of the stretch will not be able to return to their original length. This is tissue deformation, shown in the lower right of Figure 6.1. The simplest analogy of this process occurs when a plastic ring, such as those that hold multiple beverage cans, is stretched until the ring rapidly lengthens, thinning the plastic, but does not break. While the polymers in plastic are different, the proteins in load-bearing tissues can be deformed without the tissue fully rupturing. Over time, the damage of this deformation can heal in many cases.

6.2   Teeth and bone

The stress–strain curves of mineral approach Hooke's law. The upper and lower bounds of Young's modulus for the dental composite dentin, a mixture of hydroxyapatite and collagen, are shown in Figure 6.2. The equations governing the upper and lower limits of the mixture elasticity E_{den} (Kinney *et al.*, 2003) are:

$$E_{den} \leq V_A E_A + V_C E_C = V_A(E_A - E_C) + E_C \qquad (6.5)$$

$$E_{den} \geq \frac{E_A E_C}{V_A E_C + V_C E_A} \qquad (6.6)$$

where V_A and V_C are the volume fractions of apatite and collagen, respectively, and E_A and E_C are the Young's moduli for apatite and collagen, respectively. Since E_A will always be much larger than E_C, the $V_A E_A$ term will dominate the upper limit. The lower limit is curvilinear because the denominator will decrease as V_A goes up, reflecting the increasing mathematical influence of E_C. The important element here is that a composite can have a very wide range of potential elasticity, and the measurements of Young's modulus using a variety of methods at physiological apatite concentration, shown by the vertical bar in Figure 6.2, show a very wide range of values. Conceptually, while it is difficult to predict elasticity based on composition alone, the actual composite elasticity is not dominated by either of the two materials in the composite. In both a very hard composite, dentin, and a very soft composite, the wall of a blood vessel (Figure 6.5 in the

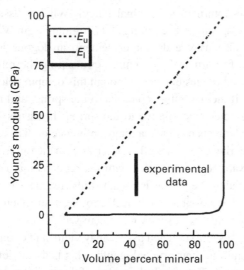

Figure 6.2   Upper and lower bounds for Young's modulus in dentin as a function of the percentage of hydroxyapatite. The dashed line represents the modulus of hydroxyapatite, and the lower solid line the modulus of collagen fibers. The vertical bar represents the range of experimentally determined Young's modulus values at a dentin mineral composition of 45%. The wide range of dentin modulus values falls between those of hydroxyapatite and collagen. Dentin will have a modulus that is a combination of the composite materials, without one dominating the other in the physiological range. (Reproduced from Kinney *et al.*, 2003, by permission of SAGE Publications.)

next section), the modulus of elasticity for the mixed material falls between the elasticity of the two primary materials in the mixture.

Given the composite nature not only of biological material, but also their elasticity, the range of elasticity of common biophysical materials is of interest. Table 6.1 shows the range of Young's modulus values from hydroxyapatite to elastin. Enamel, dentin and bone are all composites, with their high moduli reflecting in part their mineral content, hydroxyapatite (calcium phosphate), in enamel, dentin and bone. Collagen is a component of all three mixtures, as well as being present along with elastin in the connective tissue surrounding blood vessels. Tendons and ligaments are almost entirely collagen. Cartilage is a mixture of collagen, elastin, and proteoglycans, which have a core protein with glycosaminoglycan carbohydrates attached. The absence of blood vessels in cartilage is reflected in their slow growth and repair rates relative to other load-bearing tissues. Table 6.1 shows that the range of different load-bearing substances has a range of over five orders of magnitude, providing a great deal of flexibility in biophysical function.

Collagen serves as a good example of the complexity in estimating the elastic modulus (An *et al.*, 2004). The elastic modulus can be calculated based on a polymer's radius and its persistence length. The persistence length P can be thought of as a fundamental length of a polymer unit, such that

$$\langle \cos \theta \rangle = e^{-(L/P)} \tag{6.7}$$

Table 6.1 Young's modulus in different physiological materials

Material	Young's modulus	Source
Hydroxyapatite	110 GPa	Kinney *et al.*, 2003
Enamel	40–100 GPa	He and Swain, 2008
Dentin	12–25 GPa	Kinney *et al.*, 2003
Bone	10–30 GPa	Thurner, 2009
Tendon	1.3 GPa	Reeves, 2006
Cartilage ECM	5–20 MPa	Guilak *et al.*, 1999
Collagen	6.5 MPa–12.2 GPa	An *et al.*, 2004
Ligament	5 MPa	Frank *et al.*, 1999
Elastin	0.6 MPa	Glaser, 2001

where L is the length of the polymer and $\langle \cos \theta \rangle$ is the possible angle between position zero and position L. For lengths shorter than P, the polymer behaves like a stiff rod with the cosine approaching 1 and the angle zero; for lengths much greater than P, the polymer behaves statistically, with the cosine approaching zero and the angle spanning the range 0–90°. The equation predicting the elastic modulus E then is (An *et al.*, 2004):

$$E = 4kTP/\pi r^4 \tag{6.8}$$

where k is Boltzmann's constant, T is the absolute temperature, and r is the polymer radius. Estimates of the radius vary from 0.28 nm to 1.23 nm, giving a range of moduli from 12.2 GPa down to 32.7 MPa, almost a 1000-fold difference. Further, tests using light and x-ray scattering yield modulus estimates in the 3–9 GPa range, while stretching collagen-dominated tissues, as in Figure 6.5 in the next section, gives much smaller estimates. This is an area in which how the measurement is made, including the size of the sample tested, is a significant factor in the magnitude of the measurement. For investigation at the molecular, not tissue level, nanoindentation has become a common method.

Nanoindentation is used to measure elasticity in materials using atomic force microscopy (AFM) (Oliver and Pharr, 2004). AFM uses a nanoscale, computer-controlled cantilever silicon tip. When the tip approaches a surface, there is a Hooke's law relation between the force of the tip on the surface and the distance of the tip from the surface. Lasers detect the tip position and feedback systems maintain the tip distance from the surface as the tip rasters across the material surface. AFM has multiple uses, including the three-dimensional imaging of a surface at the nanometer level, shown in Figure 6.3. Another application is measurement of the amount of force necessary to indent the material by a given distance. The force, converted to the stiffness of the contact, and the size of the indentation can be used to measure Young's modulus at nanometer distances. The stiffness has to be corrected because both the material and the tip will move during the contact, and the indentation has to be corrected for the geometry of the tip. The initial calculation, which would have elements of both the material and the tip elastic moduli, is the reduced modulus of elasticity. The reduced modulus of elasticity E_r is

$$E_r = \frac{1}{\beta} \frac{\sqrt{\pi}}{2} \frac{S}{\sqrt{A(h_c)}} \tag{6.9}$$

where β is a geometrical constant varying between 1.0055 and 1.14 depending on the shape of the indenter tip, S is the stiffness of the contact (the measured force variable), and $A(h_c)$ is the area of indentation, which can be calculated from the polynomial fit

$$A(h_c) = 24.5h_c^2 + C_1 h_c^1 + C_2 h_c^{1/2} + C_3 h_c^{1/4} + \ldots + C_8 h_c^{1/128} \tag{6.10}$$

where h_c is the contact depth (the measured length variable) and the C_i values are the fitted constants. The relation between the reduced modulus of elasticity and the modulus of elasticity E_s is

$$1/E_r = (1 - v_i^2)/E_i + (1 - v_s^2)/E_s \tag{6.11}$$

where v_i is the Poisson ratio of the indenter tip material and v_s is the Poisson ratio of the test material, and E_i is the modulus of elasticity of the indenter material, which will usually be much higher than the modulus of the test material. For an isotropic material, the Poisson ratio is

$$\nu = -\frac{\varepsilon_{trans}}{\varepsilon_{axial}} \tag{6.12}$$

where for the compression of a material in the axial direction, the material will expand in transverse direction in most (but not all) materials. Under these conditions, ε_{trans} is positive and ε_{axial} is negative, so that the Poisson ratio v is positive, usually between 0.0 and 0.5. This method allows calculation of the modulus of elasticity on a very small scale, but is still subject to error from the estimates of the constants at this level. Since in most cases biological materials are not isotropic, variations in Young's modulus based on the orientation of the material are possible.

The anisotropic nature of bone is clearly present in Figure 6.3. The hydroxyapatite layer apparent in (a) overlays the collagen layer in (b). EDTA is a chelator of heavy metal ions. In biophysical systems, it is used to chelate calcium and magnesium. When bone is

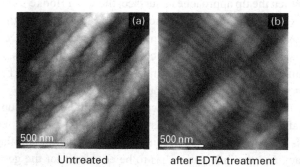

Untreated   after EDTA treatment

Figure 6.3   Atomic force microscopy scans of bone. (a) The hydroxyapatite surface and (b) after EDTA removal of the mineral phase, the banding pattern of the collagen fibers from the same sample. (Reproduced from Thurner, 2009, with the permission of John Wiley & Sons, Inc.)

exposed to EDTA solutions, the hydroxyapatite, a form of calcium phosphate, dissolves as the calcium binds to the EDTA. Removal of the mineral layer of bone would of course change its modulus of elasticity in a non-physiological manner, as the rigidity of bone depends on the presence of hydroxyapatite. Its removal would leave the strong, flexible collagen, which while not easily broken, would be unable to bear weight, a condition that occurs in osteoporosis. Conversely, removal of the collagen component, leaving only the mineral component, would result in a stiff but brittle condition which would chip or shatter easily. This condition can occur in mutations that lead to osteogenesis imperfecta (OI).

Osteogenesis imperfecta, also called brittle bone disease, is caused by genetic mutations that weaken collagen, leading to catastrophic tissue breakdown (Gautieri *et al.*, 2009). The weakened collagen produces pathological impairment in bones, joints, auditory apparatus, neurological and pulmonary functions. There are several known mutations that result in a range of impairments, from mild to severe. One of the amino acids involved in stabilizing the collagen triple helix structure is glycine. There are eight known mutations of this glycine residue, leading to the range of pathologies in OI. The general shape of the energy curves of the mutations relative to the normal collagen is shown in Figure 6.4. If you recall from Chapter 2, the formation of the secondary minimum produced a balance point between the London attraction forces and the van der Waals repulsion forces between two atoms. A similar relation occurs between collagen strands, yielding the intermolecular adhesion profile. Under normal circumstances, with the glycine residue present, there is an energy minimum between 1.9 and 2.0 nm. The energy minimum has a free energy of −11.7 kJ/mol, much more negative,

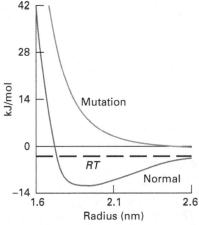

Figure 6.4   Intermolecular adhesion profiles for normal and mutated collagen. There are a number of genetic mutations that cause osteogenesis imperfecta, or brittle bone disease. All of these mutations minimize or eliminate the energy well between the collagen strands, where the negative free energy enhances strand crosslink formation. Positive adhesion energies cause strand repulsion. The changes in energy profiles are consistent with weaknesses in collagen found in osteogenesis imperfecta. (Redrawn from Guatieri *et al.*, 2009, with permission from Elsevier.)

and therefore providing more stability, than the ambient free energy of 2.58 kJ/mol. This stability puts the collagen in a position to form both enzymatically and non-enzymatically generated crosslinks between the collagen strands. These crosslinks provide much of the strength of collagen in tissues. Several of the mutations, as seen in Figure 6.4, have an adhesion profile with no minimum. With positive energies, these mutations will repel the adjacent strands, decreasing the formation of crosslinks and weakening the collagen. Even as the distance between the strands increases, the energy approaches zero, and the ambient energy will create a local environment that will further decrease binding between adjacent strands. Thus, the mutations produce a physical alteration at the molecular level that is manifested by the crippling effects at the tissue and organ level. At this time, there are relatively few mutations, such as the OI and sickle cell mutations, where the molecular changes that lead to the pathological state are understood.

6.3   Blood vessels

In contrast to bones and teeth, where collagen provides structural support to tissues whose stiffness defines their physiological function, collagen and its parallel protein elastin provide the range of displacement for blood vessels, protecting them from overextension. In Figure 6.5, we see the control curve for the stretch of a passive (i.e., non-contracting) blood vessel. We saw this curve in the previous chapter, in the length–

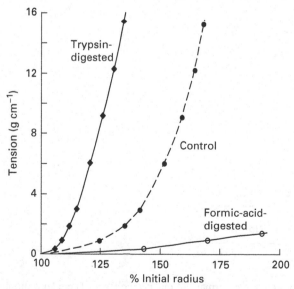

Figure 6.5   Contributions of collagen and elastin to the passive tension curve of iliac arteries. The control curve is a combination of collagen and elastin. Trypsin digestion of elastin reveals the collagen curve. Formic acid digestion of collagen reveals the elastin curve. As in the dentin curve in Figure 6.2, the composite material has characteristics of both components in the physiological range. (Reproduced from Shadwick, 1999, with permission of the *Journal of Experimental Biology*.)

tension curve from smooth muscle. Note that the extension leading to increased tension does not start at zero, and that there is no linear phase initially. Blood vessels do not obey the linear part of Hooke's law, and there is no single defined value for their modulus of elasticity at the whole tissue level. Any definition of the modulus for a blood vessel, or for other tissues with a curvilinear passive curve, such as the passive tension in skeletal or cardiac muscle, or any of the smooth muscle dominated tissues such as the bladder, uterus, GI tract or pulmonary bronchioles, depends on where the measurement is made. As the tissue is slowly stretched from a zero tension length, it will have a low modulus, but as the tissue reaches greater lengths, the stress–strain slope will increase. Without a single defined value, the circumstances of measurement become important. For the stretch away from a completely relaxed state, the low modulus will be appropriate. For conditions where the connective tissue is under constant high tension, as in the case of an aneurysm, the higher modulus at the stretched length is appropriate. This is a case in which the circumstances determine the appropriate measurement value.

Connective tissue in blood vessels primarily consists of two proteins, collagen and elastin. Both are flexible, but they have very different distension properties. Trypsin will digest elastin much more rapidly than collagen. Following trypsin exposure, the remaining collagen with its high modulus of elasticity dominates the stress–strain relation. Formic acid digests collagen at a higher rate than elastin, so that following formic acid exposure, elastin dominates. Estimation of Young's modulus in the stiffer regions of the two proteins shows that collagen is about 11 times stiffer than elastin. Collagen will behave more like rope, while elastin behaves like a rubber band. As with dentin, neither protein dominates the tissue modulus of elasticity. The tissue behaves midway between the two protein moduli. It is sometimes said that elastin dominates the low tension region of the passive curve, and collagen dominates the high tension region. While this may be true as a first approximation, both proteins contribute to the total tissue behavior at all lengths. It appears that at the microscopic level elastin limits the stiffness of collagen, possibly by linking collagen anchor points. This, or some related mechanism, must be in play to explain the intermediate whole tissue behavior.

Earlier, we showed that the hysteresis loop defines a region that is the energy lost during the loop cycle. At the time, we showed that the smaller area occurred during the return phase. If the return phase went on the other side of the Hooke's law line, the dashed line in Figure 6.6, the return phase would have the larger area, and the area of the loop would be the energy gained by the hysteresis. This would violate the laws of thermodynamics, which prevent getting more energy out of a system than you put in: this would be a perpetual motion machine. A clever student might contend, however, that if you just change the axes, putting the stress on the x-axis and the strain on the y-axis, then the dashed line would define a smaller area than the Hooke line, and there would no problem with the thermodynamic laws, which is essentially asking about the equivalence of

$$P\mathrm{d}l = l\mathrm{d}P. \tag{6.13}$$

This analysis should trouble you; can the laws of thermodynamics be so easily circumvented just by plotting the data differently? Of course they can't. Here, the concept of the independent and dependent variables is important. In a relation that produces a hysteresis

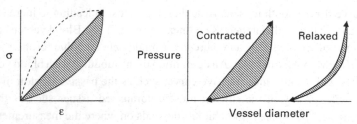

Figure 6.6   Hysteresis curves of activated and relaxed blood vessels. (Left) The area of the hysteresis curve is determined by the driving element, the independent variable in the relation. (Right) There is a large hysteresis loop in activated smooth muscle, indicating a large loss of energy during the stretch–release cycle as it moves up and down the muscle length–tension curve. The relaxed vessels have a rightward shift, reflecting the position of the passive curve. There is no large energy loss during a stretch release cycle of a relaxed blood vessel.

loop, the return phase must begin by approaching the independent variable axis, the variable that drives the other factor. Here, changes in stress do not cause the changes in length: it is changes in length that cause the increase in pressure, so strain will always be the independent variable, regardless of how the data is plotted. Never bet against the laws of thermodynamics.

There is another aspect of this behavior that reflects an earlier topic. Strain is a vector; it has both magnitude and direction. Pressure is a scalar, with magnitude but without a particular direction. The product of a vector and a scalar is a vector, so that the energy of the loop, the heat lost or entropy gained, will be directed toward an energy minimum, as are all vector processes. Also, strain is an extensive property, and is additive like mass, while pressure is an intensive property, like density, and is not additive. These concepts are all interconnected, and understanding the aspects of different properties allows you to avoid untenable situations, such as concluding that you can gain energy by putting a tissue through a stress–strain hysteresis cycle.

The right side of Figure 6.6 shows the hysteresis curves for activated and relaxed blood vessels. They start at different diameters because the unrestrained active muscle will slide down its length–tension curve until force, or pressure, is zero. The relaxed blood vessel will reach zero pressure when its passive curve reaches zero, which, as was seen in the last chapter, occurs at a longer length, a larger diameter, than the zero active force length. Stretching an active vessel produces a large hysteresis curve, indicating that the vessel has a significant energy loss during the cycle (Bauer *et al.*, 1983). This reflects the energetic inefficiency of the myosin ATPase, forming crossbridges broken by both the stretching process and during the active shortening on the return phase. In contrast, the passive curve has virtually no area in a hysteresis loop, following an essentially curved Hooke's law path. There is almost no energy lost by stretching and releasing the connective tissue composed of collagen and elastin. This difference shows that the size of a hysteresis loop is proportional to the amount of inefficiency in the enzymatic (or chemical) activity in the material. Since the area under the stretch phase, the triangle area from Figure 6.1, defines the total energy of the system, an approximation of the efficiency of the total system can be made. Since this total energy would include both passive and active elements, the inefficiency only defines an upper limit for the active processes. If the passive elements contribute significantly to the stretch and return, the active phase efficiency may be very much lower than the system as a whole.

An important aspect of passive tension stiffness is its contribution to systolic blood pressure. When blood is pumped from the left ventricle, it enters the aorta and the large arteries. These vessels distend slightly, then rebound during diastole to maintain blood flow throughout the cardiac cycle. Blood flow is pulsatile in these large arteries. The pulsatile flow is damped out as blood flows through the arterioles. While we will cover blood flow extensively in the next chapter, the load-bearing function of the large arteries is related to the systolic pressure. If arteries were made of stainless steel, they would not expand when blood entered. The systolic pressure would skyrocket. Conversely, if the large arteries were very elastic, balloon-like, they would easily expand as blood entered without a significant rise in blood pressure. The systolic pressure therefore is a function of the stiffness of the large elastic arteries. Figure 6.7 shows that as people age there is a leftward shift in the elastic artery stress–strain curve, with a higher modulus of elasticity at every vessel diameter (Nichols and Edwards, 2001). Since, as we shall see, the amount of blood flow is a function of the vessel diameter and the blood pressure, as people age the heart must develop increasing pressure in order to reach the same diameter and deliver the same amount of blood. This causes an increase in systolic pressure over decades, in people with both normotensive and hypertensive blood pressure (Ganong, 2005). Having a higher average systolic pressure means that the probability of reaching a pressure where the tension of the blood vessel cannot resist the distending pressure increases, and so the risk of a stroke increases.

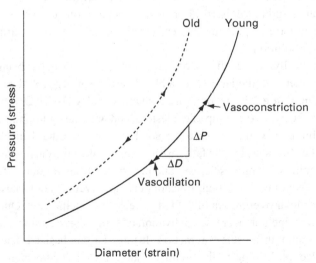

Figure 6.7 Stress–strain relation for elastic arteries. Vasoactive substances do not shift the pressure–diameter curve, but do alter the initial diameter, pressure and modulus of the curve. This indicates that the stress–strain relation in elastic arteries is dominated by the connective tissue. With age, the curve shifts to the left, increasing the modulus, the slope, at all diameters. (Redrawn from Nichols and Edwards, 2001, by permission of SAGE Publications.)

This brings us to the biophysical effect of alcohol on the cardiovascular system. Ethanol, the most common alcohol consumed, is an amphipathic molecule, with a hydrophobic methyl group and a hydrophilic hydroxyl group. While ethanol is highly soluble in water, it is also soluble in hydrophobic phases of membranes. Moderate consumption of ethanol has been shown to have positive effects on the cardiovascular system (Gunzerath *et al.*, 2004; Lakshman *et al.*, 2010) in comparison to abstinence or heavy ethanol consumption. The effects of ethanol have a U- or J-shaped curve, with the most positive effect occurring at modest intake levels, while heavy intake negates any benefits and results in significant deleterious effects. Ethanol, in contrast to other mood-altering substances, does not bind to a specific receptor, but has its effect by binding to proteins and altering their activity. One of the positive effects of ethanol consumption is a reduction in coronary artery disease. This positive benefit is due in part to ethanol's biophysical effect on the first stage of atherosclerosis. In this stage, low density lipoprotein (LDL) deposits cholesterol as a fatty streak on the intima of blood vessels. These lipid patches are unlikely to dissolve in plasma, given the partition constant of cholesterol. The partition constant K_D^o, formerly called the partition coefficient (IUPAC Gold Book), is the ratio of the distribution of a molecule at equilibrium between two immiscible phases, an organic hydrophobic phase and an aqueous hydrophilic phase:

$$(K_D^o)_A = \frac{a_{A,org}}{a_{A,aq}}. \tag{6.14}$$

The molecule activity a is used in this definition. From Chapter 2, a molecule's activity indicates the fraction of that molecule's concentration c that is effectively available in a solution. In dilute phases, the value of a is approximately c, but in highly concentrated phases, a is a reduced fraction of c. For cholesterol, the partition fraction between red blood cell membranes and buffer is 10^7. In plasma, this partition constant is reflected in the binding of cholesterol to proteins, as very small amounts of cholesterol will exist freely in hydrophilic solution.

Shown diagramatically on the left side of Figure 6.8, a fatty streak has been deposited on the intima, the inner endothelial layer of a blood vessel. Cholesterol, the structure of which was shown in Figure 2.12, is much larger than ethanol, CH_3CH_2OH. Ethanol can associate with both the hydroxyl group of cholesterol and with the hydrophobic rings and tail region. This binding would increase the solubility of cholesterol in plasma, and gradually reduce the cholesterol in the fatty streak. This would have the positive effect of slowing the development of atherosclerosis, and may be the basis for one of the positive effects of modest ethanol consumption. Ethanol does not reverse later stages of atherosclerosis, such as cellular overgrowth of the fatty streak and calcification. Other positive effects of ethanol may include decreased activation of sympathetic neurons, resulting in decreased vascular smooth muscle contraction. It must be emphasized that the heavy consumption of alcohol, leading to liver damage and the well-documented deleterious effects of alcoholism, has negative effects that far outweigh any positive benefits on the cardiovascular system.

As discussed in Chapter 2, cholesterol is not covalently bound to membranes. Red blood cell membranes lose cholesterol over time, increasing the stiffness of their

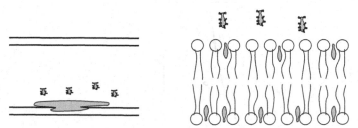

Figure 6.8 Effect of ethanol on the early stages of atherosclerosis and on membranes. Modest alcohol consumption has positive cardiovascular effects. Ethanol binding, shown as small additions on the larger cholesterol molecules, will solubilize cholesterol in plasma, decreasing cholesterol in fatty streaks. Ethanol will also extract small amounts of cholesterol from membranes. Over time, this will stiffen the membrane, increasing its elastic modulus.

membranes and the likelihood of rupture. The right side of Figure 6.8 shows that in the presence of ethanol, there is an increased potential for cholesterol to leave the cell membrane and bind to ethanol molecules in plasma, by a mechanism similar to that which reduces the cholesterol presence in fatty streaks. While modest alcohol consumption in general benefits cardiovascular function, in the elderly alcohol consumption is associated with increased systolic pressure in every racial group (Perissonotto *et al.*, 2010; Gu *et al.*, 2007; Curtis *et al.*, 1997). The mechanism behind this increase is over and above the increase in systolic pressure with age associated with connective tissue changes (Ganong, 2005). The increased removal of cholesterol from membranes of the elastic arteries could play a role. The mechanism, as shown in Figure 6.8, would be the same as that of the reduction in atherosclerosis development associated with ethanol. From Figures 6.2 and 6.5 above, it is clear that in a composite material, the modulus of elasticity is function of both of the elements in the composite. While it is tempting to assume that the majority of the stiffness in large elastic blood vessels is borne by the collagen, the cellular elements should also be important in the overall stiffness of the vessels. Making these structures more rigid by the removal of cholesterol would shift the overall curve to the left, consistent with Figure 6.7, even if there were no change in collagen or elastin, which of course ethanol-induced membrane changes do not rule out.

One of the benefits of the passive tension in large blood vessels, regardless of its relative composition of cellular and connective tissue, is its monotonic increase in tension as a function of inflation, shown in Figure 6.9 (Shadwick, 1999). Radiating from zero length and tension, the Laplace pressure lines are the slope of tension/radius. In a relaxed artery, the radius will be stable at the intersection point of the passive tension and the Laplace pressure line. As the pressure increases, the stability point is progressively higher on the passive tension curve. For some materials, the tension-inflation curve is not monotonic, but has a plateau phase with a virtually constant tension over a range of diameters. In this case, there will be a jump in inflation as the Laplace line passes the shoulder of the plateau. While this will not happen in normal arteries, materials with this type of plateau should not be used as artificial vessels because of the potential for uncontrolled inflation over a small pressure range. In the previous chapter on muscle

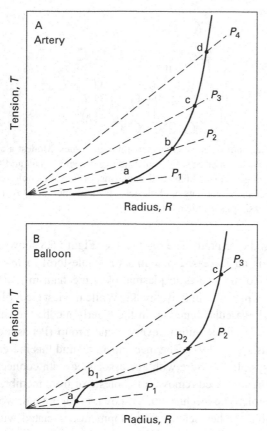

Figure 6.9   Laplace stability of inflated arteries. An artery will have an increasing wall pressure during inflation. For a monotonically increasing curvature, the Laplace pressure lines will intersect the tension curve at unique points. This results in stable inflation. A sigmoidal tension curve, present in non-physiological materials such as rubber, can have unstable regions when the tension curve parallels the pressure line. Activated blood vessels can also have sigmoidal tension curves. (Reproduced from Shadwick, 1999, with permission of the *Journal of Experimental Biology*.)

we saw that activated smooth muscle in blood vessels can produce a shoulder region near the peak of the force generation, a condition that can produce the same large increase in diameter over a small pressure range. Since blood vessels and the heart usually work on the short end of the length–tension curve, this sudden expansion over a small pressure range is not common.

For a relaxed blood vessel at the end of diastole, the stability point will be the starting point for inflation during systole. While the effects of chronic hypertension on the energetics of the heart are well known as a potential cause of myocardial infarction, this extension of the starting point for vessel inflation also defines how close a vessel will be to potentially rupturing during systole, resulting in a stroke. Efforts to reduce diastolic pressure are widespread, but education on the risk of stroke in addition to the risk of heart attack may be lost on much of the general public. Still, the combination of vessel stiffness

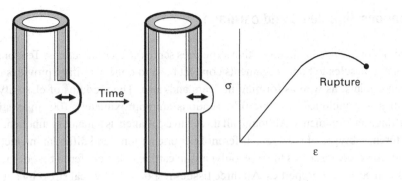

Figure 6.10 Pressure cycles effect on aneurysms and the stress–strain rupture curve. (Left) Aneurysms in large arteries will not be static structures, but will fluctuate with the pressure changes during the cardiac cycle. Over time, the connective tissue of the aneurysm will weaken as its elastic fibers decrease, indicated by the thinner wall thickness. (Right) All materials have a rupture or tear length. A tissue will not return from this position. Rupture of an aneurysm is often fatal.

and diastolic inflation are both very real risk factors for stroke, and should be taken into account by both health care professionals and the general public.

Among the most dangerous conditions associated with load-bearing tissue is the formation of an aneurysm. An aneurysm forms when the distending blood pressure breaks through the two innermost layers of a blood vessel, the endothelial intima and the vascular smooth muscle media layers, leaving only the outer adventitia layer of connective tissue preventing a catastrophic hemorrhage. The left side of Figure 6.10 shows that during the initial phase of aneurysm formation, the intact adventitia pouches out. In experimental models of aneurysm, experiments that could not ethically be done on humans, the temporal integrity of aneurysms was tested (Novak *et al.*, 1996). Aneurysms are not static, but pulsate with the changes in blood pressure during the cardiac cycle. Aneurysms exhibited hysteresis, which accelerates the degeneration of the elastic fibers in the adventitia. Over time, if the aneurysm is not surgically corrected, the wall gradually gets thinner, increasing the risk of rupture and potential death.

The right side of Figure 6.10 completes the set of stress–strain curves first presented in Figure 6.1. Beyond the length at which tissue deformation occurs, there will be a fall in wall stress as the tissue is stretched. As the stretched tissue is less and less able to resist distension, the tissue will eventually tear, and not return to the initial zero force length or a deformed length, but will be catastrophically changed. The length at which the rupture occurs is also called the tear length. For an isotropic tissue, the site of rupture cannot be predicted. Biological tissues only have isotropic regions over a very small strain range, if such a region is present at all. In both an anisotropic tissue and a complex system in which several tissue types, such as a limb in which muscle, tendon, bone and ligament are in series with one another, the weakest point has the highest probability of rupture. Eccentric contractions increase the possibility of muscle damage. For any contraction, either concentric or eccentric, if one of the structures in series with the contraction is not able to withstand the force generated, the overall structure will rupture at that weakest point.

6.4   Tendons, ligaments and cartilage

The major extracellular connective structures serve different functions. Tendons connect skeletal muscles to bone. Ligaments connect bone to bone. Cartilage provides resiliency around joints, as well as forming the ears and nose. The modulus of elasticity of these tissues was included in Table 6.1. Tendons are approximately 100 times stiffer than cartilage and ligaments. Although all three have collagen as a major component, there are different subtypes of collagen, different fiber orientations, and different interactions with other molecules, such as the proteoglycans in cartilage that confer in tissues the different tissue mechanical properties. All three tissues respond to physical stress by altering their structure over time to better withstand the external forces. Cartilage, which is avascular, will respond more slowly than either tendons or ligaments, which do have blood vessels.

An example of the response of tendons to exercise is shown in Figure 6.11. In contrast to arteries, where age is associated with an increase in the modulus of activity (Figure 6.7 above), the modulus decreases with age in tendons. This appears to be associated with weaker connections between the collagen fibers that make up the bulk of tendons. Collagen molecules form a triple helix with a glycine residue repeating every third amino acid (Braunwald *et al.*, 2001). Glycine, the smallest amino acid, allows the

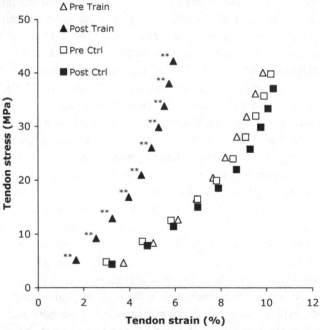

Figure 6.11   Effect of exercise on the patellar tendon stress–strain curve in older adults. Aging results in a decrease in the modulus of elasticity. The stress–strain curves for control and pre-training groups were similar, but following an exercise regimen the modulus increased significantly in the exercise group. The post-control group remained unchanged. Exercise can play a significant role in countering the effect of aging on tendons. (Reproduced from Reeves, 2006, with permission.)

entwining collagen fibers to pack very closely, while the other amino acids, a large number of which are either proline or hydroxyproline, have either charged or hydrophobic side groups projecting out from the sides of the helix. Both enzymatic and non-enzymatic crosslinks form between collagen fibers, giving the collagen, and the tendon, great strength. The relation between connective tissue stiffness and strength is complex, for while exercise increases tendon strength, stretching during a warm-up before competing reduces injury and improves performance by giving greater elasticity to connective tissue for a modest duration. This greater elasticity makes the connective tissue more likely to stretch and less likely to rupture. It is interesting to note that while stretching immediately prior to high performance athletics, like track sprinting, is routine, in other sports in which players might not enter a game for up to an hour or more after pre-game warm-ups, like basketball, players do not stretch just before entry. It would be interesting to find out if stretching just prior to entering basketball games had any effect on the incidence of connective tissue injuries.

There is a decrease in the binding within collagen in tendons as people age. This is not, however, an irreversible process. As shown in Figure 6.11, there is a leftward shift in the stress–strain relation of the patellar tendons of older people in an exercise program. It is important for people to be aware of the relation of the tendon strength compared with the strength of their muscles. While both weaken with age, skeletal muscles do not lose significant strength in men until after the age of 50. In some men in their 40s, engaging in vigorous physical exercise with a high degree of twisting and stretching (handball is a major culprit) will rupture their Achilles tendon. The tendon must have been the weak link in the force conduction chain. Regular exercise specifically aimed at strengthening the tendons can reduce the incidence of this type of injury. As people age further, weakening muscles also means less stress on the tendons, and therefore less chance of injury.

Ligament damage occurs when force on a joint is sufficiently large to produce deformation of the ligament. Since a ligament is designed to stabilize a joint, deformation will reduce the stability of the joint, increasing the possibility of further injury as both the injured ligament and other healthy tissues are moved into unusual positions. Ligament injuries are often referred to as sprains, and the extent of the injury is dependent on the extent of the deformation. A mild sprain occurs when the deformation decreases the joint stability but not to the extent that the joint position cannot be maintained. In more severe sprains, the deformed ligament is partially torn, and is unable to adequately maintain the joint position. The most severe sprains have an entirely severed ligament, which does not contribute to the joint stability. Rehabilitation of milder sprains involves rest, immobilization, compression and elevation (components of the RICE acronym), and exercises to strengthen the other structures of the joint, such as the muscles and tendons. Strengthening these other structures helps to enhance joint stability lost to ligament deformation. Rejoining torn pieces of ligament is often unsuccessful in restoring function, in which case complete ligament replacement is necessary. Tommy John surgery, named for the Los Angeles Dodger pitcher on whom it was first performed, replaces the torn ulnar collateral ligament with a tendon, surgery often performed on a baseball pitcher. Recovery takes up to a year. Studies of young pitchers (Lyman *et al.*, 2002) show a strong correlation of the risk of elbow injury with the number of pitches, and have led to limitations on pitch number in young players. This is consistent with relatively small sites of damage from repetitive motion gradually weakening the overall strength of the ligament until a single catastrophic event produces a major tear.

Cartilage in joints does not bear weight to the degree that tendons and ligaments do, but mechanically buffers the joint, limiting its range of motion in directions that would produce damage. Tears in cartilage result in deformations that impair the joint's normal function in two ways: not preventing movements that can result in joint injury, or moving into a non-normal position that restricts the range of motion in normal directions. Because cartilage is avascular, it often does not heal well following injury, and has to be surgically removed. While the load the joint can bear is not necessarily limited by cartilage removal, a greater range of motion can increase the risk of injury to load-bearing structures in the joint.

One of the consequences of joint injury is local swelling. This edema plays an important physiological function. Injuries occur when a joint moves rapidly in an unnatural direction, or exceeds its normal range of motion in the normal direction. The load-bearing elements in the joint can produce and transmit a force resisting the forces trying to change the angle of the joint. Water is virtually incompressible, so increasing the hydration in a joint increases the resistive force. The swelling helps the joint remain immobile, allowing the healing process to occur. Anti-inflammatory treatments reduce hydration in the damaged joint, allow faster movement, but also increase the risk of further damage. Glucocorticoid steroids, cortisol or cortisone, are the most powerful anti-inflammatory compounds, and are used by athletes to reduce swelling in injured joints and allow them to compete. They are balancing the desire to compete with the increased risk of catastrophic damage by competing on a partially damaged, non-swollen joint. Leukocyte apoptosis induced by glucocorticoids is also a factor in slowing joint healing. There are serious ethical issues involved when coaches of non-professional athletes encourage, or demand, the use of anti-inflammatory steroids in order to have them compete.

There has been no area of athletics in which load-bearing structures are affected more than the increase in body weight by American football players. Figure 6.12 shows that between 1972 and 2002, the self-reported average weight of football linemen at Michigan State University has increased by over 25 kg or 60 pounds. This is typical of other university football teams, as the self-reported linemen weights for 362 other players from 11 other Midwestern universities in 2002 was 127.3 kg, or over 280 pounds. Much of this involves increased training efforts on the part of the athletes, but there have been reports that some athletes have used anabolic steroids to increase muscle hypertrophy. These changes have several biophysical consequences. As with the concerns over elbow injuries in young athletes, increased weight borne by the joints of young football players should also be a concern. Damage to the musculoskeletal system during football occurs when some action, either by the individual himself or by an opponent, causes the weakest element in series with applied force to break. The elements involved include the muscles, tendons, bones, ligaments and the attachment sites between the different parts. While cartilage is not technically in series with the force/movement parts, increased force on joints will also demand more from cartilage that cushions joint movements.

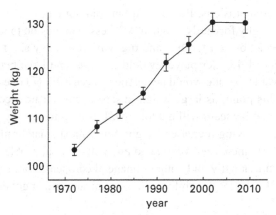

Figure 6.12   Change in body weight of football linemen from Michigan State University 1972–2009. Over a 30 year period 1972–2002, football linemen exhibited a more than 25 kg increase in average weight, plateauing between 2002 and 2009. This increase in weight occurred across both college and professional football. If there is no corresponding decrease in foot speed, the weight increase will produce an increase in muscle momentum.

The different parts of the force-generating pathway respond to training at different rates. Muscles are cellular, and can hypertrophy relatively rapidly. Tendons and ligaments have blood supplies, and the cells that extrude the proteins that bear loads in these connective tissues can increase the filament strength over time, but probably not as quickly as muscles. Cartilage, being avascular, will respond most slowly. The use of anabolic steroids by professional football players has been associated with ligament and joint injuries, but not muscle and tendon injuries (Horn *et al.*, 2009). The increased hypertrophy of muscles appears to be paralleled by increases in tendon strength, but ligaments are apparently unable to adapt to their greater demands, or do not adapt quickly enough. There are still many remaining questions in this field. For example, entheses are the attachment sites of tendons or ligaments onto bones. For muscles to move bones, the force must pass through entheses. In a very under-researched field, there is evidence based on endurance exercises that the structure of attachment sites does not reflect muscle size or activity, evidence that contradicts preconceived notions (Zumwalt, 2006). Further research on the effects of muscle hypertrophy, including any steroid effects, on both tendon-bone and ligament-bone entheses, is still needed.

Football injuries are not only acquired by internal stress on connective tissue. Opposing players can also cause serious damage to players. Both professional football players (Horn *et al.*, 2009) and front row rugby players (Hogan *et al.*, 2009) have significant neck injuries, injuries that occur during contact with opponents. The increase in player weight creates an increased risk, as having a 130 kg opponent fall across your knee or neck is more likely to cause injury than the damage a 100 kg opponent may cause. Playing contact sports comes with an innate element of risk, and if players know and understand the risks, they and the people who watch them can derive great enjoyment. Professional players understand these risks, and are willing to trade them for a lucrative living. These factors are less clear at amateur levels, particularly where both player size and speed create high-energy collisions. Players and fans did not enjoy games any less when football linemen and rugby forwards weighed 20% less than they do today. If there were weight-for-height restrictions on football and rugby players many short and long-term injuries could be reduced. Lighter players would be faster and have better endurance, and those with the best technique would be the most successful. No one coach can institute such changes at this point, as large weight discrepancies create a competitive disadvantage that a coach and his team will not tolerate. It is hard to see any long-term advantage in player health to having increased weight. Since most amateur players will never play professionally, this increased weight and any pathological problems associated with it will be borne individually and without financial compensation. Restrictions on the size of players must be the responsibility of the organizations that govern weight-dependent sports.

Physical quantitation of load-bearing tissues is made by measuring the stress (the force/cross-sectional area) produced in response to an applied strain (the fractional length change), the slope of which is the modulus of elasticity. Depending on the amount of strain, the physical responses go through a progressive series of changes: linear, hysteresis loop, viscoelastic, permanent deformation, and rupture. Compound tissues, such as bone and teeth, have moduli of elasticity intermediate between the moduli of their individual components. The elastic modulus of a tissue can be measured at the microscopic level, but at the tissue and organ level the modulus can be non-linear, dependent on the displacement and activation of the tissue. Pathological conditions can weaken connective tissue, such as the strength of mutated collagen. Conditions can alter connective tissue, collagen strengthening in response to exercise and weakening during prolonged blood pressure cycling of an aneurysm. Changes in sports training, both appropriate and not, alter connective tissue in ways that are yet to be fully understood, particularly in our understanding of the long-term consequences.

References

An K-N, Sun Y-L and Luo Z-P. *Biorheology*. **41**:239–46, 2004.

Bauer R D, Busse R and Wetterer E. In *Biophysics*, ed. Hoppe W, Lohmann W, Markl H and Ziegler H, 2nd edn. New York: Springer-Verlag, pp. 618–30, 1983.

Braunwald E, *et al. Harrison's Principles of Internal Medicine*, 15th edn. New York: McGraw-Hill, 2001.

Curtis A B, James S A, Strogatz D S, Raghunathan T E and Harlow S. *Am J Epidemiol*. **146**:727–33, 1997.

Frank C B, Hart D A and Shrive N G. *Osteoarthr Cartilage* 7:130–140, 1999.

Ganong W. *Review of Medical Physiology*. New York: Lange, 2005.

Glaser R. *Biophysics*. Berlin: Springer-Verlag, 2001.

Gu D, Wildman R P, Wu X, *et al. J Hypertens*. **25**:517–23, 2007.

Guatieri A, Uzel S, Vesentini S, Redaelli A and Buehler M J. *Biophys. J*. **97**:857–65, 2009.

Guilak F, Jones W R, Ting-Beall H P and Lee G M. *Osteoarthr Cartilage*. **7**:59–70, 1999.

Gunzerath L, Faden V, Zakhari S and Warren K. *Alcohol Clin Exp Res.* **28**:829–47, 2004.

He L H and Swain M V. *J Mech Behav Biomed Mater.* **1**:18–29, 2008.

Hogan B A, Hogan N A, Vos P M, Eustace S J and Kenny P J. *Ir J Med Sci.* **179**:259–63, 2009.

Horn S, Gregory P and Guskiewicz K M. *Am J Phys Med Rehabil.* **88**:192–200, 2009.

Kinney J H, Marshall S J and Marshall G W. *Crit Rev Oral Biol Med.* **14**:13–29, 2003.

Lakshman R, Garige M, Gong M and Leckeyl L. *Genes Nutr.* **5**:111–20, 2010.

Lyman S, Fleisig G S, Andrews J R and Osinski E D. *Am J Sports Med* **30**:463–8, 2002.

Nichols W W and Edwards D G. *J Cardiovasc Pharmacol Ther.* **6**:5–21, 2001.

Novak P, Glikstein R and Mohr G. *Neurol Res.* **18**:377–82, 1996.

Oliver W C and Pharr G M. *J. Mater Res.* **19**:3–20, 2004.

Perissinotto E, Buja A, Maggi S, *et al. Nutr Metab Cardiovasc Dis.* **20**:647–55, 2010.

Reeves N D. *J Musculoskelet Neuronal Interact.* **6**:174–80, 2006.

Shadwick R E. *J Exper Biol.* **202**(23):3305–13, 1999.

Thurner P J. *Wiley Interdiscip Rev Nanomed Nanobiotechnol.* **1**:624–49, 2009.

Zumwalt A. *J Exp Biol.* **209**:444–54, 2006.

7 Fluid and air flow

The flow of material is central to many biophysical functions. The flow of blood supplies the tissues with the energy and raw materials they need and removes the waste products they generate. Air flow to the lungs is critical in supplying the oxygen needed for energy production. In large joints, synovial fluid lubricates surfaces under both high pressure and a wide range of motion. Disruptions of flow often have a physical basis, such as the kinking of the great veins during pneumothorax. Others, such as the buildup of atherosclerotic plaques, place additional burdens on flow-generating systems. Understanding both normal and pathological conditions affecting flow starts with understanding the properties of physiological fluids.

7.1 Fluid properties

The fixed surface a fluid is passing imparts resistance to the flow, and at the fluid–surface interface the velocity will be zero. At distances farther from the surface, the velocity will increase. The change in velocity dv as a function of the distance from the fixed surface dz is the shear rate γ

$$\gamma = \frac{dv}{dz} \tag{7.1}$$

which has units of s^{-1}. A common way in which a shear rate is measured is by having two parallel surfaces pass one another at a given distance (Figure 7.1).

This produces a constant shear rate between the plates (i.e., laminar flow) and is often used to study the characteristics of a particular fluid (Dörr, 1983). The force F on the moving plate producing the laminar flow is

$$F = \eta \gamma A \tag{7.2}$$

where η is the viscosity of the fluid between the plates and A is the area of the moving plate. The sheer stress τ is

$$\tau = \eta \gamma. \tag{7.3}$$

The changes in velocity in the z-direction create an energy gradient. In Figure 7.1, the energy gradient is constant, and therefore there will not be any preferred position relative to the distance from the fixed surface; that is, the change in velocity, the shear rate, is constant across the fluid profile.

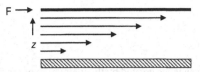

Figure 7.1 Laminar flow produced by a plate (thick black line) moving across a fixed surface (cross-hatched). The arrows represent the velocities of different lamina, the zones of different velocity.

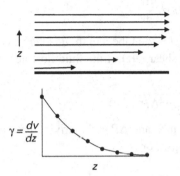

Figure 7.2 Laminar flow: velocity gradient of flow along a surface. The upper panel arrows represent the velocities of different flow layers adjacent to a fixed surface. At the fluid–surface interface, the velocity is zero. The velocity increases in the z-direction perpendicular to the fixed surface until the maximum velocity is reached. The change in velocity dv over the change in distance dz creates a velocity gradient, the relative magnitude of which is shown in the lower panel. This gradient is the shear rate, and is highest at the interface and approaches zero well away from the fixed surface.

When a fluid moving at a fixed velocity crosses a fixed surface, different velocities occur because the fixed surface creates friction against the moving fluid (Figure 7.2). This friction decreases with the distance from the fixed surface, and the fluid will approach a maximum velocity at a sufficient distance from the fixed surface. This produces a reduction in shear rate along the z-direction. The shear stress is greatest at the surface, and reaches zero when the different lamina have the same, maximum velocity sufficiently far from the surface.

This change in shear stress occurs in blood vessels. The sheer stress plays an important role in blood flow. The Prigogine principle of minimal entropy production has solid materials in fluid move to regions of minimal sheer stress. In blood vessels, minimum sheer stress will occur at the center of a vessel, and blood cells will have a higher concentration there compared with areas of high sheer stress near the vessel walls. This behavior, the Fahraeus–Lindqvist effect, also plays important roles in pathological conditions. To understand this, consider flow in a tube. With fixed surfaces all around the tube, minimum velocity will occur around the circumference. The flow pattern will be similar to that in Figure 7.3.

There are two forces at work during flow in a tube: the frictional force of the fluid against the vessel and the different laminae, and the driving force propelling the fluid down the tube. The frictional force F_f is

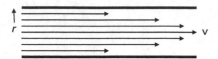

Figure 7.3   Velocity profile of fluid flow in a tube. The resistance created by the interaction of the fluid with the walls of the tube will reduce the velocity there compared with regions farther from the walls.

$$F_f = 2\pi r l \eta \frac{dv}{dr} \tag{7.4}$$

where $2\pi r$ is the circumference of the tube, l is the tube length, η is the viscosity of the fluid, and dv/dr is the velocity gradient (or shear rate) of the fluid flow across the tube. The driving force F_D is

$$F_D = \pi r^2 \Delta P \tag{7.5}$$

where πr^2 is the cross-sectional area of the tube and ΔP is the driving pressure. During stationary flow, these two forces must be equal, or

$$2\pi r l \eta \frac{dv}{dr} = F_f = F_D = \pi r^2 \Delta P \tag{7.6}$$

$$2l\eta \frac{dv}{dr} = r\Delta P. \tag{7.7}$$

The tube has a radius r_t with $r = 0$ at the center of the tube. At the wall, the velocity must be zero, and the velocity will reach its maximum v_o at the center of the tube. With these parameters, rearranging and taking the integral of both sides of the above equation yields

$$\int_0^{v_o} dv = \int_0^{r_t} \frac{\Delta P}{2l\eta} r dr \tag{7.8}$$

which is solved for the velocity v_r at a given radius r as

$$v_r = \frac{\Delta P}{4l\eta}(r_t^2 - r^2). \tag{7.9}$$

Under constant pressure conditions, this equation has the shape of a parabola, as seen in Figure 7.3, with its highest velocity at $r = 0$, the center of the tube. Flow profiles that fit a parabola are termed Newtonian, as they can be derived from Newton's equation $F = ma$, without the need to take friction into account. Newton was of course aware of the effect of friction, but the terminology has evolved such that non-parabolic flow profiles are termed non-Newtonian. They are not non-Newtonian in the sense that was discussed in Chapter 3, where a very large mass can bend a light wave.

Newtonian flow can be modeled as a series of lamina. The flow J_r in each lamina will be the velocity of that lamina times the incremental cross-sectional area of that lamina $2\pi r dr$, yielding

$$J_r = \frac{\Delta P}{4l\eta}(r_t^2 - r^2)2\pi r \mathrm{d}r = \frac{\pi\Delta P}{2l\eta}(r_t^2 r - r^3)\mathrm{d}r. \tag{7.10}$$

This equation is integrated over all r to give the Poiseuille equation, also called the Hagen–Poiseuille equation, which measures the total flow J_t in the tube:

$$J_t = \int \frac{\pi\Delta P}{2l\eta}(r_t^2 r - r^3)\mathrm{d}r = \frac{\pi\Delta P}{8l\eta}r^4. \tag{7.11}$$

The Poiseuille equation demonstrates the importance of vessel radius in controlling flow. With a fourth-power relation between radius and flow, even small changes in radius result in very, very significant changes in blood flow. This minute level of control is important, as there are about five liters of blood in the body, and about ten liters of space that blood can occupy when all the vessels are dilated. Most vessels have to be constricted most of the time, or the blood pressure would fall so low that the brain would not be perfused, followed shortly by unconsciousness, coma and death. The complexities of blood flow will be explored further in this chapter.

The Poiseuille equation only applies to laminar flow. Common experience allows us to relate smooth laminar and turbulent non-laminar flow. Water rushing at high speeds from a fire hydrant or down the rapids of a river appears wild and uncontrolled. At lower speeds, we see the smooth flow of a quiet river or the water from a faucet. At still lower speeds, turbulence reasserts itself in a babbling brook or a dripping pipe. These different behaviors are controlled by the Reynolds number, named for nineteenth-century engineer Osborne Reynolds. If the Reynolds number is too high, the flow is turbulent and non-laminar, like the river rapids. Within an intermediate range of Re, flow is smooth and laminar. At low Re values, non-laminar flow returns in the brook or dripping pipe. Flow is laminar when the fluid has a range of velocities, with adjacent regions of different velocities, laminae, passing one another with little disturbance, creating a smooth flow. If there is disruption in the smooth passage of different fluid layers, non-laminar flow results and turbulence is produced. Reynolds number Re is defined (Ganong, 2005) as

$$\mathrm{Re} = \frac{\rho v d}{\eta} \tag{7.12}$$

where ρ is the density of the fluid, d is the tube diameter, v is the velocity of flow and η is the fluid viscosity. Low values of Re only occur under non-laminar conditions, where the physical surroundings play a relatively high role in causing resistance to flow, as in a low volume brook or the occasional drop dripping through a pipe opening. For tubes with laminar flow, the transition from laminar to turbulent flow will occur with increasing probability as the value of Re rises. The constriction of blood vessels during blood pressure measurements or the narrowing of blood vessels by atherosclerotic stenosis narrows the vessel, but is compensated by a rise in velocity, leading to an increase in Re. Laminar flow is nearly silent, while turbulence generates sound waves. Thus, turbulent flow is used to detect partially open blood vessels during blood pressure measurement or when detecting the bruits (pathological sounds) associated with atherosclerotic plaques.

The viscosity is a function of the composition of the fluid. Fluid can be composed of only a solvent, or contain both a solvent and additional material, which may be dissolved in the solvent or be present as particulate matter. In physiological systems, water is the only solvent of consequence, and is never without additional salts, metabolites or even cells. That means that both factors, the properties of the solvent water and the materials present in it, influence the fluid's viscosity. For a pure solvent, the kinematic viscosity ν is

$$\nu = \frac{\eta}{\rho} \qquad (7.13)$$

where ρ is the density of the solvent. While this factor would be important if there were a range of physiological solvents, the sole presence of water as a solvent limits us to water's properties. A major factor in the viscosity of water is the number of H-bonds forming water clusters. As expected, the kinematic viscosity of water is temperature dependent, as there are many more H-bonds near $0\,°C$ than at higher temperatures. In mammals, of course, there are no significant variations in temperature (at least, not significant in terms of H-bond formation), so that water at $37\,°C$ is our primary concern. Further, pure water is never present in biophysical systems: there are always ions and molecules dissolved in water. The relative viscosity η_{rel} is

$$\eta_{rel} = \frac{\eta}{\eta_{solv}}, \qquad (7.14)$$

the ratio of the measured viscosity of a solution η to the viscosity of the pure solvent η_{solv}. Other useful viscosity measurements (Glaser, 2001) are

$$\text{Specific viscosity: } \eta_{sp} = \eta_{rel} - 1 \qquad (7.15)$$

$$\text{Reduced viscosity: } \eta_{red} = \frac{\eta_{sp}}{c} \qquad (7.16)$$

$$\text{Intrinsic viscosity: } [\eta] = \lim_{c \to 0} \eta_{red} \qquad (7.17)$$

where c is the concentration of a dissolved solute. These equations are useful in providing the framework for measuring the effect of molecules (and in the case of blood, cells) present in the solution. When used to study molecules, the shape of different molecules, from spherical to oblong, will alter the viscosity in different ways, so that viscosity measurements are used to infer the physical shape of molecules in physiological solutions. Also, for molecules that may form complexes, the concentration c may be replaced with the activity a. Variations in expected viscosity as the concentration is changed are one of the ways of inferring autocomplexation. In more complex solutions, the equivalent to the solvent is the solution that serves as the starting solution. For example, it would not be appropriate to measure the viscosity effect of red blood cells added to water, as these cells would osmotically lyse. In this case, the starting solution would be plasma. Then, measuring the viscosity of plasma in the presence and absence of the cells would be appropriate.

Viscosity is not constant in all fluids. Newtonian fluids have constant viscosity, independent of changes in shear rate (or velocity gradient). Non-Newtonian fluids change as the shear rate changes. In biological fluids, the presence of cells, proteins and macromolecules significantly affects viscosity. Cells and macromolecules can aggregate at low flow rates. These aggregations dissipate as the shear rate increases. Additionally, some proteins and carbohydrates have elongated shapes. These molecules align with the direction of flow. Further, cells can change their shape, becoming more oval, as shear stress increases, orienting the cell's long axis in the direction of flow (Figure 7.4)

Disaggregation and alignment decrease the viscosity, a characteristic called pseudo-plastic flow. Both blood and synovial fluid have non-Newtonian, pseudoplastic behavior exhibiting decreased viscosity as the shear rate increases (Figure 7.5). Blood (Schmid-Schönbein *et al.*, 1968) has a higher viscosity at low shear rates than does synovial fluid (Wright and Dowson, 1976). Over the range of shear rates measured, both show a continuing decrease in viscosity. Synovial fluid in particular exhibits a continuing decrease in viscosity of more than three orders of magnitude over four orders of magnitude of shear rate without approaching an asymptote (Wright and Dowson, 1976). This demonstrates that for biological fluids, the Poiseuille equation has to be modified for changes in viscosity when calculating fluid flow.

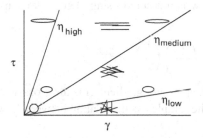

Figure 7.4   Deformation by shear stress. Shear stress τ is the product of shear rate γ and viscosity η. Cells change shape and molecules change orientation as the shear stress increases. Cells, the round and oval structures, become increasingly oval as the shear stress increases, a process that is enhanced in high viscosity solutions. Rod-shaped or oblong molecules, shown as clusters of lines, will orient with their long axis in the direction the flow. Both processes minimize the free energy of the system.

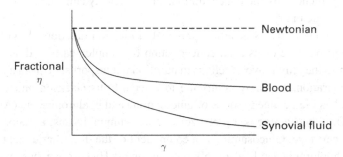

Figure 7.5   Pseudoplastic behavior of blood and synovial fluid. The viscosity of Newtonian fluids is independent of shear rate. Both blood and synovial fluid exhibit pseudoplastic flow. At low shear rates, blood has a viscosity about eight times that of synovial fluid. Over similar ranges of shear rate, synovial fluid has a greater change in viscosity than blood.

7.2   Synovial fluid flow

Cartilage is an avascular tissue that must supply the cells that form its extracellular matrix, the chondrocytes, by a combination of diffusion and convective flow (Jackson and Gu, 2009) from the capillaries on the edge of the tissue. The relative importance of these two processes is measured by the Peclet number Pe, which is defined as

$$Pe = \frac{UL}{D} \tag{7.18}$$

where U is the fluid flow rate, L is the diffusion distance, and D is the diffusivity, related to the diffusion constant. The higher the Peclet number, the greater the fraction of transport borne by convective flow. In cartilage, small molecules are primarily carried by diffusion, while large molecules move primarily by convective flow. For molecules with a Peclet number near 1, both diffusion and flow carry a significant amount of the transported substance.

The hydraulic permeability of a fluid is the ease with which it can flow through a soft tissue (Jackson and Gu, 2009). The flow is measured using Darcy's law, in which the hydraulic permeability k is

$$k = \left(\frac{Q}{\Delta P}\right)\left(\frac{h}{A}\right) \tag{7.19}$$

where Q is the volumetric flow, ΔP is the pressure gradient, A is the cross-sectional area of the sample, and h is the thickness of the sample. As the relation between flow, pressure and resistance R is

$$P = QR \tag{7.20}$$

then the resistance to flow through a soft tissue must be

$$R = \frac{h}{kA}. \tag{7.21}$$

The hydraulic permeability will be a function of the interactions of the fluid with its surroundings, influenced by the concentration, charge density and friction of its components with the tissue matrix.

In joints, synovial fluid serves both supply and lubrication functions. It will carry the metabolites, and because of its high concentration of complex carbohydrates will also reduce friction as the joint moves. Deficits in nutritional transport are a primary source of cartilage degeneration, and may contribute to osteoarthritis (Jackson and Gu, 2009). Synovial fluid has high concentrations of glucosamine and hyaluronic acid (Altman and Dittmer, 1964). Glucosamine is a precursor to glucosaminoglycans, a major structural component of cartilage. Glucosamine is a component of the disaccharide that polymerizes to form hyaluronic acid, also called hyaluronan. Hyaluronan has over 10 000 monomers and its high molecular weight is responsible for the high viscosity in synovial fluid. Synovial fluid also contains the smaller molecular weight protein–glucosaminoglycan molecule lubricin (Swann et al., 1981), whose relatively smaller size compared with hyaluronan may contribute to lubrication at the edges of joints where hyaluronan may be unable to reach effectively.

The composition of synovial fluid allows it to lubricate joints over a wide range of force and velocity. Muscle force and velocity, as we saw in Chapter 5, are inversely related. When a muscle, and the joints it is connected to, bears a large load, the movement of that load will be slow. In contrast, when a muscle is bearing a light load, it can contract quickly, moving the joint through its range of motion rapidly. Rapid joint movement means that the shear velocity of the synovial fluid in the joint will also be high. From Figure 7.5, as the shear velocity increases, the slope of the viscosity of synovial fluid approaches zero; that is, its viscosity does not change greatly as the velocity changes, and the viscosity is low. Low-viscosity fluid flows easily, and as the joint changes its position, the fluid can disperse across the joint surfaces to prevent friction. At high loads, the shear velocity will be low, and the viscosity of the synovial fluid will dramatically increase by orders of magnitude, reaching a gel-like consistency. This gel-like state will prevent the fluid from flowing away from the joint when large loads are being borne, and again preventing friction from damaging articular surfaces.

The gel nature of synovial fluid is a consequence of the hyaluronan. Under high pressure, the water, electrolytes and metabolites are squeezed away from the high molecular weight hyaluronan, significantly increasing its concentration at high pressure points in the joint (Laurent *et al.*, 1996). Normal synovial fluid both flows into the cartilage in the joint, and creates a barrier between the different surfaces in the joint, the gel-fluid being thickest in the center of the joint (Hlavácek, 1995). Synovial fluid from an inflamed joint has a much lower viscosity (Hlavácek, 1995), and will not form as thick a gel under high pressure conditions. Under low force, high speed conditions, the viscosity of synovial fluid in an inflamed joint is similar to that of normal synovial fluid. This adds a dimension to the use of anti-inflammatory treatments of athletic injuries. For activities where the force generation is low, synovial fluid present in the injured joint will not be as compromised as in cases where high force must be generated. High force injuries, where the connective tissue may be less able to stabilize a joint, will be further complicated by the thinner synovial fluid preventing contact between damaged surfaces. The synovial fluid thickness at an intact joint with sealed edges is only about 100 nm, and is homogeneously distributed (Hlavácek, 2002). If the joint capsule is damaged and some synovial fluid can escape through the edges, the remaining fluid will be non-homogeneous and, being under lower pressure and higher velocity, will not increase its viscosity as much, lessening the protection of the damaged joint. The range of load-bearing ability, lubrication and viscosity of synovial fluid exhibits remarkable static and dynamic behavior.

7.3 Arterial blood flow

Blood flow follows the same principles as other types of physical flow. The relation between flow Q, resistance R and pressure P

$$P = QR \qquad (7.22)$$

is the same as that for current, resistance and voltage in electrical systems. The driving force generated by the heart propels the blood forward. The friction of the blood against the vessel walls imparts resistance to the flow. In large vessels, blood flow is laminar, obeying the Poiseuille equation

$$Q = \frac{\pi \Delta \Pr^4}{8l\eta}.$$

(7.23)

Combining these two equations, the resistance R is

$$R = \frac{8l\eta}{\pi r^4}.$$

(7.24)

Thus, resistance is a function of both the blood viscosity and the fourth power of the radius. It is also a function of the vessel length, but within any individual, the length will be a constant at a given age, and as such is not a physiological variable for that individual. Both the viscosity and the radius can change. As discussed above, blood is a non-Newtonian solution. At very high shear rates, the viscosity will flatten out and approach Newtonian behavior. It is at very low shear rates that blood is non-Newtonian. In a vessel with laminar flow, the regions near the vessel walls will have a high shear rate, and blood flow is Newtonian in these regions. Mathematically, this means that the velocity profile of blood will have a parabolic shape as it moves away from the wall. In the center of the vessel, with the shear stress at zero in the center of the vessel, blood would be non-Newtonian. Near the walls, there will be a lateral difference in shear rate and shear stress on the wall side (high shear) and axial sides (low shear) of the blood cells; the larger the cells, the greater the shear difference. This difference creates a force driving the cells toward the low shear side, the center of the vessel. This follows the Prigogine principle of minimal entropy production, as the least energy would be needed to keep the cells along the axis of the vessel. This is the basis for the Fahraeus–Lindqvist effect, in which particulate mass will migrate to the center of a stream of flow. This effect is applicable to any smoothly flowing system, be it the bloodstream or a river. With a low shear rate, the lateral forces on the blood cells are minimized in the center of the vessel. Random collisions that propel the cells toward the walls will move them into areas of higher shear stress, creating a force driving them back toward the center. Under stationary flow conditions, there will be a balance between the two forces of dispersion and shear.

The other important variable in the Poiseuille equation is the vessel radius. In humans there are about five liters of blood in the circulatory system. If the entire system were fully dilated, there would be space for about ten liters of blood. Rapid dilation would cause blood to pool to the lowest point (gravity always wins), normally the feet, and the person faints from lack of blood to the brain. To maintain blood flow to the brain, the blood pressure must be maintained by partially constricting vascular smooth muscle (VSM). VSM has tone, partial contraction in the absence of a stimulus. Tone is caused by calcium leaking into VSM and is generally about 10–15% of maximum VSM contraction force. It is enough to reduce the radius of vessels. As resistance is inversely related to the fourth power of the radius, tone plays a major role in generating resistance and maintaining blood pressure. By being partially contracted, blood vessels are poised to either relax or constrict further to meet the blood flow needs of the body, with those tissues that most need the blood dilating their vessels and increasing their blood flow.

On the arterial side of the circulation, blood is moving from larger vessels into smaller vessels. The Fahraeus–Lindqvist effect complicates analysis of blood entry into smaller vessels, as the relative size of the smaller vessel and the angle off the larger vessel make it difficult to calculate the fraction of cells and plasma entering the smaller vessel. For a vessel that narrows in the direction of blood flow, a uniform reduction in radius means that the blood in the center of the vessel, with the lowest shear stress, would enter the narrowed region (Glaser, 2001). Upon entry, the velocity profile will be flatter than in the larger vessel, but as the blood flows along the smaller vessel, friction with the walls will generate shear stress and the laminar profile will reappear.

Blood flow in the arteries is pulsatile. The laminar flow conditions above assumed that the driving force and the frictional force were equal, conditions that do not apply to pulsatile flow. The stiffness of the vessel walls becomes an important variable during pulsatile flow. The Moens–Korteweg equation for the velocity v_{pw} of pulse wave propagation in thin-walled tubes is

$$v_{pw} = \sqrt{\frac{Ed}{2r\rho}} \tag{7.25}$$

where E is the modulus of elasticity, d is the wall thickness, r is the radius and ρ is the density of the medium. The elasticity is difficult to measure, particularly if some atherosclerosis may be present. Bramwell and Hill (1922) converted the Moens–Korteweg equation into a form with measurable hemodynamic variables pressure P and volume V:

$$\frac{Ed}{2r} = \frac{V\Delta P}{\Delta V} \tag{7.26}$$

and

$$v_{pw} = \sqrt{\frac{V\Delta P}{\rho \Delta V}}. \tag{7.27}$$

The two forms of the pulse velocity equation allow changes to either the wall elasticity, the radius or the blood flow to be assessed quantitatively. It is immediately apparent that atherosclerotic plaques will significantly increase the propagation velocity, as the plaque will increase both modulus of elasticity E and the wall thickness d. Pulsatile flow persists through the length of the arterial tree, as you can easily check your pulse at your ankle. Pulsatile flow is rapidly damped out in arterioles (20–100 μm diameter), which are about 1% of the diameter of the average artery (3–6 mm). Conditions that control damping of pulsatile flow are quantified by the similarity α parameter (Glaser, 2001):

$$\alpha = r\sqrt{\frac{\omega\rho}{\eta}} \tag{7.28}$$

where in addition to the density the pulse frequency ω and the viscosity η are taken into account. Changes in frequency, viscosity and density will be small relative to the changes in radius. When $\alpha \geq 3$, the pulse propagation velocity is the streaming velocity, and the pulse will not damp out. The aorta has a similarity of 15, and the femoral artery, with a much smaller radius, has a similarity of 3, the limit of pulsatile flow. With the much smaller radii of arterioles, the similarity factor is well below 3, and the pulsatile blood flow quickly damps out.

7.4   Arteriole blood flow

Arterioles are the control sites in the circulatory system. About 60% of the total drop in blood pressure occurs in the arterioles, so that systemic changes in the arterioles have major effects on the overall blood pressure. The friction caused by the contact of blood cells with the arteriolar walls is a major source of resistance to blood flow. Tone is present in all arterioles, and sensitivity to neural, humoral and pharmaceutical agents controls the dilation and constriction of the arterioles, and therefore controls where in the body the limited supply of blood will go. Arterioles get progressively smaller as they approach the capillaries, with the smallest arterioles having a diameter of about 20 μm, about three times the diameter of a red blood cell.

The decrease in arteriole diameter affects the velocity profile. Figure 7.6 shows velocity profiles for four different velocity conditions: one Newtonian, fitting an x^2 function, and three non-Newtonian profiles fitting x^3, x^5 and x^6 functions. Over a wide range of diameters and velocities in the 45–102 μm arterioles, the velocity profile as a function of the radius is remarkably similar. The mathematical best fit of these profiles is parabolic, fitting a Newtonian flow profile (Mamisashvili *et al.*, 2006). In intermediate size 30–32 μm arterioles, the velocity profiles are velocity dependent. At a low velocity, the profile is Newtonian; at twice this velocity, the profile becomes non-Newtonian.

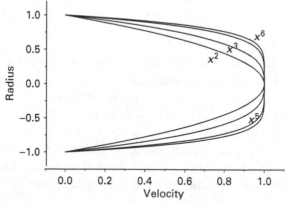

Figure 7.6   Velocity profiles of Newtonian and non-Newtonian flow in arterioles. The graphs plot relative velocity as a function of the distance from the center of the arteriole at $r = 0$. A Newtonian profile, like that shown in Figure 7.3, has a parabolic profile fitting the x^2 function. Non-Newtonian profiles have progressively flatter profiles and narrower high shear stress regions fitting higher order functions such as x^3, x^5, or x^6.

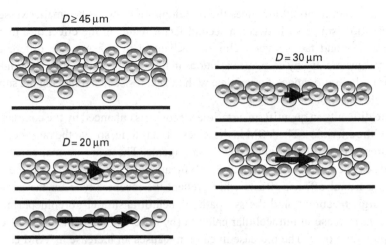

Fahraeus–Lindqvist effect in arterioles. For vessels of diameter D that is 45 μm or greater, the velocity gradient profile is constant with red blood cells clustering in the center of the vessel. The distribution is also velocity independent. Smaller vessels produce variable behavior. The arrows in 30 μm and 20 μm vessels indicate the relative velocity. Lower speed velocity in 30 μm vessels produces Newtonian behavior with strong center clustering of cells, while higher velocity has non-Newtonian flow with cells closer to the vessel wall. In 20 μm vessels, high velocity produces Newtonian flow, while low velocity produces non-Newtonian flow. These results show that both vessel size relative to the cells and velocity can alter flow conditions.

The non-Newtonian profile fits a higher order model and the profile is flatter in the middle of the vessel, with a narrower high shear stress region near the vessel wall, fitting an x^5 function. The non-Newtonian flow will not concentrate the cells in the center of the vessels as much as the Newtonian flow will. Small arterioles have Newtonian flow at high velocity, but non-Newtonian flow profiles at low velocity can fit functions up to x^6 (Mamisashvili *et al.*, 2006).

Figure 7.7 shows the relationship of the cells to the arteriole diameter in 45, 30 and 20 μm vessels drawn to scale. The Newtonian 45 μm arteriole has most of the red cells toward the center of the vessel, with a few cells shown closer to the cell wall. The balance of shear stress and random collisions will produce this distribution. The 30 μm arterioles are shown at low (short arrow) and high (long arrow) velocities. Here, the Newtonian high shear region is thicker for the low velocity flow, so that more cells will be concentrated in the center of the vessel. As the vessel size is reduced to 20 μm, an interesting reversal of flow occurs: the high velocity flow is Newtonian, while the low velocity flow is non-Newtonian and exhibiting pseudoplastic behavior. This could only occur if there is more than one factor controlling the blood cell distribution. In the 30 μm vessels the high flow must create turbulence that disrupts laminar flow, while in the 20 μm vessel the narrowing of the vessel to within three times the red cell diameter must play a role in altering the velocity profiles. There are two logical paths to follow here. First, it is the increase in velocity in going from 45 to 30 that produces the pseudoplastic flow, and then the decrease in vessel size from 30 to 20 that counters pseudoplastic flow at high velocity and returns the flow to Newtonian. Alternatively, the narrowing of the vessel at low flow from 30 to 20 produces the pseudoplastic flow in the smaller vessel. It is not possible to distinguish between a second factor reducing the effect of a first factor, or having a second factor overwhelm an unchanged first factor. The difference may be purely semantic. In any case, the variations in flow profile under different conditions may occur in the entire range of arterioles, with the net resistance reflecting the sum of all the arteriole behaviors.

The difficulty in quantifying arteriole viscosity is enhanced by the contractile properties of the arterioles. The VSM in these vessels has tone, so that the radius can change in the presence of dilating and constricting agents. These are normal changes, and the agents can be neural or hormonal. The sympathetic nervous system has the most wide-ranging neural effects, effects which reflect the receptor type on the vessels. Alpha adrenergic receptors bind the sympathetic neurotransmitter norepinephrine (NE) and cause an increase in intracellular calcium (by different mechanisms in different alpha receptor subtypes). The increase in calcium causes an increase in VSM contraction, a decrease in vessel diameter, an increase in resistance and a decrease in blood flow. Beta-2 adrenergic receptors also bind NE, but produce a decrease in calcium, leading to dilation, decreased resistance and increased flow. There are a wide range of vasoactive hormones, including: epinephrine, which in general mimics NE effects; histamine, which constricts large arteries and dilates arterioles; angiotensin II, which is a general vasoconstrictor; and the endothelial paracrines nitric oxide (dilation) and endothelin (constriction).

The adrenergic and histaminergic effects are particularly instructive: the same molecule can have opposite effects in different tissues. This illustrates a general problem for the pharmaceutical industry. It is easy to find a molecule that can do what you want; it is very hard to find a molecule that will do what you want that does not have serious side effects in other tissues. There are only a limited number of direct and second messenger systems in biology. It is the redundancy of these systems in different tissues that causes side effects. An example of this is the effect of phenylpropanolamine (PPA) on VSM. PPA is a psychotropic drug that causes the release of NE and epinephrine. It is used as a diet drug. PPA does not cause contractions of VSM directly, but enhances adrenergic contractions of VSM (Dillon *et al.*, 2004), and has been linked to strokes in young women. The mechanism of PPA enhancement of adrenergic activity is consistent with binding to the adrenergic receptor simultaneously with adrenergic binding, increasing the sensitivity and duration of adrenergic activity. Figure 7.8 shows the effect of PPA enhancement of NE-induced VSM contractions on blood flow. Over a range of initial lengths, corresponding to different radii, and with a range of zero VSM force lengths, PPA enhancement that increases NE force by as little as 20% causes more than a 50% decrease in blood flow under constant pressure conditions. Compensatory mechanisms in the body would increase pressure to return flow to normal levels, increases in pressure that would increase the risk of stroke. Similar to the cases of creatine, erythropoietin (EPO) and muscle hypertrophic drugs, people who are inclined to take PPA as a diet drug might think that if one pill helps me lose 5 pounds, maybe taking more pills will help me lose even more. Increased PPA intake enhances adrenergic contractions, with greater cardiovascular consequences. Nobody takes PPA to have a stroke, but the side effect of

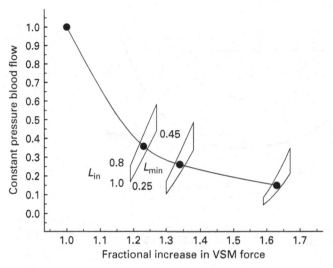

Figure 7.8 Dependence of blood flow on the fractional increase in vascular smooth muscle (VSM) activation under constant pressure conditions. The line indicates the changes that occur starting at an initial length L_{in} of 0.9 L_o and a zero force length L_{min} of 0.35 L_o. The boxes encompass the changes over a range of 0.8–1.0 L_{in} and 0.25–0.45 L_{min}. The decreases in blood flow are substantial under all conditions.

enhanced VSM contraction can be catastrophic. This is not a unique case: even a casual look at the possible side effects of virtually every drug hammers this point home.

7.5   Viscosity and hematocrit

The viscosity of blood is strongly dependent on the hematocrit, the volume percent of blood occupied by cells and other formed elements. Hematocrit is dominated by the red blood cell volume, which dwarfs the volume of white blood cells and platelets. The Fahraeus–Lindqvist effect determines the distribution of the cells during flow. The larger a formed element, the more it is affected by the shear stress profile, and the more likely that it will be found in the center of a vessel. The distribution of cells is also affected by any turbulence in the flow and the friction of cells colliding with the vessel walls, as well as specific adherence to the vessel walls. The variations in velocity profiles in Figure 7.6 reflect these different factors. In particular, the transposition of velocity effects in 30 and 20 μm vessels could reflect increased contact with the wall as the vessel diameter approaches the cell diameter, contacts that could be decreased (i.e., decreasing the entropy production) at a higher flow rate in the 20 μm vessels.

Figure 7.9 shows that the blood pressure changes in a biphasic manner as the hematocrit increases above the normal (Martini *et al.*, 2006). The initial drop in blood pressure is attributed to an increase in the release of the vasodilator nitric oxide from the endothelium in response to the increased hematocrit. Beyond a 10% increase in hematocrit above normal (i.e., above 50% in a person with a normal hematocrit of 45%), the

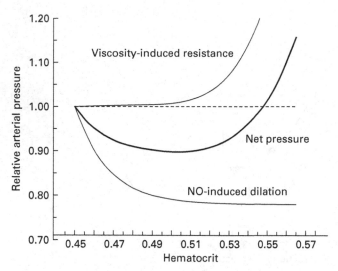

Figure 7.9   Biphasic response of blood pressure to increases in hematocrit. The decrease in BP with small increases in hematocrit is attributed to nitric oxide release. The increase in BP with large increases in hematocrit is attributed to increased viscosity and resistance to flow.

effect of increased hematocrit is reversed. The blood pressure increases above normal when the increased hematocrit is above 55%, and continues to rise thereafter. The severe medical consequences associated with athletes taking exogenous erythropoietin have been well documented (Hainline, 1993), with hematocrit values reaching up to 80% (Dhar *et al.*, 2005). The consequences associated with the increased resistance caused by the increased number of contacts of red blood cells with the blood vessel walls correlates with an increase in blood viscosity (Alkhamis *et al.*, 1990). The increased hematocrit is able to overcome the Fahraeus–Lindqvist effect, as the cells are so concentrated that they are impeded from moving to the central axis of the vessels. This increased cell-wall contact is consistent with the increase in hemoglobin release, as the increased friction would increase the probability of red cell lysis. Figure 7.10 shows a model of the circulatory system in which the non-linear increase in viscosity extends across all vessel diameters. The exponentially increasing viscosity in the smallest vessels reflects the cells having to be forced through the vessels with extended duration of contact between the cells and the endothelial wall. Experimental data imbedded in the diagram shows good agreement with the model.

An increase in hematocrit has a direct effect on hemodynamics. The increased viscosity will increase the total peripheral resistance. To maintain constant blood flow in the presence of increased resistance, the blood pressure must increase. The inflow rate to the coronary arterial system decreases as the hematocrit goes up, delivering blood to the heart capillaries at a slower rate (Huo and Kassab, 2009). This slower delivery is in part offset by the increased oxygen content per unit volume present as the hematocrit goes up. Despite this, the increased resistance to flow as the hematocrit rises must be overcome to pump blood, and the oxygen consumption of the cardiac muscle cells will rise.

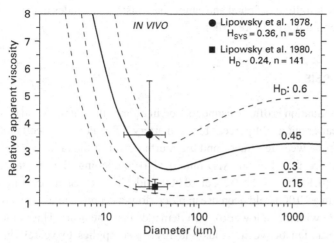

Figure 7.10   Network analysis model hematocrit dependence of blood flow in the mesentery. The bars are from experimental data (Lipowsky *et al.*, 1978, 1980). The non-linear increase of viscosity at elevated hematocrit occurs across all vessel sizes. (Redrawn from Pries *et al.*, 1996 by permission of Oxford University Press.)

One of the triumphs of recombinant DNA research is the development of exogenous sources of erythropoietin production. Erythropoietin (EPO) is a hormone produced by the juxtaglomerular (JG) cells of the kidney. When the dissolved oxygen content of the blood decreases, the JG cells release EPO into the plasma. EPO migrates to the bone marrow and stimulates the stem cells there to increase the production of red blood cells. The increase in RBCs boosts the oxygen content reaching the JG cells and through a negative feedback mechanism reduces the release of EPO. Under steady-state conditions, the amount of EPO released daily is just sufficient to replace the number of red cells lysed per day. People with chronic renal disease, decreased EPO release, or anemia secondary to cancer therapies can be given recombinant EPO to increase their red cell production and improve their quality of life. The dose of EPO has to be individually adjusted to compensate for the red cell deficiency in each person, striking a balance between anemia and the cardiovascular consequences of viscosity, pressure and flow at a high hematocrit.

The use of EPO to increase oxygen delivery has also been used by high-level athletes in whom a slight oxygen advantage may be what they need to win. No sport has been more affected by illicit EPO use than cycling. A large number of these incredibly fit athletes have died suddenly from apparent heart attacks. EPO use is the suspected cause of these deaths. The sequence of events involves injecting EPO, followed by an increased hematocrit response. During bicycle racing, traversing several mountains in single day, the loss of water in sweat further increases the hematocrit to the point where the heart cannot develop enough pressure to overcome the increased viscosity and resistance, oxygen delivery falls and the heart fails. EPO injection has been banned, and teams have been removed from the Tour de France when team support members have been found with EPO in their possession. Unfair competition and potential injury and death balanced by a better life for millions of renal and cancer patients: the two sides of recombinant EPO development could not be more striking.

7.6   Arterial stenosis

The Poiseuille equation is often misunderstood in a particular pathological condition. It is common to hear about a "90-percent blocked artery." The Poiseuille equation shows an inverse relation between resistance and the fourth power of the radius. A 90% blockage implies a 10% opening. For an axisymmetrical 10% opening, the resistance would be increased by $1/(0.1 \times 0.1 \times 0.1 \times 0.1)$, or 10 000-fold. Under constant pressure conditions, this would mean a 10 000-fold decrease in blood flow, a decrease fatal to any downstream tissue, even one with much lower oxygen demands than the heart. This is the misunderstanding: the relation between radius and resistance applies to vessels in which the reduction runs along the entire length of the vessel. Atherosclerotic plaques can form along the length of a vessel, but the stenotic narrowing in a small region, perhaps with a stenotic length down the vessel that is 10% of the diameter of the vessel (Spencer and Reid, 1979), is more common. A limited-length narrow opening will restrict blood flow, but not to the degree that a complete narrowing of the vessel along its entire length would. The effects of a focal stenosis are quite different from those predicted by the Poiseuille equation narrowing.

The paper by Spencer and Reid (1979) represented a significant step forward in the analysis of stenotic blood vessels. They used Doppler ultrasound images to measure the velocity of blood flow through a stenotic area, calculating the amount of blood available downstream from the stenosis. Recall that ultrasound works by measuring the reflection of sound waves back toward the source. If a tissue is exposed to high frequency sound waves, and the tissue moves before the waves are reflected back, there will be a shift in the frequency of the reflected sound waves. If the tissue is moving toward the sound source, the frequency would increase; if the tissue is moving away from the sound source, the frequency would decrease. The signals can be collected at angles and trigonometrically corrected for the change in signal. Using 5 MHz waves, Spencer and Reid determined that a 1 KHz shift was proportional to a blood velocity of 30 Hz. Concluding from x-ray analysis that a short, axisymmetrical stenosis was the best model of stenosis shape, they applied Bernoulli's constant energy equation (at constant gravity) for a change in vessel velocity and pressure as the vessel diameter changes:

$$\frac{V_o^2}{2} + \frac{P_o}{\rho} = \frac{V_s^2}{2} + \frac{P_s}{\rho} \tag{7.29}$$

where V_o and V_s are the open and stenotic velocities, P_o and P_s are the open and stenotic pressures, and ρ is the blood viscosity, which is assumed to be constant. They calculated that for a given blood pressure ΔP condition, a decrease in the cross-sectional area of the artery would be proportional to the increase in velocity:

$$V_o A_o = V_s A_s \tag{7.30}$$

$$\frac{V_{\mathrm{s}}}{V_{\mathrm{o}}} = \frac{A_{\mathrm{o}}}{A_{\mathrm{s}}} = \frac{D_{\mathrm{o}}^2}{D_{\mathrm{s}}^2} \tag{7.31}$$

where V_{o} and V_{s} are the open and stenotic velocities, respectively, A_{o} and A_{s} are the areas, D_{o} and D_{s} are the diameters. The resistance R to flow in the stenotic area would be

$$R = 8\eta L/\pi r^2 = 8\eta L/A_{\mathrm{s}} \tag{7.32}$$

where η is the blood viscosity, L is the axial length of the stenosis, and A_{s} is the cross-sectional area of the narrowed region. Since flow Q is

$$Q = \frac{\Delta P}{R} \tag{7.33}$$

they could determine flow through the stenotic region by measuring velocity using the Doppler ultrasound and the diameter of the vessel using x-ray. They produced what is now called Spencer's curve (Spencer and Reid, 1979), which has been used for both research and clinical applications, particularly in cerebrovascular hemodynamics (Alexandrov, 2007).

Figure 7.11 shows the dramatic blood flow changes predicted by the Spencer model. In an area of stenosis, there will be a large increase in the velocity of blood flow. This increase in velocity can compensate for the decrease in cross-sectional area caused by the stenosis. A reduction in vessel diameter to 40% leaves only 16% of the cross-section, yet the increase in velocity balances this, and 99% of the blood flow remains. Even when only 20% of the diameter remains, or a cross-section of 4%, 83% of the blood flow remains. Below this, however, further narrowing of the vessel cannot maintain nearly normal blood flow, and there will be a dramatic decrease in blood flow, which would

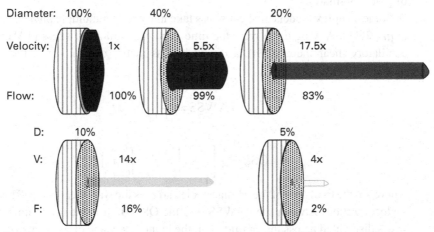

Figure 7.11   Effect of arterial stenosis on blood velocity and flow from the Spencer model. The model stenosis is smooth, of uniform length, and axisymmetrical with no turbulence. The blood arrow length is proportional to normal velocity, and the arrow intensity is proportional to blood flow within the stenosis. Over a wide range of stenosis, increased blood velocity compensates for the decrease in diameter. Below 20% diameter, velocity decreases and blood flow decreases dramatically.

severely limit downstream supplies of blood. These great differences in downstream blood availability are crucial in determining a course of treatment for vascular stenosis.

7.7   Arterial asymmetry: atherosclerosis and buckling

Biophysical processes contribute to two distinct pathologies in large blood vessels, the formation of atherosclerotic plaques and the buckling, twisting, of large vessels. Plaque formation is dependent on the shear rate, and the medical observations, experimental data and theoretical models of the carotid bifurcation all converge in predicting the site of plaque formation in this region. An extensive series of studies on buckling of large vessels gives us insight into this potentially catastrophic event. Both of these are included below.

The Fahraeus–Lindqvist effect above showed us that blood cells will not congregate in areas of high shear. At arterial bifurcations, the disruption of the flow pattern leaves the highest velocity flow near the inside walls of the bifurcation, with a sharp drop in velocity, and therefore a high shear rate, along the insides of the bifurcation. On the outsides of both branches the velocity falls off more slowly. For the internal carotid artery, both the velocity and the shear rates are low on the outside of the carotid sinus (Wells and Archie, 1996).

A variety of different metrics have been used to model the flow patterns of the carotid bifurcation. Models of oscillatory and steady-state shear stress have shown that areas of high shear stress are associated with increased risk of aneurysm, while areas of low shear stress are associated with increased risk of atherosclerotic plaque formation (Chatziprodromoua *et al.*, 2007). The bifurcation of the carotid artery has regions of low shear stress located on the lateral surface just prior to the bifurcation (Figure 7.12). There is good agreement between the models of low shear areas and the clinical findings of plaque formation.

Some complex models of shear stress take into account multiple physical factors (Lee *et al.*, 2009). Among these were the time averaged wall shear stress (TAWSS) and the oscillatory shear index (OSI), a dimensionless metric of changes in the shear stress direction:

$$\text{TAWSS} = \frac{1}{T} \int_0^T |\tau_\text{w}| \, \mathrm{d}t \tag{7.34}$$

$$\text{OSI} = \frac{1}{2} \left[1 - \left| \int_0^T \tau_\text{w} \right| \bigg/ \int_0^T |\tau_\text{w}| \, \mathrm{d}t \right] \tag{7.35}$$

where T is the total time of the cardiac cycle and τ_w is the wall shear stress. There is clearly a close relation between the TAWSS and the OSI. Both relate to the amount of shear stress distributed across the carotid wall: the higher the value of either metric, the greater the shear stress. The relative residence time (RRT) is

$$\text{RRT} = \frac{1}{(1 - 2 \times \text{OSI}) \times \text{TAWSS}} \tag{7.36}$$

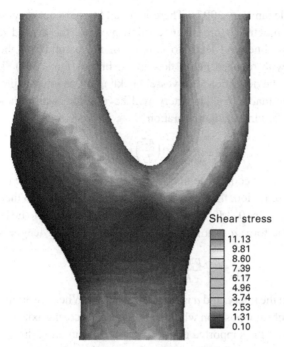

Shear stress

11.13
9.81
8.60
7.39
6.17
4.96
3.74
2.53
1.31
0.10

Figure 7.12   Shear stress pattern at the carotid bifurcation. The dark areas are the sites of lowest shear stress, and are likely sites of atherosclerotic plaque and thrombus formation. (Reproduced from Chatziprodromoua *et al.*, 2007, with permission from Elsevier.)

and is a measurement of how long particles will stay near the wall. The RRT was found to be the best metric for measuring low and oscillating shear at the carotid bifurcation. This complex analysis is in general agreement with the model in Figure 7.12 (Chatziprodromoua *et al.*, 2007)

Experimental measurements of different areas of the carotid bifurcation found that those areas on the outside of the proximal internal carotid and the midpoint of the carotid sinus had intimal thickness (0.49–0.63 mm) far greater than any part of the common carotid (0.10–0.15 mm), the distal internal carotid (0.06–0.09 mm) or the external carotid (0.08–0.27 mm) (Ku *et al.*, 1985). In addition to the intimal thickening, areas of low shear stress are associated with both atherosclerotic plaque formation (Cunningham and Gotlieb, 2005) and thrombus formation (Runyon *et al.*, 2008). The initial formation of a plaque or thrombus can be exacerbated by the changes in flow pattern they produce. By increasing the velocity through the narrowed area, as shown in Figure 7.11, there will be areas of low shear behind the plaque or thrombus, increasing the probability of its enlargement. Maintaining conditions of high shear stress minimizes the formation of these deleterious structures, but high shear stress does coincide with areas prone to form aneurysms.

The shear stress models and anatomical findings show that atheroma and thrombus formation is not axisymmetrical. The blood flow changes predicted by Spencer's curve assumed a smooth, axisymmetrical stenosis, a model which has proven to be useful for clinical applications (Alexandrov, 2007). There is an additional consideration in large blood vessels with asymmetrical forces: under high pressure the carotid (and other) arteries can buckle. This buckling leads to kinking of the carotid and the expected pathologies associated with reduced blood flow to the brain (Han, 2007). Theory and experiment have shown the dependence of vessel buckling on its internal pressure. The simplest model assumes that a buckled artery will have a sine wave shape, with its deformed central axis x_c having the configuration

$$x_c = C \sin\left(\frac{\pi z}{L}\right) \qquad (7.37)$$

where C is a constant, z is the coordinate in the axial direction, and L is the length of the artery. Being a sine configuration, the artery is tethered at its two ends, so that the model extends over one-half cycle for 0–180°. The buckling force would have to be lateral to the axial direction, so that the force q at point z along the buckled artery length is

$$q(z) = \frac{p\pi^3 a^2}{L^2} C \sin\left(\frac{\pi z}{L}\right) \qquad (7.38)$$

where p is the pressure in the artery, and a is the arterial radius. When the artery buckles, it will move to a new equilibrium position where the buckling force, the axial tension along the artery, and the force on the supported ends of the artery countering the buckling will all balance one another, producing a net force of zero and no further movement. The critical pressure p_{cr} at which an artery will buckle is

$$p_{cr} = E\frac{\pi^2 at}{L^2} + 2E(\lambda_z - 1)\frac{t}{a} \qquad (7.39)$$

where E is the modulus of elasticity of the wall, t is the wall thickness, and λz is stretch ratio of the artery, the observed length of the artery divided by its slack length. Several elements can be derived from this equation. The higher the modulus, that is, the stiffer the artery, the higher the pressure needed for buckling. The bigger the radius and thicker the wall, the higher the buckling pressure. These points are applicable to large veins, often used for grafts, which have a larger radius, lower modulus, and thinner wall than arteries, and have been shown to have much lower critical buckling pressures than arteries (Martinez *et al.*, 2010). The longer the length of vessel, the lower the pressure needed for buckling. For a given artery, there is a linear relation between the stretch ratio and the critical pressure. When the stretch ratio approaches its minimum of 1.0 (where it would go slack), the pressure needed for buckling would approach 0 (Han, 2007). This has implication for vessel transplants, where the insertion of a graft with a thinner wall and less axial stretch than the original vessel would decrease the critical buckling pressure. Insertion of grafts under axial pressure would increase the critical pressure, making the graft less likely to buckle. For vessels within an elastic substrate, the buckling takes on higher buckling modes (i.e., several sine waves), producing a tortuosity similar to that seen in vivo (Han, 2009).

Given the effect on blood flow, plaque formation and mechanical stability, it is likely that vessel mechanics will continue to be a fruitful area of research.

7.8   Lung air flow

The reciprocal flow of inspiration and expiration has elements of both synovial fluid flow in its reciprocal nature, and blood flow in its flow through tubes. While air is a fluid, the equations governing aerodynamics are slightly different from those for hydrodynamics. Among the simplest but most important analytical tools in respiration is the anatomic dead space. During inspiration, fresh air must pass through the mouth, pharynx, trachea, bronchi and bronchioles, all non-exchanging sites, before reaching the alveoli, where oxygen and carbon dioxide are exchanged with the blood capillaries. This means that the tidal volume V_t is

$$V_t = V_{ADS} + V_{alv} \tag{7.40}$$

where V_{ADS} is the anatomical dead space volume and V_{alv} is the alveolar volume. The anatomical dead space volume must be filled before the alveoli receive any fresh air, so that the dead space volume is a price that must be paid on every breath. For high respiratory rates with low tidal volumes, the amount of fresh air reaching the alveoli is limited. For a normal tidal volume of 500 ml, the dead space is 150 ml. That means that if the tidal volume fell to 150 ml, essentially no fresh air would reach the alveoli. During severe asthma attacks, this rapid breathing can occur, and bronchodilators must be introduced quickly to alleviate the attack or the person would suffocate. Thus, the factors controlling the delivery of aerosol medicines are important.

   In the case of inhaled drug delivery, practical considerations are more important than theoretical models. The key determinant is what fraction of an aerosol is deposited in the respiratory tract (Kim and Hu, 2006). Testing the total deposition fraction (TDF) in men and women, a sigmoidal model was empirically fitted to the experimental data. The sigmoidal equation is

$$\text{TDF} = 1 - \{a/(1 + b\omega)\} \tag{7.41}$$

where a and b are fitted constants, ω is a composite factor relating aerosol particle size D_a, tidal volume V_t and respiratory flow rate Q:

$$\omega = D_a^x V_t^y Q^z \tag{7.42}$$

where x, y and z are fitted to the data. Increases in ω increase the denominator in the brackets, lowering the bracket value, and therefore increasing the amount of medicine deposited. Note from the TDF equation that TDF will vary in a sigmoidal fashion as a function of ω, varying between the limits of $(1 - a)$ when ω is zero and 1 as ω approaches infinity. The net result is that increases in ω increase medicine delivery. The values of x and y are positive values, indicating that increased particle size and tidal volume increase delivery. The flow rate exponent z is negative, so that slower flow rates increase delivery. These have practical benefits: sufferers should use an inhaler that generates large size particles and inhale deeply and slowly to maximize delivery. Gender-specific analysis shows that the fractional delivery is more efficient for women when using larger particles.

The flow dynamics of forced expiration are quite complex. The increase in transmural pressure that propels air out of the alveoli also puts pressure on the conducting airways. When the intraluminal pressure in the airway drops below the intrapleural pressure during expiration, the transmural pressure will collapse the airway, creating a narrowing or choke point (Hayes and Kraman, 2009). This creates a "waterfall" effect, where flow downstream, toward the mouth, from the choke point is independent of the downstream pressure, as its resistance is so much lower than at the choke point. Choke point resistance creates downstream oscillations that cause snoring during sleep apnea. High speed flow creates central choke points, while lower flow rates create choke point in the periphery. This difficulty in identifying the physical location of a choke point is one of the difficulties in analyzing expiratory air flow.

Models of expiration show that at a critical air wave velocity V_c the expiration reaches a maximum (Dawson and Elliot, 1977), with V_c being

$$V_c = \left(\frac{A^3 \mathrm{d}B/\mathrm{d}A}{q\rho} \right)^{1/2} \tag{7.43}$$

where A is the area of the choke point, $\mathrm{d}B$ is the transmural pressure, $\mathrm{d}B/\mathrm{d}A$ is the stiffness of the choke point, q is a correction factor for Poiseuille flow and ρ is the density of the air. The two largest variables are the area of the choke point (the exact position of which is unknown) and the stiffness. Higher velocities can be tolerated if the choke point area is larger (less choke) and if the airways are stiffer, better able to resist collapse. The analytical difficulties lie in the anatomical unknowns, although multiple experiments confirm the general applicability of the model (Hayes and Kraman, 2009).

7.9   Aqueous humor and cerebrospinal fluid flow

A buildup of aqueous humor pressure in the eye can lead to glaucoma (Nilsson and Bill, 1994). Aqueous humor is produced by the ciliary processes posterior to the iris. It then circulates around the iris, bathing the avascular lens and cornea, and is absorbed by the trabecular meshwork posterior to the iris and into Schlemm's canal. If the aqueous humor drainage is blocked, pressure builds up in the eye, damaging the axons of the ganglion cells at the optic nerve. The drainage of aqueous humor is pressure dependent: if there is an increase in resistance at the drainage sites, pressure will increase until the outflow matches the inflow rate from the ciliary processes. Inflow is caused by Na–K ATPase activity increasing the sodium concentration in the spaces between the cilary cells, raising the osmotic pressure. Water follows the sodium until the osmolarity matches that of the posterior chamber. The interesting point here is that the inflow rate is not dependent on the internal pressure. This system does not use normal flow mechanics, which have a strong relation between pressure, flow and resistance, since aqueous humor entry into the eye is not from a pool of fluid, but across the cell membrane. Since aqueous humor entry is not strongly pressure dependent, disruption of the normal outflow results in the increase in pressure until the inflow and outflow rates match. The pathological consequences of glaucoma that follow possibly would not occur if both inflow and outflow were both pressure dependent, as a reduction in drainage would be accompanied by a decrease in fluid inflow. Cerebrospinal fluid (CSF) is also produced by ion-flow-induced osmosis (Ransom, 2005). Reduced drainage of CSF can lead to an increase in CSF pressure, following the same principles as those leading to glaucoma. These processes lead us naturally into the next sections on membrane function.

Fluid flow across a surface produces laminar flow, with areas of high shear stress near the surface. Particulate matter will be driven from the surface by a differential shear gradient. Flow in a vessel is described by the Poiseuille equation, which includes pressure, vessel length, fluid viscosity and the most important variable, vessel radius. In joints, synovial fluid exhibits viscosity changes over orders of magnitude, protecting joint movement during both high speed, low force movement and high force, low speed movement. In a blood vessel the formed elements in the blood will congregate in the center of flow, the site of lowest shear. Disruptions in flow by atherosclerotic plaques will increase the Reynolds number of the flowing material, increasing the probability of transition to turbulent flow. The flow rate is partially dependent on the viscosity of the solution, which in blood is determined by the hematocrit, the percentage of red blood cells in the blood. High hematocrits increase the interaction of blood cells with the vessel walls, significantly increasing resistance to blood flow and putting increased demands on the heart. Changing vessel diameters throughout the circulatory system alter the influence of different factors in different sized vessels. Pathological stenoses will decrease vessel diameter, but increased flow velocity will compensate significantly until the stenosis closes more than 80% of the vessel. Atherosclerotic plaques will tend to form in areas of low shear stress, and blood vessels will buckle at a critical pressure related to their elasticity, wall thickness, and axial stretch, with veins having much lower critical pressures than arteries, an element important in vein grafts. Air flow in the lungs during asthma attacks affects the delivery of dilating medicine, and is controlled by particle size, tidal volume and flow rate. The production of aqueous humor in the eye is independent of fluid pressure, which has important consequences in the development of glaucoma. There continues to be both theoretical and practical applications of fluid and air flow in physiological systems.

References

Alexandrov A V. *J Neuroimaging*. **17**:6–10, 2007.
Alkhamis T M, Beissinger R L and Chediak J R. *Blood*. **75**:1568–75, 1990.
Altman P L and Dittmer D S. *Biology Data Book*. FASEB, Washington, DC: 1964.
Bramwell J C and Hill A V. *Proc R Soc Lond B*. **93**:298–306, 1922.
Chatziprodromoua I, Poulikakosa D and Ventikos Y. *J Biomech*. **40**:3626–40, 2007.
Cunningham K S and Gotlieb A I. *Lab Invest*. **85**:9–23, 2005.
Dawson S V and Elliott E A. *J Appl Physiol*. **43**:498–515, 1977.

Dhar R, Stout C W, Link M S, *et al. Mayo Clin Proc.* **80**:1307–15, 2005.

Dillon P F, Root-Bernstein R S and Lieder C M. *Am J Physiol Heart Circ Physiol.* **286**:H2353–60, 2004.

Dörr F. In *Biophysics*, ed. Hoppe W, Lohmann W, Markl H, and Ziegler H. Berlin: Springer-Verlag, pp. 42–50, 1983.

Ganong, W. *Review of Medical Physiology.* New York: Lange, 2005.

Glaser, R. *Biophysics.* Berlin: Springer-Verlag, 2001.

Hainline B. In *Cardiovascular Evaluation of Athletes: Toward Recognizing Athletes at Risk of Sudden Death*, ed. Waller B F and Harvey W P. Newton, NJ: Laennec Publishing, pp. 129–37, 1993.

Han H C. *J Biomech.* **40**:3672–78, 2007.

Han H C. *J Biomech.* **42**:2797–801, 2009.

Hayes D Jr and Kraman S S. *Respir Care.* **54**:1717–26, 2009.

Hlavácek M. *J Biomech.* **28**:1199–205, 1995.

Hlavácek M. *J Biomech.* **35**:1325–35, 2002.

Huo Y and Kassab G S. *J Appl Physiol.* **107**:500–5, 2009.

Jackson A and Gu W. *Curr Rheumatol Rev.* **5**:40–50, 2009.

Kim C S and Hu S C. *J Appl Physiol.* **101**:401–12, 2006.

Ku D N, Giddens D P Zarins, C K and Giagov S. *Arteriosclerosis.* **5**: 293–302, 1985.

Laurent T C, Laurent U B and Fraser J R. *Immunol Cell Biol.* **74**:A1–7, 1996.

Lee S W, Antiga L and Steinman D A. *J Biomech Eng.* **131**:061013, 2009.

Lipowsky H H, Kavalcheck S and Zweifach B W. *Circ Res.* **43**:738–49, 1978.

Lipowsky H H, Usami S and Chien S. *Microvasc Res.* **19**:297–319, 1980.

Mamisashvili V A, McHedlishvili N T, Chachanidze E T, Urotadze K N and Gongadze M V. *Bull Exp Biol Med.* **142**:748–50, 2006.

Martinez R, Fierro C A, Shireman P K and Han H C. *Ann Biomed Eng.* **38**:1345–53, 2010.

Martini J, Carpentier B, Negrete A C, Frangos J A and Intaglietta M. *Am J Physiol Heart Circ Physiol.* **289**:H2136–43, 2006.

Nilsson S F E and Bill A. In *Glaucoma*, ed. Kaufman P L and Mittag T W. London: Mosby, pp. 1.17–1.34, 1994.

Pries A R, Secomb T W and Gaehtgens P. *Cardiovasc Res.* **32**:654–67, 1996.

Ransom B R. In *Medical Physiology*, ed. Boron W F and Boulpaep E L Philadelphia, PA: Elsevier Saunders, p. 403, 2005.

Runyon M K, Kastrup C J, Johnson-Kerner B L, Ha T G and Ismagilov R F. *J Am Chem Soc.* **30**:3458–64, 2008.

Schmid-Schönbein H, Gaehtgens P and Hirsch H. *J Clin Invest.* **47**:1447–54, 1968.

Spencer M P and Reid J M. *Stroke.* **10**:326–30, 1979.

Swann D A, Slayter H S and Silver F H. *J Biol Chem.* **256**:5921–25, 1981.

Wells D R and Archie Jr J P. *J Vasc Surg.* **23**:667–78, 1996.

Wright V and Dowson D. *J Anat.* **121**:107–18, 1976.

8 Biophysical interfaces: surface tension and membrane structural properties

Phase transitions always indicate that something interesting is happening. In physiological systems, physical phase transitions occur in two areas, at the air–tissue surface in the lungs, and at membranes. In the lungs, the surface tension of water and its alteration by surfactant controls the ease of alveolar inflation. At membranes, there are multiple factors: the fluidity of molecules within the membrane; the stability and organization of integral proteins; and the maintenance of membrane integrity under changing external conditions. In this chapter, we will cover the principles of surface tension and the inflation of alveoli; the composition of the membrane and the factors that control molecular activity; and the response of the membrane to ultrasonic waves, deformation and ethanol.

8.1 Surface tension

The common solvent in physiological systems is water. As we have seen, water molecules form hydrogen bonds with other water molecules, with electrolytes, and with proteins. At equilibrium, there will be no net force on any molecule in the system. Any perturbations in the system will, given enough time, return to an equilibrium state if no additional energy is put into the system. Recalling the Boltzmann energy distribution from Chapter 1, over a long time the energy of a system will have an average energy of kT, although almost no molecules will have exactly the energy of kT. The energy of any physical part of the system dl at this sufficiently long time dt is

$$\iint E \, dl \, dt = C \tag{8.1}$$

where the energy E is integrated over the space and time intervals, and C is a constant. These conditions must apply to molecules at a gas–liquid interface. Not all the molecules in the vicinity of the interface will be under the same conditions, however. The different equilibrium conditions for the different molecules are shown in Figure 8.1.

Four different solvent molecules are considered in Figure 8.1: a molecule in the bulk solution (M1); a molecule at the interface (M2); a molecule at the interface, but separated from the gas phase by a non-associating surface structure or immiscible liquid (M3); and a molecule at the interface that has an ionic bond with an ion in solution (M4). For molecule 1 (M1) in Figure 8.1, with the selection of coordinates such that the energy profile is centered on M1, $C = 0$, as the force in all directions will be canceled out. This is

Figure 8.1   Solvent interactions and surface tension. Solvent molecules distribute their interaction energy with the surroundings. Molecule 1 has its bond energy distributed equally over all the molecules around it, producing weak individual interactions (light gray). Molecule 2, on the gas–liquid surface, has fewer interactions, and therefore more energy (mid gray) in its individual bonds, but can potentially enter the gas phase. Molecule 3, at a solvent–solvent interface, cannot bond with or enter the external solvent, and has higher bond energies (dark gray) than M1 or M2. Molecule 4 can form strong bonds with a dissolved ion (square structure), has a strong polarity, and has increased energy in its solvent–solvent bonds, increasing its surface tension.

not the case for molecule M2. M2 is on the surface, and the center of its energy profile (where $C = 0$) will not be centered on M2, but will be below the surface of the liquid at $M2_0$. At M2, there will be a force in the direction of $M2_0$. M2 can only maintain its position if there is a balancing force applied to it. It cannot form significant bonds above the solvent surface (essentially, the activation energy for bond formation in this direction is very high), so the bonds it does form with other surface molecules and with molecules just below the surface must be stronger than the individual bonds of M1. These stronger intermolecular bonds of surface molecules are the basis for surface tension. Because of the net inward force on surface molecules, a system will try to minimize this force, minimizing the surface area of the system. With no structural restrictions, liquid droplets will form spheres, the shape that minimizes the surface-to-volume ratio.

The stronger bonds between molecules at the surface means that expansion of the surface requires more energy than that required to move a molecule between two other molecules in the bulk phase. The surface tension γ is the energy needed to expand the surface, in J/m^2 or N/m. The specific surface energy is the energy required to enlarge a surface by $1\ m^2$ (Glaser, 2001).

When two immiscible liquids are in contact, an interface will form between them. Molecule 3 in Figure 8.1 is at such an interface. Unable to bond with the other solvent, and also unable to have any molecules leave its surface as they could at the gas–liquid interface, the bonds between the solvent surface molecules will be stronger yet. This would not be the case if some bonding were possible between the molecules of the two different phases: in this case, the intermolecular bond energies for M3 would be lower than for M2.

The presence of solutes dissolved in a solvent, such as salts dissolved in water, can also affect surface tension. Molecule M4 is in such a solution. M4 is able to bind tightly to one of the electrolytes of the salt, ionizing the water in the water–salt bond. This will increase the relative energy between M4 and adjacent solvent molecules, enlarging the size and duration of water clusters, and increase the surface tension. Other solutes will not form such ionizing structures, and in general most solutes will lower surface tension, with ionizing salts the exception (Tinoco *et al.*, 1995).

Since surface tension is a function of hydrogen and non-polar bond formation, it is temperature dependent. Water, for example, decreases its surface tension from 72 mN/m at 25 °C to 59 mN/m at 100 °C. While this is physically important, for the range of temperature change in physiological systems it is not of great significance. Surface tension is important at the cellular and tissue level, influencing the behavior of cells and the inflation of alveoli, but less so at the organismal level. The surface tension of water is not so high that someone jumping into a pool encounters a great deal of resistance. By contrast, in falling from a great height gravity plays a far greater role, and whether one falls into water or on land, damage will result if the fall is far enough. Conversely, to a small insect, gravity is not fatal from a great height, as air resistance would slow its rate of fall so significantly that the insect would not strike the earth very hard. But to an insect, the surface tension in a pool that we can ignore is a deathtrap: caught within the grip of surface tension, the insect cannot get free, similar to the animals whose fossils are found in ancient tar pits.

The energy difference between molecules at the surface and in the bulk phase means that molecules or ions which increase surface tension would increase the energy of the system more if they concentrate at the surface, a violation of the Prigogine principle of minimal entropy production. Molecules that lower the surface tension will lower the overall energy of the system when they concentrate at the surface. The adsorption Γ of molecules to the surface is

$$\Gamma = \frac{n_i}{A} \tag{8.2}$$

where n_i is the number of molecules i and A is the area of the surface. The addition of molecules to a surface from the bulk solution is related to the surface tension by the Gibbs adsorption isotherm Γ (Tinoco *et al.*, 1995):

$$\Gamma = -\frac{1}{RT}\frac{d\gamma}{d(\ln a)} \cong -\frac{1}{RT}\frac{d\gamma}{d(\ln c)} \tag{8.3}$$

where $d\gamma$ is the change in surface tension and $d(\ln a)$ and $d(\ln c)$ are the changes in the natural logarithms of activity and concentration in the bulk solution, respectively. In physiological systems, hydrophobic molecules will lower the surface tension of a water solvent system, and will therefore gather at the surface. Amphiphilic molecules will gather at this new interface between hydrophilic and hydrophobic molecules.

In our systems, phospholipids are amphiphilic, having both a hydrophilic and hydrophobic component. In the absence of hydrophobic molecules, phospholipid molecules will collect at the surface of water, extending their hydrophobic tail outward from the water surface (Figure 8.2).

Because these molecules collect at the surface, interspersing between the water molecules, they will lower the surface tension of the system. The change in surface tension produced by adding surface active molecules can be measured using a Langmuir film balance (Figure 8.3). The mechanical compression of the surface layer reduces the surface area in which the surface active molecules are confined. Since it requires a force to confine the molecules, there will be an equal and opposite force pushing back on the film balance, resulting in an increased surface tension.

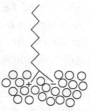

Figure 8.2   An amphiphilic molecule at a water–air surface.

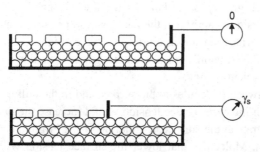

Figure 8.3   Surfactant surface tension measured with a Langmuir film balance. Compression of the surface layer increases its surface tension γ_s.

The addition of amphiphilic surfactant molecules to the surface of water lowers overall surface tension of the system. The molecules are mobile, and will move around the surface driven by Brownian motion. When the surface molecules are compressed by a film balance, the surface pressure, measured in N/m just as for surface tension, will increase. The surface pressure $P_{surface}$ is compared with the surface tension of the solvent, which in physiological systems is water:

$$P_{surface} = \gamma_{water} - \gamma_s \tag{8.4}$$

where γ_s is the surface tension of the surfactant measured by the film balance. The lower limiting case would occur as the concentration of the surfactant and therefore its surface tension approach zero, where the surface pressure would then be the surface tension of water. When the surface molecules are compressed by the film balance into a confined space, the surface pressure will rise exponentially. There will be a pressure at which the surfactant molecules will no longer maintain a single molecule layer and the surface pressure will no longer increase as the area decreases (Figure 8.4).

Fluorescence microscopy of monolayers of dipalmitoylphosphatidylcholine (DPPC) in a Langmuir apparatus is shown in Figure 8.4(A). The lipids are not in a continuous layer, but exhibit distinct structures. The spiral asymmetry of the DPPC cluster is due to unequal addition of the R- and S-forms of the molecules, resulting in the chiral-dependent structures (Stottrup *et al.*, 2010). Figure 8.4(B) shows the change in surface pressure as the area is compressed. Viewing the area axis from right to left, as the film balance decreases the surface area, the lipids will transition from a liquid-expanded phase to a solid, condensed phase in which the lipids are crowded together. The discontinuity of

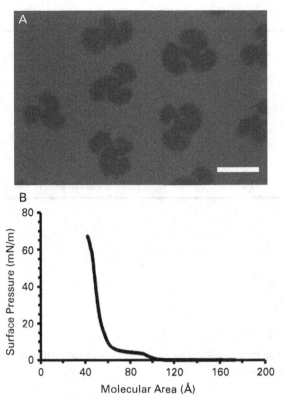

Figure 8.4   DPPC Monolayers. (A) Clusters of DPPC lipid molecules form asymmetrical structures due to the chirality of the lipid composition. The lipids form clusters, not a continuous layer, at sufficiently low concentrations. Scale bar is 50 μm. (B) As the area of the surface monolayer is compressed the lipids will transition from a liquid to a condensed phase, rupturing the monolayer when the surface pressure is sufficiently high. (Reprinted from Stottrup *et al.*, 2010, with permission from Elsevier.)

the surface pressure occurs when the compressive force is so high that the lipids no longer maintain a monolayer, essentially piling up upon each other, rupturing the surface.

The surface tension of several biophysically related materials is shown in Table 8.1. Water, present in concentration (55.4 M) 400 times greater than the next most abundant extracellular substance, sodium (140 mM), is the solvent for most biological molecules, the exception being those in the hydrophobic core of a membrane. Water has only a small variation in surface tension between its melting (75.6 mN/m) and boiling points (58.9 mN/m). The very small thermal gradients in physiological systems near 37 °C will not produce significant functional changes in the surface tension of water. Pure water has nearly the identical surface tension as plasma with its proteins and metabolites. The inclusion of blood cells, dominated by red blood cells, produces a 15–20% decrease in the overall surface tension of blood compared with plasma. There is a tight range of surface tensions covering pulmonary surfactant, ethanol, lecithin/cholesterol and phosphatidyl choline/cholesterol. These values are similar to those of organic solvents. Surfactant, as we will see below, significantly lowers the surface tension of the water

Table 8.1 Surface tensions of biophysically related materials

	Surface tension (mN/m)	Reference
Water (37 °C)	70	Weast, 1975
Plasma (adults)	70–75	Altman and Dittmer, 1964
Blood	56–61	Altman and Dittmer, 1964
Surfactant	22.2	Otis *et al.*, 1994
Ethanol (96%, 37 °C)	21.5	Weast, 1975
Lecithin/cholesterol	20	Reifenrath and Zimmerman, 1976
Phosphatidyl choline/ cholesterol	18	Brzozowska and Figaszewski, 2002
Ectodermal cells	1.1–7.7	Kalantarian *et al.*, 2009

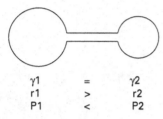

$$
\begin{array}{ccc}
\gamma 1 & = & \gamma 2 \\
r1 & > & r2 \\
P1 & < & P2
\end{array}
$$

Figure 8.5 Law of Laplace for two connected vesicles. The higher pressure in the smaller vesicle 2 will cause it to collapse into the larger vesicle 1.

molecules lining the inside of the lung alveoli, greatly reducing the effort needed to inflate the alveoli. Cells have far lower surface tension than water, a consequence of the membrane composition and its interactions with the molecules below its surface. This reduction in surface tension allows cells to deviate from spherical structures more easily than water droplets can.

8.2 Action of surfactant on lung surface tension

Surface tension plays an important role in the alveoli of the lungs. The relation between alveolar wall tension, inflation pressure, and radius is a function of the Law of Laplace, which for alveoli will be

$$P = \frac{2\gamma}{r} \qquad (8.5)$$

as we have seen in Chapter 6. Consider two vesicles connected by a tube (Figure 8.5). These two vesicles will have the same composition, and the same surface tension. The vesicle with the larger radius will have a lower pressure than the smaller vesicle, and molecules in the small vesicle will migrate into the larger vesicle, further decreasing the radius of the smaller vesicle until it collapses into the larger vesicle. The rate of this collapse is decreased by the presence of pulmonary surfactant (PS). PS is manufactured

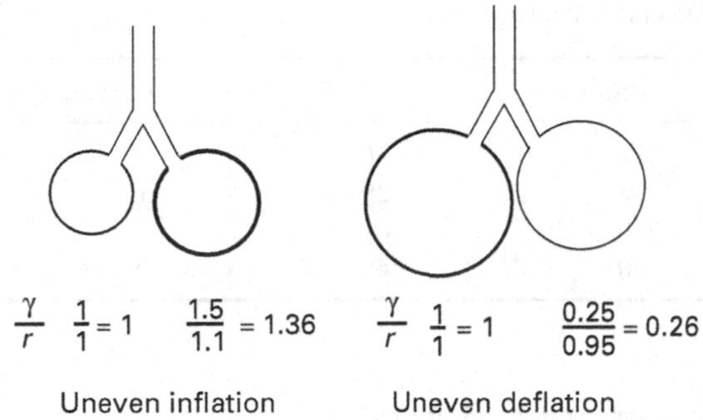

$$\frac{\gamma}{r} \quad \frac{1}{1} = 1 \qquad \frac{1.5}{1.1} = 1.36 \qquad \frac{\gamma}{r} \quad \frac{1}{1} = 1 \qquad \frac{0.25}{0.95} = 0.26$$

Uneven inflation   Uneven deflation

Figure 8.6   Balancing of uneven alveolar cycling. The alveolar thickness represents the surface tension. (Left) If one alveolus inflates faster than the others, surfactant will thin, and the ratio of the surface tension to radius will rise, decreasing its rate of inflation and allowing the other alveoli to catch up. (Right) If one of the alveoli deflates faster than the others, the surfactant will thicken, and the ratio of surface tension to radius will fall, decreasing its rate of deflation and allowing the other alveoli to catch up. This balancing during inflation and deflation maintains uniform alveolar size.

by the lungs in the last few weeks before birth. It coats the inside of the alveoli, and binds to the surface layer of water. It disrupts the hydrogen bonds between the water molecules, lowering the net surface tension of the alveoli. This decreases the transmural pressure gradient that is needed to inflate the lungs. The pressure gradient needed to inflate alveoli without PS is significantly increased, a condition that occurs in premature births. These children need positive pressure inflation to assist the inflation of their lungs as their respiratory muscles cannot produce sufficient force to inflate their lungs. The presence of PS does not negate the Law of Laplace: given enough time, our alveoli would eventually collapse into one another, greatly reducing the area available for gas exchange. This collapse is avoided by pulmonary cycling.

Figure 8.6 shows the effects of cyclic inflation and deflation on the alveoli (Otis *et al.*, 1994). As the alveoli inflate, the pulmonary surfactant thins out on the surface of the alveoli. This thinning results in an increase in the surface tension in the wall of the alveoli. If one alveolus increases more than the others, both its radius and its surface tension would increase, with the ratio of surface tension/radius being greater than in the smaller alveoli. This would make it harder to further inflate this larger alveolus, and the other alveoli could catch up. The reverse happens during deflation. An alveolus that decreases its radius more than the others will lower its surface tension/radius ratio below that of the other alveoli, so that it will not deflate as fast as the other alveoli. By cycling, the size of all the alveoli is kept constant, and the collapse that would happen without cycling does not occur.

8.3   Membrane lipids

Phospholipids form the matrix of cell membranes. They are not covalently bound to one another, associating through London–van der Waals bonds, discussed in Chapter 2. These weak bonds account for the fluidity of membranes, both the overall movements

Table 8.2 Chemical composition of cell membranes

	Protein	Lipid	Carbohydrate
Erythrocytes	54	46	
Liver cells	68	29	3
Myelin sheath	20	80	
Rod outer segment	60	40	

Table 8.3 Percent lipid content in erythrocytes and liver cells

	Erythrocytes	Liver cells
Phospholipids	50	52
Sphingomyelin	18	19
Cholesterol	32	29

that a membrane can make without rupturing, and the movements of molecules relative to one another within the plane of the membrane. There are multiple forms of phospholipids that provide different local environments for the proteins imbedded within the membrane. Different cell types show a wide range of lipid/protein ratios (Frömpter, 1983), but a remarkable similarity in the types of lipid present in erythrocyte and liver cell membranes (Tables 8.2 and 8.3).

It is not surprising that myelin has a high lipid content relative to its protein content. Many layers of the wound sheath are not in contact with either the cytoplasm or the interstitial fluid, removing both the functional need for many proteins in those areas and access to newly made proteins. The variation in relative protein content in erythrocytes, liver cells and the rod outer segment of the retina is not great, despite the very different functions of these cells. Erythrocytes function to carry gases, and the absence of organelles means that some functions, such as energy production, that are carried out internally in most cells are confined to the cell membrane. The major energetic function of erythrocytes is the maintenance of their ionic gradients and, as mentioned in Chapter 4, the glycolytic enzymes in the RBC membrane can contribute their newly produced ATP directly to ion ATPases. The rod outer segment also has a limited function, but here that function has the membrane heavily involved. The multiple folds of membrane that hold retinal, a hydrophobic molecule, actually have more protein than lipids. Liver cells, perhaps the most biochemically complex cells in the body, have the highest concentration of protein in their membranes, reflecting the many reaction sequences these cells carry out. While ion pumps make up a significant portion of the proteins in any cell, the diversity of cellular activity means that many other cell-specific proteins will also be present.

As they are perhaps the two cell types with the greatest difference in biochemical complexity, the similarity in lipid distribution between erythrocytes and liver cells is unexpected. The phospholipids can be further separated into phosphatidylcholine, phosphatidylinositol, phosphatidylserine, and phosphatidylethanolamine, all of which are present in similar fractions in erythrocytes and liver cells. The chemical non-differences probably represent the functions supported by these molecules: the general fluidity required of any cell membrane is going to be driven in great part by its lipid composition. Any radical change in content could produce membranes that had extremes in either rigidity or instability that compromise the integrity of the cell. Since one of the key functions of any membrane is always to maintain the chemical separation of the inside and outside chemical phases, it is perhaps not surprising that there is a similarity in the total lipid membrane profile of very different cells.

There is considerable asymmetry in the distribution of glycerophospholipids on the inside and outside of the cell membrane (Daleke, 2003). This group includes phosphatidylcholine (PC), phosphatidylinositol (PI), phosphatidylserine (PS), and phosphatidylethanolamine (PE). Initial membrane synthesis is asymmetric, with all four of these lipids oriented toward the cytoplasmic side of the membrane. Each of these lipids has a hydrophilic head group, making it thermodynamically unfavorable for them to spontaneously reorient to the other side of the membrane (i.e., the energy barrier for transition is much higher than the ambient energy), in contrast to the lateral diffusion that occurs easily in the plane of the membrane. The addition of these lipids to one side of a membrane creates bending and instability in the membrane. During membrane synthesis in the endoplasmic reticulum (ER), a lipid transporter redistributes the lipids, creating an essentially random distribution. Asymmetry is re-established by the action of ATP-dependent lipid transporters. Flippases transport PS from the outer leaflet to the cytoplasmic side. Floppases transport PC and cholesterol to the outer side. A non-specific cell membrane calcium-dependent scramblase moves lipids in both directions. The low inherent transport rates of flippase and floppase activity serve to produce and maintain phospholipid asymmetry, with PS, PI and PE on the inner side and PC on the outer side. Scramblases will degrade the asymmetry. Being calcium dependent, scramblase activity is not continuous in healthy cells with low resting calcium, but can occur during cell activation. Damaged cells often have increased calcium, and loss of asymmetry is often a sign of apoptosis.

Transfer of phospholipids across the membrane is one cause of membrane bending, discussed below. A second effect of asymmetry is differential charge on the two sides of a membrane. The phosphate group of all phospholipids has a negative charge at physiological pH. PC has a positively charged choline group, resulting in a net neutral zwitterionic molecule. PS does not have a charged side group, so that it has an overall negative charge. The net effect is an inner membrane with mostly negative charges on its surface and an outer membrane with both positive and negative local charges.

8.4   Membrane curvature

Membrane asymmetry is one contributor to membrane curvature. The degree of curvature of a membrane is determined by its two radii of curvature, perpendicular to each other and to the plane of the membrane. Each radius has a curvature R and a principal curvature C (Graham and Kozlov, 2010), which are related as

$$C = \frac{1}{R}. \tag{8.6}$$

Whether curvature is small or large is determined by the ratio of the membrane thickness h divided by the radius, or h/R. For the 7 nm plasma membrane in a cell with a 7 μm radius, the ratio would be 10^{-3}, and the curvature would be small. For an intracellular vesicle with a radius of 70 nm, the ratio would be 10^{-1}, and the curvature would be large. Mechanisms that produce curvature must operate most effectively in small, intracellular vesicles.

Curvature that deforms membranes into the cytosol is considered positive curvature; curvature into the lumen of a vesicle or the outside of a cell would be negative curvature. There are multiple mechanisms that produce curvature of a membrane. A membrane that is symmetrical in both its leaflets will remain flat: since there is no preferred direction in a symmetrical structure, there is no energy gradient to induce curvature. The mechanisms that can induce curvature include: (1) unidirectional transfer of lipids from one leaflet to the other; (2) asymmetrical distribution of lipids with different bending characteristics; and (3) shallow insertion of proteins into one side of the membrane, inducing local curvature. All three of these mechanisms are known to actively contribute to membrane curvature (Graham and Kozlov, 2010), and although other mechanisms have also been proposed, these appear to be the dominant ones.

Unidirectional transfers resulting in a differential distribution of lipids is a direct consequence of flippase activity. Translocating PS molecules in an initially symmetrical membrane results in a numerical disparity. This mechanism would have minimal effect over a large area such as the entire cell membrane, as the lateral diffusion of PS would minimize the numerical difference. In a small vesicle, however, this difference can be substantial.

There is a variation in the physical structure of membrane lipids that results in differences in their intrinsic curvature. The quantitation of this effect is based on three factors: the width of the hydrophilic head (A), the volume of the hydrophobic domain (V), and the length of the hydrophobic domain (l), which yield the fractional asymmetry f (Glaser, 2001):

$$f = \frac{V}{lA}. \tag{8.7}$$

Cylindrical lipids would be symmetrical along their long axis and have $f = 1$; a monolayer of these would be flat. For lipids with a head wider than its tail, modeled as an inverted cone, $f < 1$; these monolayers would be curved with the wider head side curved around the narrower tail side, producing positive curvature. For lipids with the tail region wider than the head, modeled as an upright cone, $f > 1$; these monolayers are curved with the head on the inside curve and the tail region on the outside curve, producing negative curvature. Diglycerides with unsaturated fatty acid tails have the tails repel one another, producing negative intrinsic curvature. Lysolipids, with only one tail, form positive curvature monolayer membranes, as will monolayers in which one of the two tails of a diacylipid is enzymatically removed. Intact sphingomyelin has zero curvature, but removal of its phosphocholine head group by sphingomyelinase, producing ceramide, decreases the

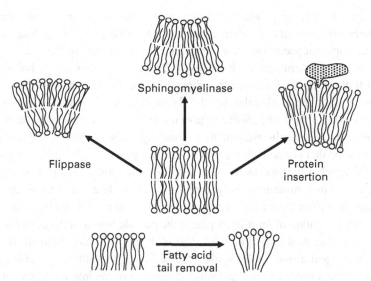

Figure 8.7   Mechanisms of membrane curvature. The uncurved membrane in the center has both diglycerides, shown with a circular head and two tails, and sphingomyelin, shown as a single line. The upper side of each structure is the cytosol side. Flippases transfer phosphatidylserine to the cytosol side, increasing the number of molecules on that side and producing positive curvature. Sphingomyelinase removes the choline head group from sphingomyelin on the cytosol side, inducing negative curvature leading to exocytosis. Shallow protein insertion on the cytosol side moves the head groups apart, producing positive curvature. The removal of one fatty acid tail from the diglyceride results in a conical molecule and a positive curvature monolayer. Lysophospholipids also have one tail and would also form a positive curvature monolayer.

size of its head, resulting in a negative curvature molecule (Graham and Kozlov, 2010). Exocytotic vesicles have a high concentration of ceramide.

The insertion of proteins into the head region of one side of a membrane will induce curvature. There are a number of proteins that exhibit this behavior. The key element is that the insertion be relatively shallow. It is immediately obvious that if a protein extended uniformly through the entire width of a membrane then there would be no change in the membrane curvature (Graham and Kozlov, 2010). The mechanisms are shown in Figure 8.7.

8.5   Membrane protein and carbohydrate environment

The proteins in membranes comprise a very large fraction of the total output of the genome, providing both structural and functional support for the cell. Many proteins have carbohydrates attached to the outside surface, carbohydrates that provide both physical stability as well as physiological effects, including antibody recognition. In this section, we will cover the structural aspects of these molecules.

As discussed in Chapter 2, all binding requires contact between molecules that produces an energy well with a depth greater than the local ambient energy. If a molecule can minimize its entropy production, it will stay in that environment. Membranes have two distinct environments: the charged surface produced by the phosphate head groups and the hydrophobic membrane core. Protein can bind to each of these areas.

Proteins in the membrane core will only bind if they have a hydrophobic surface exposed to the fatty acid tails of the phospholipids. Proteins acquire their secondary structure through intramolecular bonds. These bonds result in a three-dimensional structure with the lowest free energy. One of the guiding principles in the study of protein evolution is that a cytosolic enzyme most naturally folds into the lowest energy configuration, a shape that is also its most functional configuration. Any other structure would require the input of energy to form, a condition that could result in an evolutionary disadvantage. For a protein in a hydrophilic solution, like the cytosol, the hydrogen bonds needed to form the α-helix of a protein must compete with the water molecules for binding. The formation of an α-helix places the peptide bonds in the center of the helix, leaving the amino acid side groups to deal with the local environment. If those side groups are charged or inducible dipoles, the helix can be stable as the side groups will structure the local water and prevent water disruption of the internal hydrogen bonds. If the side groups are hydrophobic, however, the overall structure will be dynamic, with the side groups continually assuming new positions to avoid the water dipoles. This condition can disrupt the helix, exposing the protein peptide bonds to the external water, yielding an unfolded structure. Beta sheet protein structures are often more stable in hydrophilic environments.

In the core of a membrane the fatty acid tails provide a hydrophobic environment. A protein in this environment would form an α-helix. The peptide bonds in the middle of the helix are shielded from the hydrophobic carbons of the fatty acid chains. If the side groups are hydrophobic, the overall structure becomes quite stable. There are many different membrane proteins. Some of these have a single helix extending into the membrane core, anchoring additional parts of the protein that are outside the membrane. Other proteins have multiple protein helices, up to the very large seven-membrane-spanning protein family which includes rhodopsin and the G-protein-coupled receptor group. It is characteristic of these proteins that their amino acid side groups are hydrophobic within the core of the membrane, and hydrophilic outside of the membrane. This will result in the most stable energetic configuration. While the number of detailed structural models of these transmembrane proteins is increasing, the hydrophobic nature of portions of their outer structure makes the formation of crystals very difficult. Improved methods have made X-ray crystallography possible for the determination of the three-dimensional structure of many of these proteins. There is a wide range of extension of transmembrane proteins into the hydrophilic cytosol or extracellular fluid, from 1 nm to more than 10 nm.

A number of basic proteins bind to the surface of membranes, including cytochrome c, myelin basic protein and protein kinase C. These proteins bind to the acidic head groups and are heavily influenced by the local ionic conditions. Increases in ionic strength decrease the binding of proteins to membranes, but given the virtually constant ionic strength inside cells, this is not a physiologically important variable. Most of the energy of binding for proteins to membrane surfaces is borne by electrostatic interactions between the negative charges on the lipid heads and positively charged amino acids such as arginine and lysine in the proteins (Ben-Tal *et al.*, 1997). There is also a contribution made by non-specific, non-polar binding as well when proteins are within 0.2 nm of the surface, an effect associated with a change in surface tension. The asymmetry of the membrane is important here as well, as PS-rich leaflets will have many negatively charged sites for binding, without the positive charges also present in PC-rich leaflets.

The negative charges on the external surface of cell membranes are not only borne by phospholipids. Many membrane proteins have carbohydrates attached to their extracellular side. The carbohydrates that form this layer are called the glycocalyx, and each individual has a series of genetically controlled enzymes that determine which carbohydrates will be attached to their cells. These carbohydrates are used by the immune system in the identification of self, and play an important role in protecting our cells from attack by our own immune system. There are also significant biophysical effects of these carbohydrates. One of the most common sugars in the glycocalyx is neuraminic acid, also called sialic acid. This sugar has a negative charge, and this negative charge extends the electric potential of the membrane several nanometers into the interstitial fluid. The exponential decay of electric potential on the outside of the cell will therefore begin several nanometers from the membrane (Glaser, 2001). This is in contrast to the falloff in potential on the inside of the cell, where the exponential decay will be at the edge of the membrane. In addition, the hydrophilic carbohydrates will structure the water near the membrane surface, creating an unstirred layer on the outside of the cell membrane. This area will have a lower diffusion coefficient for lateral diffusion, parallel to the membrane, than for diffusion normal to the membrane. These electrical and diffusional conditions will affect near membrane phenomena such as ion mobility and agonist-receptor binding.

The advent of gold-nanoparticle near-field microscopy has greatly increased the precision of membrane protein imaging. The principles of this measurement are shown in Figure 8.8. A gold nanoparticle is suspended in a focused laser beam. The molecules of interest in the membrane, usually proteins, are labeled with fluorescent tags that are

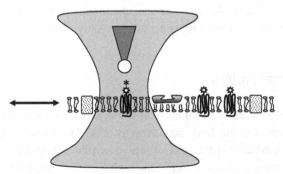

Figure 8.8   Gold-nanoparticle antenna-based near-field microscopy. The gold nanoparticle (white circle) is suspended in a focused laser beam. Fluorescent tags are chemically attached to the proteins of interest, in this case the seven transmembrane-spanning proteins with the fluorescent star-shaped head groups. Other proteins do not have tags attached. As the membrane is rastered through the laser beam, the fluorescent tag is activated (*) by the beam. The gold particle detects the emission of the tag.

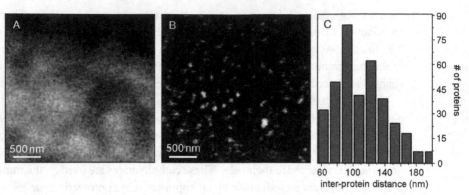

Figure 8.9   Erythrocyte membrane images. (A) Confocal image. (B) Corresponding near-field fluorescence image of PMCA4 calcium ATPases acquired using an 80 nm gold particle antenna. (C) Histogram of the nearest neighbor protein distance. (Reprinted from Höppener and Novotny, 2008, with permission from American Chemical Society.)

sensitive to the wavelength of the laser. The membrane is rastered through the laser beam by computer control, and the fluorescent tag is activated by the laser beam. When the excited tag has its energy decay, the signal is picked up by the gold nanoparticle antenna. These systems can localize radiation to 50 nm, significantly smaller than the diffraction limit of light (Höppener and Novotny, 2008) and have the potential to be even more sensitive.

An example of the increased sensitivity of the gold-nanoparticle system is shown in Figure 8.9. Part B of the figure shows a confocal fluorescence image of an erythrocyte membrane. Using a He–Ne laser with the gold bead serving as an antenna in the focused part of the beam, fluorescently labeled calcium pump proteins are shown at the same magnification. In part B individual protein can be seen using the gold particle antenna, with an 8–10-fold image enhancement. The nearest neighbor analysis in part C shows that this average interprotein distance is 90 nm. This technology is not at its limit: improved antenna geometries could enhance the identification of single membrane proteins with a resolution of 5–10 nm (Höppener and Novotny, 2008).

8.6   Membrane protein transporters

Membrane proteins are responsible for facilitated diffusion, active transport and secondary active transport across the membrane. Ions use the energy of ATP for transport against their electrochemical gradients, balancing the leak of ions down their gradients through ion channels. Facilitated diffusion allows transfer of hydrophilic substances across the membrane through hydrophilic pathways. Secondary active transport uses the energy of an ion gradient, usually sodium, to drive other hydrophilic substances against their gradients. Secondary active transporters in which the driving ion and the secondary substance move in the same direction are termed cotransporters or symports; those in which the driving ion and substance move in opposite directions are countertransporters or antiports. In this section membrane ion ATPases, facilitated transporters and secondary active transport proteins will be discussed. Ion channel function strongly depends on the electrical properties of the membrane, and will be covered in the next chapter.

Ion ATPases work by altering the dissociation constants for the transferred ions. The energy of ATP is transferred to the transporter protein, increasing the energy of that molecule, energy that is lost when the protein is dephosphorylated. The high and low affinity states are interconverted by the phosphorylation transfer. In the Na–K ATPase, sodium has a higher affinity in the non-phosphorylated E1 state than in the phosphorylated E2 state, and potassium has a higher affinity in the phosphorylated E2 state than the non-phosphorylated E1. In the Ca ATPase, the non-phosphorylated E1 state has the higher ion affinity. The consequence of the affinity differences is that ions are translocated from their high affinity side, where ions at low concentrations can still bind, to the low affinity side, where high ion concentrations will not prevent dissociation from the transporter. The sodium and calcium high affinity states communicate with the cytosol, with low affinity on the extracellular side. There is also low calcium affinity on the lumenal side of the sarcoplasmic reticulum (SR).

The reaction sequence of the Na–K ATPase is shown in Figure 8.10 (Scheiner-Bobis, 2002). In the E1 configuration at position 1, the ATPase has high affinity for sodium at three sites ($K_D = 0.19$–0.26 mM) and ATP ($K_D = 0.1$–0.2 µM), both K_Ds lower than the concentrations of either sodium or ATP in any cell. Two of the sodium binding sites are also used to bind potassium (Li *et al.*, 2005). There will be competition between sodium and potassium for these binding sites, and the preference for sodium has to reflect dissociation constants and relative ion concentrations for sodium that must be fractionally stronger than for potassium. The phosphorylation of an aspartate residue and release of ADP leads to occlusion of the bound sodium ions within the protein by reorientation of protein subunits. This new E2–P3Na$^+$ state has a K_D for sodium of 14 mM, more than 50 times less sensitive than the E1 configuration, and opens to the extracellular side. The sodium ions dissociate and are replaced by potassium ions with a $K_D = 0.1$ mM. Potassium binding causes dephosphorylation of the protein and reorientation of the protein, occluding the potassium within the center of the protein. ATP has a low affinity in this state, $K_D = 0.45$ mM, but the K_D is still lower than the cytosolic ATP concentration, so ATP will bind. Quantitation of this process is difficult because while the sodium pump is accessible to cytosolic ATP, ATP contributed by membrane-bound glycolysis is its primary source of ATP (Paul *et al.*, 1979), making estimation of the available ATP concentration questionable at best. In any case, weak ATP binding returns the protein to the E1 state with a low affinity for potassium. The potassium ions dissociate, increasing the ATP affinity. Sodium ions then bind in the ion pocket, continuing the cycle.

An important feature of all ion ATPases is the flexing of the ion binding pocket, only exposing its opening to one membrane side at a time (Figure 8.11). A continuous channel permeable by an ion would yield ion channel activity and a dissipation of the ion gradients. The general structure of the calcium ATPase and the Na–K ATPase is similar. The binding sites of calcium and sodium are equivalent, with the unoccupied calcium pump structure equivalent to the potassium binding state (Ogawa and Toyoshima, 2002), noting that the ionic radii of sodium (0.102 nm) and calcium (0.100 nm) are virtually

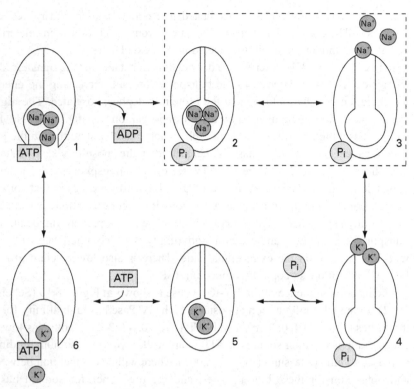

Figure 8.10   Reaction sequence of the Na–K ATPase. Sodium and ATP bind to high affinity sites in the E1 state (1). ATP hydrolysis leads to reorientation of the transmembrane helices, occluding the sodium ions (2). Subsequent helical repositioning exposing the sodium ions to the extracellular side produces a high potassium affinity and release of the sodium ions (3). Binding of potassium ions leads to dephosphorylation (4) and occlusion (5) of the potassium ions. Binding of ATP with a low affinity causes dissociation of the potassium ions (6) and reassertion of the high affinity sodium and ATP state (1). (Redrawn from Scheiner-Bobis, 2002, with permission from John Wiley and Sons.)

identical, while potassium is larger (0.138 nm). Interestingly, it is not specific negatively charged amino acid side groups that facilitate ion binding, but association of the ions with carbonyl dipoles along the transmembrane helical sequences (Scheiner-Bobis, 2002).

An important aspect of ion ATPases is the change in affinity of the ion binding sites, as noted for the Na–K ATPase, where the E1 K_D was 0.19–0.26 mM. For the Ca ATPase, the difference in affinities must be even greater in order to generate the 1000-fold or greater calcium gradient across the membrane. The E1 estimation of the binding constant is in the 0.5–1.0 μM range under most experimental conditions (Mangialavori *et al.*, 2010), much stronger binding than that of the sodium ions for the sodium pump. If the E2 calcium affinity in the Ca ATPase is similar to that of the sodium pump sodium affinity, this E1–E2 difference would be sufficient to account for the greater calcium gradient.

The SR calcium pump E1 configuration mimics that of the sodium pump (Lee, 2002), with two high affinity calcium binding sites and a high affinity ATP site. There is no clear

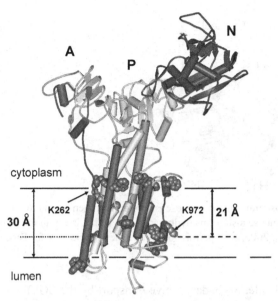

Figure 8.11 Structure of the Ca^{2+}-bound Ca ATPase. In this orientation the protein flexes to open the center region to the cytoplasm. Two calcium ions bind in the transmembrane region. One of the two calcium ions in the center of the membrane region is shown, the larger sphere on top of the lightest shaded helix. Reorientation of the protein into the calcium-free state has the N domain move laterally over the A and P segments, straightening the calcium-bound helix, decreasing the calcium affinity. The protein is now flexed to open to the SR lumen. The calcium ions dissociate into the lumenal side. (Reprinted from Lee, 2002, with permission with Elsevier.)

ion channel in any multidimensional models leading to the ion binding sites, resulting in the conclusion that rearrangement of the transmembrane helices is necessary for the ions to reach the binding sites. There is evidence that the Ca ATPase M1 helix movement may allow the ions sequential access to the binding sites. Phosphorylation of the protein closes off the path to the cytoplasm and decreases the affinity of the ions for their binding sites, the E2 state. The ions dissociate from their binding site and exit into the SR lumen, again requiring helical flex as there is no clear channel on the lumenal side. The calcium pump requires the binding of two calcium ions to binding sites on the lumenal side. Conflicting evidence using different methods has generated two alternative models, one in which the cytoplasmic ions move into the lumenal binding sites, and another in which the cytoplasmic ions move directly into the lumen. Further research may ascertain more details.

The transport of glucose uses two different families of transporters, the sodium-dependent transporters (SGLT) that move glucose against its concentration gradient using the energy of the sodium gradient, and non-sodium dependent glucose transporters (GLUT) that allow glucose to cross the membrane down its concentration gradient. GLUT transporters are present on most cell membranes. The intracellular glucose concentration is kept low by the activity of hexokinase, phophorylating glucose, keeping the free glucose concentration low, and maintaining an inward gradient. SGLT transporters are present in the mucosal cells of the GI tract and the kidney tubular cells,

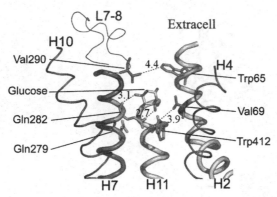

Figure 8.12   Model of the GLUT1 glucose transporter. Glucose is shown hydrogen-bound to Gln282 in the center of the transporter. Glucose also forms hydrogen bonds with Gln279 in the channel. (Reprinted from Manolescu, 2005, with permission from American Society for Biochemistry and Molecular Biology.)

localized on the lumenal side. Secondary active transport by the SGLT protein moves glucose into these cells, increasing its concentration. Glucose exits down this gradient through GLUT transporters on the basolateral side. In the mucosal and tubular cells, hexokinase activity must be low in order to maintain the outward glucose concentration gradient.

The structure of the GLUT1 transporter is shown in Figure 8.12, with an aqueous pore, through which a hexose or related substance travels, formed by transmembrane helices. Ascorbate uses SGLT transport, and its oxidized form DHA uses GLUT transport (Corti *et al.*, 2010). In addition to glucose, fructose uses the GLUT2, 5 and 7 (Manolescu *et al.*, 2005). Variations in the pore structure serve as selectivity filters, with an isoleucine substitution for valine in the pores allowing fructose transport, while those with valine do not pass fructose.

Cotransport of hydrophilic substances driven by ion gradients is a common method for molecular transport, being used for both monosaccharide and amino acid transport. As with the ion ATPases, there must be a conversion of dissociation constants between strongly and weakly bound states. A kinetic model of the SGLT1 (Wright *et al.*, 1998) shows that when nothing is bound, the transporter is oriented with its open sites facing the lumen in its low energy state. The electrochemical energy of the sodium gradient drives sodium ions into high affinity sites in the transporter pocket, increasing its local energy. Even without hexose binding, the transporter can reorient with its pocket open to the cytosol, converting the sodium binding sites to low affinity, releasing the ions. This is a sodium leak. When in the high affinity lumen-open state, there is also a high affinity hexose site to which glucose and structurally similar molecules can bind. The conversion to the cytosol-open state causes release of the glucose into the cytosol. The net effect is to transfer glucose against its concentration gradient from the jejunal lumen of the GI tract and the filtrate lumen of the kidneys into the adjacent epithelial cells. This process is so effective that no dietary glucose reaches the large intestine or filtered glucose reaches the bladder. Non-sodium dependent GLUT transporters complete the transfer process by moving glucose (and related molecules) into the interstitial fluid of the epithelium. The anatomical separation of the SGLT and GLUT processes is required for this process to work.

In addition to driving the transport of hydrophilic molecules using secondary active transport, there is also linked exchange of ions. Magnesium in smooth muscle measured by ion probe was 36 mmol/kg dry wt (Somlyo *et al.*, 1979). This value, after conversion to wet weight and for density, yields an intracellular free magnesium concentration of 7.7 mM if all the magnesium is free in the cytosol. The interstitial fluid free magnesium concentration is approximately 1 mM, yielding an outward concentration gradient and an inward electrical gradient. If magnesium were transported through a channel, it could be at equilibrium, balancing the two gradients just as chloride does. Subsequent experiments using 31P-NMR found free magnesium values of approximately 0.5 mM (Kushmerick *et al.*, 1986), resulting in both inward chemical and electrical gradients. Intracellular magnesium concentrations could not be accounted for based on equilibrium transport through a channel: there must be an energy-dependent magnesium transport mechanism moving magnesium out of the cell. The exchange of magnesium across the membrane was shown to use a sodium–magnesium antiport, transferring sodium in as magnesium is transferred out, driven by the energy of the sodium gradient. The Na–Mg antiport is present in all investigated mammalian cells (Günther, 2007). Given the ion probe measurements, most of the magnesium inside cells must be complexed and not free in the cytosol.

8.7 Membrane organization

Membrane asymmetry is not confined to differences between the two leaflets of a bilayer. There is also an uneven distribution of molecules within each leaflet of the cell membrane. Specific proteins, sphingomyelin and cholesterol, form complexes known as lipid rafts (Simons and Ikonen, 1997). Lipid rafts are not tethered, but can move laterally within the leaflet, and have been reported to exist in a wide range of sizes of 1–100 nm in width. Rafts have been found in artificial membranes, but their identification within cell membranes has been difficult. Rafts have been associated with the Src family of kinases on the inner leaflet and glycosylphosphatidylinositol (GPI) anchored proteins on the outer leaflet (Wassall and Stillwell, 2009) where phospholipase C releases proteins such as acetylcholinesterase into the interstitial fluid.

Rafts organize local membrane areas because of the affinity of raft proteins, cholesterol and sphingomyelin. Cholesterol in cell membranes is oriented perpendicularly to the plane of the membrane, with the hydroxyl group associated with the head groups as shown in Figure 2.12. In artificial membranes made entirely of polyunsaturated fatty acids (PUFA), cholesterol will not associate with the fatty acid tails (Wassall and Stillwell, 2009). Instead of its normal orientation, the cholesterol center of mass is located in the center of the bilayer, with its long axis parallel to the plane of the membrane (Figure 8.13). This represents the average position for cholesterol, which would rapidly switch between the leaflets by repulsion from the PUFA. In cell membranes phospholipids with unsaturated

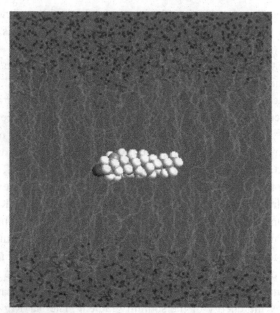

Figure 8.13   Average position of cholesterol in a membrane of polyunsaturated fatty acids. Cholesterol does not bind to the highly disordered double bonds and would rapidly flip-flop between the leaflets, generating kinetic instability and an average position in the center of the membrane. There is essentially no local environment with a significant energy minimum for cholesterol binding. (Reprinted from Wassall and Stillwell, 2009, with permission from Elsevier.)

bonds will repel cholesterol, which will then associate with saturated molecules such as sphingomyelin.

Figure 8.14 shows a model of differential molecular distribution in membranes (Wassall and Stillwell, 2009) that has generated a convergence on the health benefits of both Vitamin E and polyunsaturated fatty acids. Lipid rafts form containing proteins, sphingomyelin and cholesterol, with cholesterol being the glue which holds the raft together. This combination provides a highly structured environment: that is, an environment with a relatively long lifetime. Other proteins which do not functionally require a highly structured membrane environment will associate with polyunsaturated fatty acids, such as lipids that contain docosahexaenoic acid (DHA). These lipids will repel cholesterol, making the local physical environment more fluid. In these non-raft structures, Vitamin E is proposed to play a structural role similar to that of cholesterol in the lipid rafts. Vitamin E is able to interact with PUFA because its tail chromanol group is kept away from the disordered polyunsaturated double bonds. This model structurally links the health benefits associated with PUFA with the antioxidant behavior of Vitamin E. Vitamin E will prevent the oxidation of the proteins in the non-raft domains, prevention that has been associated with reduced cardiovascular disease. This is an intriguing and testable hypothesis.

There are a number of processes that result in membrane budding (Hurley *et al.*, 2010), including clathrin-coated vesicles, COP I and II transport systems, and cell

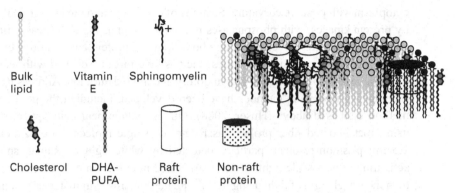

Bulk   Vitamin   Sphingomyelin
lipid   E

Cholesterol   DHA-   Raft   Non-raft
              PUFA   protein   protein

Figure 8.14   Cartoon model of raft and non-raft domains. Raft proteins, on the left, associate with the saturated fatty acid sphingomyelin and cholesterol. This produces an ordered local domain in which a number of signaling proteins function. Non-raft proteins, on the right, bind to polyunsaturated fatty acids (DHA-PUFA) and Vitamin E resulting in a more fluid domain. The linkage of the non-raft proteins and the antioxidant Vitamin E may underlie the health benefits of PUFA ingestion. (Redrawn from Wassall and Stillwell, 2009, with permission from Elsevier.)

membrane caveolae. These processes require energy use that is hundreds of times greater than ambient, meaning that ATP is necessary. On average, ATP hydrolysis has a free energy that is about 15 times greater than ambient energy (Chapter 4). Even at its most efficient, the energy of 20 or more ATPs would be needed to generate membrane budding. Some budding processes are primarily driven by lipid mechanisms, others by membrane-associated proteins.

Clathrin forms endocytotic vesicles by having its tri-armed monomer bind to adaptor proteins on the cell membrane when the adaptor protein is also bound to cargo and to inositol phospholipids on the opposite side. As more monomers bind to adaptor proteins the monomers connect to form the clathrin basket with the newly formed vesicle on the inside. After the vesicle–clathrin complex separates from the membrane, the clathrin subunits dissociate and are recycled to make new vesicles. The formation of the vesicular structure is made irreversible by the ATP-dependent clathrin dissociation using the Hsc70/auxillin complex, a process that implies that the initial vesicle formation is a near equilibrium process.

COP I and COP II form clathrin-like structures for transport from the Golgi complex to the ER (I) and from the ER to the Golgi (II). The COP II process is made irreversible using GTP hydrolysis by Sar1. An interesting aspect of COP complexes is that only protein, and not lipid, is required for the basket formation. The protein complex imposes its structure on the lipids, but is adaptable enough to form large vesicles for procollagen and chylomicron transport.

Caveolae are 60–80 nm invaginations of the cell membrane into the cytoplasm. Caveolins are pentahelical proteins with two of the helices imbedded in the lipid layer and the other three in the interfacial space (i.e., between the two leaflets). Caveolins associate with cholesterol and other raft lipids, so that caveolae have lipid rafts within them. The low energy state of caveolin structure bends the membrane toward the cytoplasm with positive curvature. Sections of caveolae can flatten using ATP hydrolysis by protein kinases, with phosphatases removing the phosphate and reasserting the low energy curved state. Caveolae require both lipid and protein contributions to form.

The study of these microstructures, such as lipid rafts, associated with membranes is difficult because of their small size, mobility and limited duration. Addressing this issue, a number of new nano methods have been developed, including the gold nanoparticle method discussed above (Huser, 2008). Others include near-field optical microscopy using metal-coated fiber probes; laser-detected single molecule localization down to 1.5 nm; plasmon-resonant particles detected by white light or Raman spectroscopy; activation of nanoscale light sources; structured illumination detecting features smaller than the wavelength of light using beat frequency patterns; stimulated emission depletion microscopy detecting red-shift emission of excited fluorophores; and nano-apertures restricting the sampling volume. This is quite a list, and the inclusions of the technical details of each of these would extend this section significantly: it is suggested that the reader seek out the specifics of methods of interest. There is one of these that has been particularly used for membrane studies, nano-apertures, and is discussed below.

Nanometer aperture methods place a window over the structure of interest and the movement of fluorescent structures is detected within a focused laser beam (Wenger *et al.*, 2007), shown in Figure 8.15. A range of apertures is used to provide a complete set of data. The diffusion rates are most accurately observed when the confining structure is near the size of the aperture. The data are fit to the fluorescence autocorrelation function $g(t)$:

$$g(t) = 1 + \frac{1}{N}\frac{1}{1 + \tau/\tau_d} \tag{8.8}$$

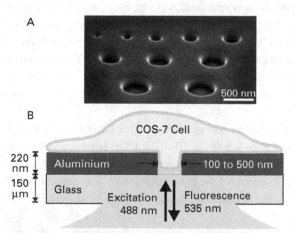

Figure 8.15   Components of nano-aperture systems. (A) Electron micrograph showing nano-apertures of different size. During experiments only a single size is used. (B) The experimental arrangement shows the cell drawn into the aperture and the excitation and emission wavelength. The lower half of the focused laser beam is shown. (Reprinted from Wenger *et al.*, 2007, with permission from Elsevier.)

where N is the number of molecules in the observation area, τ is the observation time and τ_d is the diffusion time, which is defined as

$$\tau_d = \frac{w^2}{4D} \qquad (8.9)$$

where w is the laser beam transversal waist and D is the molecular diffusion coefficient. For the observation of membranes where areas of both the membrane and the underlying cytoplasm are sampled, a multi-exponential diffusion is observed,

$$g(t) = A \cdot g_{\text{slow}}(t) + (1 - A) \cdot g_{\text{fast}}(t). \qquad (8.10)$$

The calculated τ_d of the fast phase, 0.5–1 ms, is due to movement in the cytoplasm. The slow diffusion phase has a τ_d of ~20 ms for movement within the membrane. These measurements allow estimation not only of diffusion rate, but also the size of the structure as it traverses the aperture. Experiments like these can address the question of the size and movement of lipid rafts.

8.8 Ultrasonic pore formation

External forces can be used to produce pores in vesicle and cell membranes. While both ultrasonic and electroporation methods have been employed, the consequences of applying very high voltages to intact tissue has confined most electroporation work to isolated cellular systems. In contrast, ultrasonic waves have been used for both drug and DNA delivery (Nomikou and McHale, 2010). In Chapter 1, we discussed how sound waves could be used to shatter kidney stones. Using the same principles, pores can be produced in liposomes using 20 kHz–1 MHz waves with appropriate amplitude (Schroeder *et al.*, 2009; Huang and MacDonald, 2004). The pores are produced by causing cavitations, gas bubbles, to form in the center of the lipid bilayer. The resonant radius R_r of a bubble in millimeters can be estimated by the equation

$$R_r \approx \frac{3.28}{f} \qquad (8.11)$$

where f is the frequency of irradiation in kHz (Schroeder *et al.*, 2009). A 1 MHz irradiation would then produce a bubble with a radius of 3.28 µm. Since this is much larger than the size of the membrane, the membrane will rupture when the bubble formation separates the phospholipid leaflets holding the membrane intact. The pore formation is brief, and the liposomes reseal after a short time (Figure 8.16). During the duration of the pore there will be release of the contents of the liposomes, making this method appropriate for targeted drug delivery (Huang and MacDonald, 2004).

The delivery of agents to specific sites uses microbubbles, vesicles with a diameter of less than 5 µm. Being smaller than the diameter of a capillary, microbubbles can be loaded with a molecule of interest, a contrast agent, drug, DNA, etc., and deposited into the bloodstream upstream from the target organ. The microbubble may be designed to either have a surface that will bind to the cells of interest, or have antibodies on their

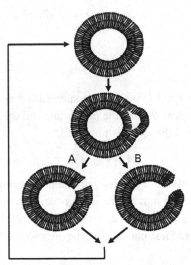

Figure 8.16   Formation of transient pores in liposome membranes. Ultrasound waves of appropriate frequency and amplitude applied to the liposome on the left produce cavitation in the center of the membrane, shown as the open white space between the leaflets in the center liposome. The cavity will expand until the pore forms. Pore formation can produce hydrophobic (A: phospholipid fatty acid tail-lined) or hydrophilic (B: phospholipid head-lined) pores. After a brief time during which the contents of the pore are released, the liposome can reseal. (Redrawn from Schroeder *et al*., 2009, with permission from Elsevier.)

surface specific for the target. As the blood containing the microbubbles passes through the target area, the region is exposed to ultrasonic waves that rupture the bubble and release the agent. The agent, such as a chemotherapy drug attacking a tumor, is delivered in a high concentration right to the site of action. Those microbubbles that escape binding and rupture do not have a long lifetime, and will soon rupture elsewhere in the circulatory system. This will release the agent in random locations throughout the body, but in concentrations far below those present in the systemic administration of a drug or contrast agent. The potential application in areas like cancer treatment are immediately obvious: minimizing exposure to chemotherapy drugs with profound side effects on healthy tissue while still attacking the cancerous site would help millions. Even if this were the only effect of microbubble delivery systems, it would be more than enough. It is expected that numerous clinical applications of microbubbles will appear.

8.9   Membrane diffusion and viscoelasticity

One of the cornerstones of the Singer–Nicholson fluid mosaic model of the membrane is the ability of unconstrained molecules to move around in the lateral surface of a membrane leaflet. This becomes a special case of the diffusion process covered in Chapter 3. The two-dimensional diffusion coefficient D_r within the membrane for a molecule of radius r is

$$D_r = \frac{k_B T}{4\pi\eta_m}\left[\ln\left(\frac{\eta_m}{\eta_w}\frac{1}{r}\right) - \gamma + \frac{1}{2}\right] \tag{8.12}$$

where γ is Euler's constant (0.5772), k_B is Boltzmann's constant, T is the absolute temperature, η_m is the membrane viscosity, and η_w is the water viscosity (Saffman and Delbrück, 1975). The ½ term applies for liquid domains in a liquid membrane (Cicuta *et al.*, 2007). This equation holds for conditions in which r is small compared with the ratio of the viscosities. For conditions in which the radius is large compared with the ratio of the viscosities, the equation reduces to

$$D_r = \frac{k_B T}{16\eta_w}\frac{1}{r} \tag{8.13}$$

with a strong dependence on the radius but independent of the membrane viscosity (Cicuta *et al.*, 2007). These studies further showed that the size-independent diffusion coefficient is

$$D_o = \frac{k_B T}{4\pi\eta_m} \tag{8.14}$$

and that the activation energy, the energy needed for a diffusing element to separate itself from its initial environment, is in the 30–80 kJ/mol range for membranes near physiological temperature. Large protein structures within the monolayer had lower activation energies.

Many membranes do not undergo extreme physical distortion in physiological systems. Perhaps the cells with the most common occurrence of membrane deformation are erythrocytes. We saw in Figure 3.6 that red blood cells have to squeeze through smaller diameter capillaries about twice a minute as they circulate. While this greatly assisted the diffusion of gases between the blood and the alveolar space, the RBCs undergo significant physical stress during this process. It is not surprising that there are many studies of the physical properties of RBCs. In addition to the membrane, they have an underlying cytoskeleton of spectrin dimers linked to actin filaments, forming a scaffolding network. This network is attached to the cell membrane by ankyrin connections between the spectrin net and transmembrane proteins. The physical properties of the membrane as a whole have to consider both the lipid bilayer and the cytoskeleton.

Measurements of the stiffness of the erythrocyte have been made using magnetic beads attached to the RBC membrane in the dimple of the biconcave disc (Puig-de-Morales-Marinkovic *et al.*, 2007) and using optical tweezers (Yoon *et al.*, 2008). Both found that the deformation of the erythrocyte does not follow simple viscoelastic models. While the different studies used both step changes in length and oscillations of over a wide frequency range, deviations from the simplest models are shown using the step change. In the chapter on load bearing (Chapter 6), the different conditions of stress–strain relation were shown. Figure 8.17 shows the behavior of a viscoelastic system with a linear Hooke's law resistive element, a spring, and a damped, viscous element. In a Maxwell model of a viscoelastic material, the spring and the viscous element are in series. In a Voigt model, the elements are in parallel. These models, and more complex

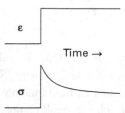

Figure 8.17   Force response to a length change in a viscoelastic system. Stress-relaxation (σ) follows the step change in length (ε).

combinations of series and parallel elements, predict the behavior of the tissue after changes in strain or stress. For example, creep is slow elongation of a tissue after a step change in stress. A Maxwell model does not predict creep but a Voigt model does. The Voigt model does not always accurately predict stress relaxation after a length step, however. While each model is useful in many cases under limited conditions, most viscoelastic materials have a more complex response than either simple model can predict.

When a viscoelastic system has a change in length (Figure 8.17) the equation describing the change in stiffness κ in the system as a function of time is the time-dependent force divided by the time-dependent change in length:

$$\kappa(t) = \frac{F(t)}{\Delta L(t)}. \tag{8.15}$$

For an exponentially decaying viscous element, the time-dependent stiffness will be

$$\kappa(t) = \kappa_\infty + \Delta\kappa(t) = \kappa_\infty + \Delta\kappa_0 (t/t_0)^{-\alpha} \tag{8.16}$$

where κ_∞ is the new stiffness due to the change in the elastic element, $\Delta\kappa_0$ is the change in stiffness of the viscous element from the initial stiffness, and α is the exponential decay constant. It is clear from the figure that given sufficient time the system will approach a new steady-state force, length and stiffness. Multiple studies (Puig-de-Morales-Marinkovic *et al.*, 2007; Yoon *et al.*, 2008) have shown that while the erythrocyte membrane does not behave as a simple elastic system, it also does not follow a model with an integral exponential decay. Analysis of the elements that could be changing when the RBC membrane is stretched showed that neither the lipid membrane alone, nor the drag of the spectrin scaffolding through the intracellular fluid, nor the viscous flow of hemoglobin are of sufficient magnitude to account for the non-linearities in the viscoelastic behavior of the erythrocyte (Yoon *et al.*, 2008). The analysis surmised that bonds connecting elements in the membrane–cytoskeleton complex break during the stretch, and that restructuring of the cytoskeleton during the recovery phase after stretch is responsible for the non-linear behavior.

A follow-up to these studies presented an interesting additional possibility (Craiem and Magin, 2010). The elastic element in a viscoelastic system behaves like a spring. The viscous element behaves like a dashpot or shock absorber. The equations defining each are below, along with a third equation describing a spring-pot.

$$\sigma(t)_{\text{elastic}} = E \cdot D^0 \varepsilon(t) \tag{8.17}$$

$$\sigma(t)_{\text{viscous}} = \eta \cdot \mathbf{D}^1 \varepsilon(t) \tag{8.18}$$

$$\sigma(t)_{\text{spring-pot}} = K_\alpha \cdot D^\alpha \varepsilon(t) \tag{8.19}$$

where $\sigma(t)$ is the time-dependent stress, $\varepsilon(t)$ is the time-dependent strain, E is the modulus of elasticity, η is the viscosity, K_α is a unit of dynamic viscosity, D are constants associated with time-dependent decay particular to a given system, and α is a non-integral, fractional exponent. In the simple models, the exponential decay has a D exponent of 1. The use of a non-unitary exponent implies that the system does not have the same probability of decay at different times, producing non-linear time-dependent rheology or tissue flow. Fractional rheology might reflect the dynamics of soft glassy materials such as foams and emulsions, and may provide more concise models of the viscoelastic properties of membranes and multilayered structures such as cartilage and blood vessels (Craiem and Magin, 2010). Since a spring-pot can be modeled as a sum of dashpots, behavioral models that link a finite number of structural elements in explaining the stress response to length change are still possible. This could lead to more complete and accurate understanding of what happens to an erythrocyte membrane as it emerges from a capillary and reasserts its biconcave disc shape.

8.10   Membrane ethanol effects

Unlike most pharmacological agents, ethanol does not bind to a specific receptor. Ethanol is an amphipathic molecule, being soluble in both membrane and plasma phases. It had long been thought that ethanol has its effects by disordering membrane lipids, but subsequent research has found that ethanol can bind directly to proteins (Harris *et al.*, 2008). In comparisons of ethanol effects on membrane lipids and proteins, the lipid effects have been found to be relatively minor, while the effects on proteins appear to be wide-ranging (Peoples *et al.*, 1996). Ethanol in sufficient amounts to induce deleterious effects will initially suppress Na–K ATPase activity (Blachley *et al.*, 1985), but with prolonged ethanol ingestion sodium pump activity and oxygen consumption increase. The cell membrane potential increases to supraphysiological levels. Cells appear to adjust to continuous alcohol exposure. Removal of chronic ethanol exposure would require return of the Na–K ATPase to its normal level over time.

The intoxicating effects of ethanol have been associated with increased GABA-receptor-induced chloride channels (Harris and Allan, 1989). It is interesting to note that GABA is the major inhibitory mediator in the brain (Ganong, 2005). It is not surprising that intoxication alters the activity of the major brain inhibitory signal. The mechanism of the ethanol effect on these channels has been determined (Harris *et al.*, 2008) with ethanol binding to a transmembrane helical site and increasing the fraction of time the channel is in the open state. The energy of the ethanol binding must be in the 10 kJ/mol range or less. The small size of the ethanol molecule limits the number of binding sites between ethanol and the protein: a hydrogen bond associated with the hydroxyl group (5 kJ/mol) and a non-polar interaction at the methyl end. The methyl group will in general have a lower binding energy than a π–π bond, which is also 5 kJ/mol, setting the upper energy limit of the ethanol binding at less than 10 kJ/mol. Bonds in this range will form and dissociate rapidly, recalling the high rate of bond turnover for water–water bonds in the 10^{11}/s range, although in a physically restricted environment the duration of ethanol binding will be longer. The structural changes caused by ethanol will not be permanent, but will alter the distribution of the molecular states, favoring the ethanol-bound state in an ethanol concentration dependent manner, in this case in the open, chloride entering channel state that would prolong hyperpolarization. Proteins will flex through a range of positions, and altering the functional state duration of an important protein will produce ethanol-dependent molecular, cellular, tissue and organ activity. The ethanol-sensitive site in the GABA receptor is an amphipathic serine residue, a small amino acid (Ueno et al., 2000). Substitution of this residue with the larger amino acid tryptophan prevents ethanol entry into the pocket between the helices, taking the place of a water molecule.

The direct binding of ethanol to proteins can alter neurotransmitter release, enzyme activity and ion channel kinetics. Because no protein with one of these functions operates in isolation, it is difficult to assign ethanol-related behavior to a specific molecule, or a system altered by that molecule (Harris et al., 2008). The continued identification of specific sites of ethanol action on proteins promises to be an area of fruitful research.

The dipolar nature of water creates surface tension at any water–gas interface. The addition of amphiphilic molecules to the surface reduces the surface tension. Pulmonary surfactant decreases the surface tension in the lungs and increases the ease of inflation. Alveolar cycling counters the collapse of small alveoli into large alveoli as they would in a static system, following the Law of Laplace. Membranes have multiple lipids distributed asymmetrically between the leaflets due to flippase and floppase activity. Lipid modification and redistribution results in positive or negative membrane curvature. There is also asymmetry within a membrane leaflet, with protein/cholesterol/sphingomyelin lipid rafts associated with many signal proteins while protein/Vitamin E/unsaturated fatty acid areas may provide antioxidant protection and cardiovascular benefits. Clathrin and COP systems create membrane-derived vesicles and caveolin forms membrane indentations. The development of nanomolecular methods has opened up new opportunities for studying membrane structures. Membrane vesicles can be ruptured using ultrasonic waves, and can be used to deliver pharmaceutical agents to targeted areas, minimizing potential side effects. Molecules can diffuse within the plane of the membrane, with analytical methods capable of measuring the energy required for molecular dissociation. Membranes exhibit viscoelastic behavior that cannot be modeled by the simplest Voigt and Maxwell models, generating a number of more complicated analyses. After long being thought to have its effects by associating with membrane lipids, ethanol is now known to bind to specific sites on membrane proteins, altering their configuration and activity. The breadth and depth of advances in our understanding of membrane structure have been great, and promise to continue.

References

Altman P and Dittmer D S. *Biology Data Book*. Washington, DC: FASEB, 1964.

Ben-Tal N, Honig B, Miller C and McLaughlin S. *Biophys J*. **73**:1717–27, 1997.

Blachley J D, Johnson J H and Knochel J P. *Am J Med Sci*. **289**:22–6, 1985.

Brzozowska I and Figaszewski Z Z. *Biophys Chem*. **95**:173–9, 2002.

Cicuta P, Keller S L and Veatch S L. *J Phys Chem B*. **111**:3328–31, 2007.

Corti A, Casini A F and Pompella A. *Arch Biochem Biophys*. **500**:107–15, 2010.

Craiem D and Magin R L. *Phys Biol*. **7**:13001, 2010.

Daleke D L. *J Lipid Res*. **44**:233–42, 2003.

Frömpter E. In *Biophysics*, ed. Hoppe W, Lohmann W, Markl H and Ziegler H, 2nd edn. New York: Springer-Verlag, pp. 465–502, 1983.

Ganong, W. *Review of Medical Physiology*. New York: Lange, 2005.

Glaser, R. *Biophysics*. Berlin: Springer-Verlag, 2001.

Graham T R and Kozlov M M. *Curr Opin Cell Biol*. **22**:430–6, 2010.

Günther T. *Magnes Res*. **20**:89–99, 2007.

Harris R A and Allan A M. *FASEB J*. **3**:1689–95, 1989.

Harris R A, Trudell J R and Mihic S J. *Sci Signal*. **1**:re7, 2008.

Höppener C and Novotny L. *Nano Lett*. **8**:642–6, 2008.

Huang S-L and MacDonald R C. *BBA* **1665**: 134–141, 2004.

Hurley J H, Evzen B, Carlson L-A and Różycki B. *Cell*. **143**:875–87, 2010.

Huser T. *Curr Opin Chem Biol*. **12**:497–504, 2008.

Kalantarian A, Ninomiya H, Saad S M I, *et al*. *Biophysical J*, **96**:1606–16, 2009.

Kushmerick M J, Dillon P F, Meyer R A, *et al*. *J Biol Chem*. **261**:14 420–9, 1986.

Lee A G. *Biochim Biophys Acta*. **1565**:246–66, 2002.

Li C, Capendeguy O, Geering K and Horisberger J D. *Proc Natl Acad Sci U S A*. **102**:12 706–11, 2005.

Mangialavori I, Ferreira-Gomes M, Pignataro M F, Strehler E E and Rossi J P. *J Biol Chem*. **285**:123–30, 2010.

Manolescu A, Salas-Burgos A M, Fischbarg J and Cheeseman C I. *J Biol Chem*. **280**:42 978–83, 2005.

Nomikou N and McHale P. *Cancer Lett*. **296**:133–43, 2010.

Ogawa H and Toyoshima C. *Proc Natl Acad Sci U S A*. **99**:15 977–82, 2002.

Otis Jr D R, Ingenito E P, Kamm R D and Johnson M. *J Appl Physiol*. **77**:2681–8, 1994.

Paul R J, Bauer M and Pease W. *Science*. **206**:1414–16, 1979.

Peoples R W, Li C and Weight F F. *Annu Rev Pharmacol Toxicol*. **36**:185–201, 1996.

Puig-de-Morales-Marinkovic M, Turner K T, Butler J P, Fredberg J J and Suresh S. *Am J Physiol Cell Physiol*. **293**: C597–605, 2007.

Reifenrath R and Zimmermann I. *Respiration*. **33**:303–14, 1976.

Saffman P G and Delbrück M. *Proc Natl Acad Sci U S A*. **72**:3111–13, 1975.

Scheiner-Bobis G. *Eur J Biochem*. **269**:2424–33, 2002.

Schroeder A, Kost J and Barenholz Y. *Chem Phys Lipids*. **162**:1–16, 2009.

Simons K and Ikonen E. *Nature*. **387**:569–72, 1997.

Somlyo A P, Somlyo A V and Shuman H. *J Cell Biol*. **81**:316–35, 1979.

Stottrup B L, Nguyen A H and Tüzel E. *Biochim Biophys Acta – Biomembr*. **1798**:1289–300, 2010.

Tinoco, JrI *et al. Physical Chemistry: Principles and Applications in the Biological Sciences*, 3rd edn. Upper Saddle River, NJ: Prentice Hall, 1995.

Ueno S, Lin A, Nikolaeva N, *et al. Br J Pharmacol.* **131**:296–302, 2000.

Wassall S R and Stillwell W. *Biochim Biophys Acta.* **1788**:24–32, 2009.

Weast R C. *Handbook of Chemistry and Physics.* Cleveland, OH: CRC Press, 1975.

Wenger J, Conchonaud F, Dintinger J, *et al. Biophys J.* **92**:913–9, 2007.

Wright E M, Loo D D, Panayotova-Heiermann M, *et al. Acta Physiol Scand Suppl.* **643**:257–64, 1998.

Yoon Y Z, Kotar J, Yoon G and Cicuta P. *Phys Biol.* **5**:036007, 2008.

9 Membrane electrical properties

The contributions of scientists measuring aspects of the membrane potential have been so well thought of that multiple Nobel prizes have been given out in this field. The field has generated quantitative findings based on the Goldman field equation and the Nernst equation that provide insight into the importance of sodium and potassium in cell signaling. The graded and action potentials that carry information within the cell and throughout the body are central in the thoughts of the brain and the movements of muscle. Local variations in cellular structure, external carbohydrates, electric fields and myelin sheaths provide a wide range of alterations in electrical function. Our understanding even extends to the structure and charges on the inside of an ion channel. At the whole body level, coordinated electrical activity in the heart provides the non-invasive electrocardiogram that allows us to easily assess cardiac health. This chapter will cover the biophysics underlying these processes.

9.1 Membrane potential

The membrane potential is a consequence of diffusible and non-diffusible ions across the membrane. The major charged substance that can never diffuse across a membrane is protein, which has a net negative charge. Start with equal concentrations of anions and cations on both sides of a cell membrane, with diffusible cation C^+ in equal concentrations on both sides of a membrane. The anions on the outside consist only of a diffusible anion $[A^-]_o$ while the inside has both protein and $[A^-]_i$, with $[A^-]_o$ greater than $[A^-]_i$. A^- will diffuse into the cell down its concentration gradient, with C^+ following into the cell down its electrical gradient. This will result in an osmotic gradient across the membrane with

$$C_i^+ + A_i^- + \text{Protein}^- > C_o^+ + A_o^-. \tag{9.1}$$

This results in the Gibbs–Donnan equilibrium, in which the ratios of the diffusible cations and anions are equal across the membrane, yielding the relation

$$C_i^+ \cdot A_i^- = C_o^+ \cdot A_o^-. \tag{9.2}$$

This produces an osmotic gradient, and water would enter the cell until the osmotic pressure either balances the inward ion flux, or the cell membrane ruptures. The process reaches equilibrium when the osmotic pressure balances the inward ionic movements.

Because protein is non-diffusible, the diffusible ions will distribute themselves based on their concentration and electrical gradients and permeabilities across the membrane. For cells, the concentration gradients are produced by the Na–K ATPase, which pumps Na^+ out of the cell and K^+ into the cell. The diffusible anion across membranes is always Cl^-. Chloride is not actively pumped, but diffuses down its concentration gradient until there is equilibrium between its concentration and electrical gradients. In all cells, K^+ is more permeable than any other cation, including Na^+, sometimes by more than 100-fold. The net effect is that the concentration of potassium builds up inside a cell because of the action of the Na–K ATPase. It will have a concentration gradient from inside to out which will be balanced by an electrical gradient, for K^+, from outside to in. If K^+ were the only diffusible cation, then it would reach equilibrium just as Cl^- does. There is, however, some diffusion of Na^+ across the membrane. The membrane potential therefore is a product of the concentration gradients and permeabilities of K^+ and Na^+. The membrane potential, concentration gradients, and permeabilities can all be measured independently. The membrane potential is measured using microelectrodes, with the experiments (1939) and theory (1952) of Hodgkin and Huxley leading to a Nobel prize. Chemical measurements of the ion concentrations combined with the membrane potential produced the Goldman field equation (1943), from which the relative permeability of K^+ and Na^+ could be calculated. The measurement of the conductance of individual ion channels using patch clamp technology (Neher *et al.*, 1978) allowed determination of the permeability of a single channel, and also led to a Nobel prize.

The membrane potential can be measured using a glass pipette microelectrode that penetrates the cell membrane (Figure 9.1). These pipettes have a very small diameter. While this produces a relatively small signal, a small diameter minimizes the change in the cell contents, as well as minimizing the leak of current around the pipette. Membrane potentials are always in the millivolt range (Table 9.1), varying between approximately −90 mV for skeletal muscle to −10 mV for red blood cells. Many neurons have a resting membrane potential of −70 mV. It would be difficult to find a cell of scientific interest that has not had its membrane potential measured.

Membrane characteristics can be measured using electrodes that set the voltage across the membrane and measure the change in current, a voltage clamp, or set the current

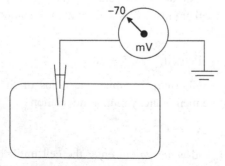

Figure 9.1   Microelectrode recording of the cell membrane potential. The electrode has a small diameter compared with the cell, minimizing chemical exchange between the cell and the electrode.

Table 9.1 Ion concentrations, equilibrium potentials, membrane potential and relative potassium/sodium permeability in skeletal muscle, neurons and red blood cells

	Skeletal muscle	Neuron	Red blood cell	Interstitial fluid
Na^+ mM	12	15	19	150
K^+ mM	155	150	136	5
Cl^- mM	3.8	9	78	120
V_{Na} mV	67.5	61.5	55.2	
V_K mV	−91.7	−90.8	−88.2	
V_{Cl} mV	−92.2	−69.2	−11.5	
V_m mV	−90	−70	−7 to −14	
P_K/P_{Na}	450	25.2	1.65	

across the membrane and measure the change in voltage, a current clamp. These methods are used to determine how ion channels operate when the cell is activated by neuro-transmitters or other molecules. Surface recording of cells in the central nervous system, called single unit recording, is used to map neural pathways in the brain. Hubel and Wiesel (1959) used this technique to map the visual cortex, work for which they also received the Nobel prize.

Progression of electrical measurement from the cell to the molecular level occurred when Neher and Sakmann (Neher et al., 1978) developed the patch clamp, which has multiple experimental configurations. The cell-attached patch isolates a small area of membrane that contains a channel of interest. Since the cell is not destroyed in this preparation, experiments testing how changes in the cell's behavior alter channel function can be performed. Also, since the contents of the pipette can be controlled, the effect of a ligand on the channel's opening can be measured, as well as the state changes in a voltage-gated channel as the voltage or current is altered. Whole cell recording uses a larger diameter pipette, rupturing the membrane at the patch site. While the larger diameter pipette gives a greater signal, this method is handicapped by the exchange of the cell contents with the pipette content, limiting the time during which experiments can be performed. This method does not study the behavior of a single channel as the other patch methods do, but considers the overall electrical response of the cell to different external conditions.

The other two patch-clamp methods involve removal of the patch from the membrane (Figure 9.2). The inside-out patch is made by rapidly removing the patch from the cell, essentially ripping the patch from the cell, which is then discarded. The concave patch remains sealed to the pipette. The pipette contents cannot be changed, but the original interior of the channel in the patch can then be exposed to different solutions. This allows direct control of intracellular ligand exposure to the channel, and other proteins that may be associated with it. The outside-out patch is made by slowly withdrawing the patch from the membrane, leaving a convex bleb of membrane extending from the pipette. This procedure flips the membrane, leaving the original outside of the membrane now on the outside of the bleb, allowing direct access of the original outside of the cell channel to a variety of external solutions while the pipette solution remains constant.

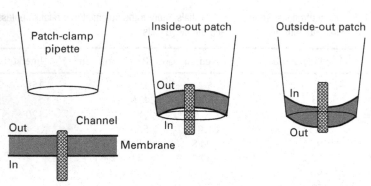

Figure 9.2   Patch clamp. The micropipette approaches an area of membrane. The pipette is applied to a small area of membrane and a high resistance seal is created with gentle suction. Four types of preparations can then be used. (1) Cell-attached patch, where the ion flux through the attached patch does not alter the remainder of the cell. (2) Whole-cell recording, in which increased suction ruptures the membrane below the pipette and the contents of the pipette are in communication with the cell interior. (3) Inside-out patch, shown in the figure, in which the pipette is rapidly withdrawn and the patch ripped from the membrane. (4) Outside-out patch, in which the pipette is slowly withdrawn from the membrane, leaving a bleb extending from the pipette with the outside of the cell membrane in communication with the external solution.

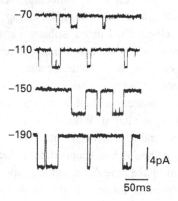

Figure 9.3   Patch clamp recording of inward Na^+ current through a single channel from a cutaneous pectoris muscle fiber. The pipette solution had Ringer's solution with 100 mM Na^+. Voltage clamps of different amplitudes produce differential channel opening, shown as downward deflections. (Reproduced from Hamill *et al.*, 1981, with permission from Springer.)

Single channel recordings from a Na^+ channel are shown in Figure 9.3 (Hamill *et al.*, 1981). The fiber had a membrane potential of -89 mV. The equilibrium potential for Na^+ is approximately $+60$ mV. The patch has a 60 GΩ seal. When the voltage is negative to the equilibrium potential, the inward current increases in a voltage dependent manner. Channels exhibit virtual square-wave behavior, jumping from the closed non-conducting state at the top of each recording to the open conducting state at the bottom of each recording. The more negative the potential, the greater the current carried by the channel, and the deeper the well when it is open. One of the striking characteristics of patch-clamp recordings is the variable duration of channel opening, a variability that can change with the applied voltage. These experiments showed a Na^+ single channel conductance of 32 pS and an average current of 2.8 pA at $-90\,mV$. This wealth of information about ion channel behavior could not have been obtained prior to the development of patch-clamp technology. It has revolutionized our understanding of membrane channels.

9.2 Goldman and Nernst equations

The membrane potential is a consequence of the concentration difference of the ions across the membranes and the permeability of those ions. If the concentrations of the ions are very small compared with other ions of similar permeability, the small concentration ions will not significantly contribute to the membrane potential. Similarly, if an ion has a very small permeability compared with other ions of similar concentration, their contribution to the membrane potential will be much smaller. In mammalian systems potassium, sodium and chloride have sufficient concentrations and permeabilities to be included in calculations of the membrane potential.

Ion differences across a membrane are a consequence of the ATPase-dependent ion pump generation of the concentration gradients and the dissipation of the electrical and concentration gradients through ion channels. The membrane potential requires differences in ionic concentration across the membrane, differences generated by the activity of ion ATPases and other ion transport systems that move ions across the membrane against their energy gradient. These processes will be covered in the next chapter. Here, the membrane potential of a cell which already has concentration differences is discussed.

The movement of an ion through ion channels across the membrane (Goldman, 1943) is a consequence of two forces: diffusion down a concentration gradient and migration down an electrical gradient. Diffusion movement is measured using the Fick equation, and electrical movement is measured using the electrical form of the Stokes equation. The total flux J_i of an ion across the membrane is

$$J_i = -D_i\left(\frac{dC_i}{dl} - \frac{V_m z_i F C_i}{RTL}\right) \tag{9.3}$$

where D_i is the diffusion constant of the ion, C_i is the local concentration of the ion, dC_i is the concentration gradient, L is the width of the membrane, dl is the change in distance over which a concentration change occurs, V_m is the membrane potential, z_i is the valence of the ion, and F is Faraday's constant. This equation can be rearranged to separate the variables dC_i and dl (from 0 to L across the membrane) and integrated to yield

$$J_i = \beta z_i P_i \frac{C_{out} - C_{in}e^{\beta z_i}}{1 - e^{\beta z_i}} \tag{9.4}$$

where

$$\beta = \frac{V_m F}{RT} \tag{9.5}$$

and P_i is the ionic permeability

$$P_i = \frac{D_i}{L}. \tag{9.6}$$

In the steady state, the membrane potential will not be changing. Thus, the net flux, the sum of the inward and outward fluxes, will be zero for the combination of all the relevant ions:

$$J_{net} = \sum_i J_i = 0 \tag{9.7}$$

which Goldman solved for monovalent ions, yielding

$$\frac{V_m F}{RT} = \ln \frac{\sum\limits_{cat} P_{cat} C^+_{out} + \sum\limits_{an} P_{an} A^-_{in}}{\sum\limits_{cat} P_{cat} C^+_{in} + \sum\limits_{an} P_{an} A^-_{out}} \tag{9.8}$$

where P_{cat} and P_{an} are the permeabilities of cations and anions, and C^+ and A^- are the concentrations of cations and anions on the inside and outside of the cell. The equation can be rearranged to solve for V_m and applied to the ions sodium, potassium and chloride

$$V_m = \frac{RT}{F} \ln \left(\frac{P_{Na}[Na]_{out} + P_K[K]_{out} + P_{Cl}[Cl]_{in}}{P_{Na}[Na]_{in} + P_K[K]_{in} + P_{Cl}[Cl]_{out}} \right). \tag{9.9}$$

The membrane potential can be measured using a microelectrode. The conductance c_i of a channel and duty cycle f_i (i.e., the fraction of time the channel is open) for a single channel can be measured using a patch clamp, shown in the recording in Figure 9.3. The permeability of an ion across the cell membrane will be

$$P_i = \sum_1^{n_i} c_i f_i \tag{9.10}$$

where n_i is the number of ion channels, each with its own conductance and duty cycle. If an ion only traveled through a single type of channel, its permeability would simplify to

$$P_i = n_i c_i f_i. \tag{9.11}$$

In the chapter on membrane structure (Chapter 8) the near-field gold-nanoparticle method demonstrated how to visualize individual proteins on the membrane surface. Every element of the Goldman equation is now subject to experimental measurement.

The Nernst equation is a reduced form of the Goldman equation applied to a single ion. The Nernst equation assumes the ion is at equilibrium, which is not true in a cell where the membrane potential is in a steady-state balance between the ion pumps and the ion channels. The utility of the Nernst equation lies in its identification of the equilibrium potential, mentioned above in describing the patch-clamp recordings. Voltage-clamp experiments, in which the sodium and potassium currents are measured at a given voltage to yield the current-voltage curves shown in Figure 9.4, can also be used to measure the

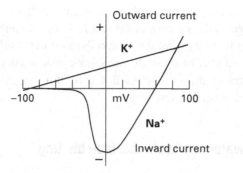

Figure 9.4 Current-voltage relations for potassium and sodium. Current measurements made during voltage-clamp experiments. Current measurements are made with positive outward current leaving the cell. The equilibrium potential for potassium occurs at the zero current voltage of −90 mV. The sodium equilibrium potential occurs at +60 mV.

equilibrium potential. The applied voltage at which the current is zero is the equilibrium potential.

The equilibrium potential is the theoretical potential at which an ion will be at equilibrium with its concentration gradient, balancing the electrical and chemical forces. The Nernst equation for a cation potential V_i at 37 °C is

$$V_i = \frac{RT}{F} \ln\left(\frac{P_i C_{out}}{P_i C_{in}}\right) = 2.303 \frac{RT}{F} \log\left(\frac{P_i C_{out}}{P_i C_{in}}\right) = 2.303 \frac{RT}{F} \log\left(\frac{C_{out}}{C_{in}}\right)$$
$$= 61.5 \, \text{mV} \left(\frac{C_{out}}{C_{in}}\right). \tag{9.12}$$

Table 9.1 shows the ion concentrations for skeletal muscle and neurons (Ganong, 2005), for red blood cells (Selkurt, 1971), and for those of interstitial fluid. The interstitial fluid concentrations are the outside ion concentrations (C_{out}) in the Goldman and Nernst equations. The ion potentials predicted by the Nernst equation show a positive intracellular value for sodium in the range of +60 mV. This is the theoretical membrane potential that would exactly counter the sodium concentration gradients across the membrane: the concentration driving sodium into the cell would be countered by the electrical gradient driving sodium out of the cell. For potassium, the equilibrium potential is near −90 mV: the outward potassium concentration gradient would be balanced by an inward electrical gradient. The chloride gradient is variable, and tracks with the membrane potential V_m. This indicates that chloride is at equilibrium across the cell membrane. This also means that chloride does not contribute to the cell membrane potential, as chloride ions will simply follow the potential generated by the other ion systems. Because of this, the Goldman equation can be rewritten as

$$V_m = \frac{RT}{F} \ln\left(\frac{[Na]_{out} + (P_K/P_{Na})[K]_{out}}{[Na]_{in} + (P_K/P_{Na})[K]_{in}}\right). \tag{9.13}$$

In this form, all the components of the equation are known except the P_K/P_{Na} ratio. The calculation of this ratio in different cells is shown in Table 9.1. Since the concentrations of sodium and potassium are similar in the three different cells, most of the difference in permeability must reside in the number of open channels, their conductance and their duty cycle. The large differences in the ratio make this a fruitful research field, as significant differences in channel behavior must occur.

9.3   The dielectric constant of water and water surface binding

The cell membrane potential separates charge across the bilayer, with net negative charge on the inside and net positive charge on the outside. The membrane potential charge separation is in addition to the local charges on the lipids and proteins: the negative charges on the phosphates of the phospholipids, the positive charge on choline head groups in some lipids, and the dissociated side groups of protein amino acids residues, such as the negative charges on aspartate and glutamate and the positive charge on lysine. All charges will affect the association of water and electrolytes with the membrane.

As a first approximation, the dipoles of water will distribute themselves in relation to the local charge on the membrane. The net positive charge on the hydrogen of water will associate with negative charges on the membrane and the negative charge on the oxygen of water will associate with the positive charges. This will structure the local water in a manner similar to the structuring by ions of sufficient charge density. Since the electric field associated with the charges on the membrane will fall off rapidly with distance, the structured water will only extend for a few molecular diameters before the ambient energy will disrupt any local ordering.

Quantitation of water binding to membrane surface features is complicated by the local non-uniformity of the surface. Many mathematical models need the simplification of a uniform surface to generate a deterministic solution. Non-uniformity makes the models too complex for exact solutions, but simplifications of non-linear models assuming functional continuities can generate iterative numerical solutions. These have been used to model the association of water to membrane proteins (Aguilella-Arzo et al., 2009).

Using the known structure of the OmpF bacterial porin, the positions and charges on the hourglass-shaped pore surface were modeled. This allowed estimation of the electric field E generated within the pore,

$$E = \frac{dV}{dx} \tag{9.14}$$

where the electric field is the length derivative of the voltage. For a flat membrane, dx is in the direction normal to the membrane surface. The membrane will act as a capacitor. For two parallel plates of a capacitor, the charge on both plates will be σ^+ and σ^-, respectively (Moore, 1972). The electric field E_o normal to the plates is

$$E_o = \frac{\sigma}{\varepsilon_o} \tag{9.15}$$

where ε_o is the permittivity in a vacuum, with a value of $\varepsilon_o = 8.854 \times 10^{-12}$ F/m, and the capacitance is

$$C_o = \frac{Q}{V_o} = \frac{\sigma A}{\sigma d/\varepsilon_o} = \frac{\varepsilon_o A}{d} \tag{9.16}$$

where Q is the charge on the plates, V_o is the voltage across the plates, A is the area of the plates and d is the distance between the plates. If a substance is introduced between the plates, that substance will have its positive and negative charges separated by the electric field, increasing the capacitance of the system by a factor ε_i,

$$C_i = \varepsilon_i C_o \tag{9.17}$$

where C_i is the new, increased capacitance of the system and ε_i is the absolute permittivity of the substance. The relative permittivity of the dimensionless dielectric constant ε_r is

$$\varepsilon_r = \frac{\varepsilon_i}{\varepsilon_o}. \tag{9.18}$$

The dielectric constant for water ε_w in interstitial fluid is 78. The electric field between the plates is reduced by a presence of the substance,

$$E_i = \frac{E_o}{\varepsilon_i} \tag{9.19}$$

due to the alignment of the dipoles in the medium substance. A uniform dielectric constant between the plates presumes that the medium is not interacting with the plates, altering the position of the dipoles. Since it is the orientation of the dipoles that alters the capacitance and yields the dielectric constant, direct contact of the media with the surface that increases (or decreases) the dipole orientation will increase (or decrease) the dielectric constant. Conversely, change in the local dielectric constant can be used to infer changes in the orientation of the medium molecules.

For a complex surface like the protein pore, the local electric field will be a function of the net distance from all local membrane structures. The dielectric constant of the medium, in this case water in the interstitial fluid, will vary inversely with the electric field. The change in the dielectric constant of water as a function of the electric field in the vicinity of a protein pore was made (Aguilella-Arzo *et al.*, 2009) by solving the Poisson equation

$$\vec{\nabla}\left(\varepsilon \vec{\nabla} V\right) = -\frac{\rho}{\varepsilon_o} \tag{9.20}$$

where V is the local voltage, ε is the dielectric constant of water and ρ is the volume charge density, a term that includes both the permanent dipoles in the protein and the orientations made by the water dipoles in the presence of the protein, recalling that an increase in orientation coincides with a decrease in the dielectric constant. The $\vec{\nabla}$ terms in the Poisson equation are divergence operators, the position-dependent gradients. The gradient within the parenthesis is the position-dependent voltage gradient, or the electric field, and the entire left side term is the position-dependent electric field. Taken

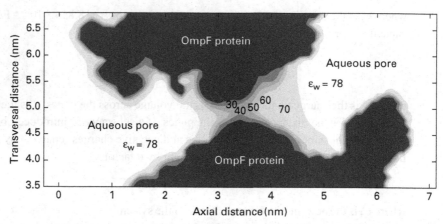

Figure 9.5   Contour plot of the water dielectric constant within the OmpF channel. The water adjacent to the protein and within the pore has a lower dielectric constant than bulk water, indicating physical structuring and orientation of the local water molecules. (Redrawn from Aguilella-Arzo *et al.*, 2009, by permission of the PCCP Owner Societies.)

together, the divergence operators in the Poisson equation are the Laplace operator. Combining the Poisson equation for the local electric field and dielectric constant with the Boltzmann distribution for the solute yields the non-linear Poisson–Boltzmann equation

$$\vec{\nabla}\left(\varepsilon \vec{\nabla} V\right) = \frac{2Fc}{\varepsilon_o} \sinh\left(\frac{eV}{kT}\right) \tag{9.21}$$

where c is the solute concentration, F is the force on the solute molecule and e is the elementary charge. This equation can be solved numerically for regions on or near proteins using computer programs that yield the potential distribution and/or the electrostatic energies in a predefined grid. For electric fields greater than 1 V/nm, the equation approaches a plateau, indicating that the water molecules have very limited degrees of freedom, limited to bond stretching or complete dissociation. Essentially, at very high electric fields the water molecules have very little mobility and therefore a low dielectric constant. In Figure 9.5 from Aguilella-Arzo *et al.* (2009), the water in the center of the pore has its dielectric constant reduced to 40, with still lower values adjacent to the protein surface. The width of the pore at its narrowest point is approximately three water molecule widths, with the low dielectric constant indicating specific orientation of the water molecules within the pore. Proteins are able to structure the adjacent environment of small molecules using the local electric fields generated by their atoms. This analysis and similar methods brings quantitation to the general concepts of water binding to charged surfaces.

9.4   Induced dipole orientation in solution

Having established the orientation of water molecules by the local protein electric fields, we now consider the degree to which the membrane electric field orients a permanent or induced dipole in the surrounding solution. For an agonist binding to a membrane receptor, the influence of the electric field on the orientation of the molecule could impact its binding to the receptor. The electric field, as noted above, is the length derivative of the voltage. The membrane separates positive and negative charges, producing an electrical double layer with the net positive charge on the external surface and the net negative charge on the internal surface.

There is a long history of models describing the electrical double layer across the membrane (Moore, 1972). The Helmholtz model of 1853 had a rigid layer of opposing charges at the solid surface, the equivalent to a plate capacitor. This is inadequate in a solution system, as the ambient energy will cause thermal motion of the ions away from a rigid structure at the membrane. The Gouy–Chapman model (1910, 1913) takes into account a statistical distribution of ions based on the electric field extending away from the membrane. This model was in turn modified by Stern (1924), to include a slowly diffusing unstirred layer of approximately one ion thickness adsorbed to the membrane charges, followed by a Gouy–Chapman distribution at farther distances from the membrane. For membranes with a glycocalyx, the membrane potential extends externally with little decrease, followed by an exponential decay (Glaser, 2001). Taking into account the many factors that influence the double layer, the change in voltage as a function of distance from the membrane is mathematically complex. Despite this, the voltage decay approximates an exponential falloff (Moore, 1972) in the region outside of the unstirred layer.

Quantitation of the falloff in potential with distance from an ionic source was addressed by Debye and Hückel (1923). Their work defined the activity coefficient γ discussed earlier, with

$$a_i = \gamma c_i \tag{9.22}$$

where a_i is the activity of an ion in solution and c_i is the concentration of the ion in solution. For very dilute solutions, γ approaches 1. At higher concentrations, ions will affect nearby ions, reducing the apparent concentration of the solution, and γ will decrease, decreasing significantly at very high concentrations. The Debye–Hückel limiting law has

$$\log \gamma = A|z_+ z_-|\sqrt{I} \tag{9.23}$$

where z_+ and z_- are the charges on the ions in solutions, I is the ionic strength, and A is a constant specific for a given solution. The potential Φ at a given distance x from a source is

$$\Phi = \frac{A}{x} e^{-x/\lambda} \tag{9.24}$$

where A/x is an ordinary coulomb potential at the source and λ is the exponential length constant, also called the Debye length (Moore, 1972). The voltage V_x at a given distance from the membrane will be

$$V_x = V_0 e^{-x/\lambda} \tag{9.25}$$

where V_o is the voltage at the external edge of the unstirred layer adjacent to the membrane. The Debye length can be calculated as

$$\frac{1}{\lambda^2} = \frac{e^2}{\varepsilon\varepsilon_o kT}\Sigma C_i z_i \qquad (9.26)$$

where C_i is the average number of i ions in a unit volume of solution (Moore, 1972). At sufficient distance from the membrane both the potential voltage and its electric field derivative will approach zero. The length constant for the fall in voltage in the exponential region of extracellular fluid is 1 nm (Delahay, 1966). The length constant will change significantly as a function of the ionic strength with the Debye length increasing approximately threefold, $\sqrt{10}$, for a tenfold increase in ionic strength. The ionic strength in physiological systems is virtually constant at 300 mOsm, so that the length constant of 1 nm is also invariant. Severe pathological increases in ionic strength in conditions such as uncontrolled diabetes would only increase the length constant by 4–5%. Only processes sensitive to small changes in the external electric field will be altered by this increase in the length constant.

Just as the voltage will decrease with distance from the membrane, the membrane electric field E_x will also decay exponentially with distance from the membrane (Dillon et al., 2000):

$$E_x = \frac{1}{\lambda} V_o e^{-x/\lambda}. \qquad (9.27)$$

As discussed above, the orientation of water near membrane proteins approaches a plateau for electric fields above 1 V/nm, which is 10^7 V/cm. Membranes have voltages in the tens of millivolts, falling off over nanometers. For a 30 mV potential exponentially decaying with a length constant of 1 nm (Tien, 1974), the initial electric field at the membrane will be 1.8×10^5 V/cm (Dillon et al., 2000), a very high electric field, but much smaller by two orders of magnitude than that needed to significantly structure water.

Having established the magnitude of the electric field near a membrane, we now consider the ability of the membrane electric field to orient molecules in nearby solutions. The important distinction is the change in the molar electric reaction moment ΔM° as a function of the electric field compared with the effect of thermal energy RT. The molar electric reaction moment is

$$\Delta M^{\circ} = \Delta\alpha^{\circ} \bullet E \qquad (9.28)$$

where $\Delta\alpha^{\circ}$ is the molar polarizability and E is the electric field. The change in equilibrium of the molecule within the electric field, dK/K, is

$$\frac{dK}{K} = \frac{\Delta\alpha^{\circ}}{RT} E \bullet dE \qquad (9.29)$$

which can be integrated to

$$\frac{\Delta K}{K(o)} = \frac{\Delta\alpha^{\circ}}{RT} \bullet \frac{E^2}{2} \qquad (9.30)$$

where $K(o)$ is the initial orientation of the molecule. High values of $\Delta\alpha°$ measure 50 pFcm^2/mol (Rüppel, 1982) but are much smaller for many molecules, water being 1.2 pFcm^2/mol (Putintsev and Putintsev, 2006). The highest electric fields near membranes are approximately 10^5 V/cm^2. Noting that

$$\text{farad} = \frac{\text{joule}}{\text{volt}^2} \tag{9.31}$$

$$J = FV^2 = FE^2\frac{\text{cm}^2}{2} \tag{9.32}$$

$$\frac{J}{\text{mol}} = \frac{\text{Fcm}^2}{\text{mol}}\frac{E^2}{2} = \Delta\alpha° \bullet \frac{E^2}{2}, \tag{9.33}$$

the energy of $\Delta\alpha°E^2/2$ for high values of $\Delta\alpha°$ near membranes is 0.25 J/mol, far less than the 2.58 kJ/mol (2580 J/mol) of ambient thermal energy in the body, resulting in an alteration of 0.01%, and much less for most other molecules, and exponentially less orientation with increasing distance from the membrane. Thus, although there is a calculable measurement of the influence of a membrane electric field on molecular orientation in solutions, it is miniscule compared with the buffeting effects of random thermal motion, and can therefore be ignored. Recalling, however, the effect of electric fields on the dissociation of molecular complexes in Chapter 2, the membrane electric field is sufficient to dissociate nearby molecular complexes.

9.5   Membrane electric field complex dissociation

Using capillary electrophoresis to measure dissociation constants, high electric fields were found to dissociate complexes of molecules. Dissociation occurred at electric fields of 100–500 V/cm, far lower than those generated at the surface of cell membranes (Dillon et al., 2000). From the equation for the exponential decay of the electric field above, electric fields that produce dissociation of molecular complexes occur between 7 nm and 8 nm from the membrane (Figure 9.6). Small variations in dissociation constants will be insignificant in altering this range, as the exponential rise in the electric field as a complex approaches the membrane will rapidly overcome any non-covalent association.

Complex formation reduces the rate of molecular oxidation (Dillon et al., 2004), a process that has been proposed as an important mechanism in molecular evolution (Root-Bernstein and Dillon, 1997). Complexes of small molecules and complexes of small proteins such as insulin/glucagon can enter the interstitial space, exposing the complex to the membrane electric field in the gap between cells. One of the common elements in all of biology is the 20 nm gap between cell membranes. For such a common occurrence, there is a dearth of research as to the reason for the consistency of this gap. Since adjacent cells are overwhelmingly of the same cell type, they will have the same

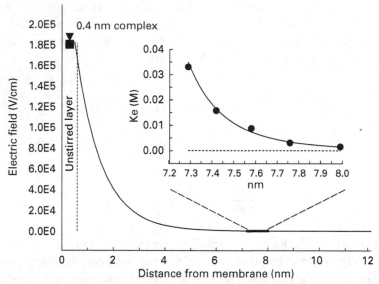

Figure 9.6   The exponential decay of membrane electric field and the electric field dissociation constants (K_e) of norepinephrine and ascorbate. Using the Stern modification of the Gouy–Chapman model of the double layer, the electric field decays exponentially with distance from the membrane unstirred layer. The range of electric field values that produce dissociation of the NE–Asc complex are shown as a function of the distance from the membrane. Complex dissociation requires 100–500 V/cm. The electric field is greater than 10^5 V/cm near the membrane. (Reproduced from Dillon *et al.*, 2000, with permission from Elsevier.)

electrical characteristics, and repel each other if they get too close. Why the gap settles at 20 nm is another question. While cells do have proteins that extend a significant distance from the membrane, there are not structures present that tether cells at 20 nm. Just as the minimum of the London–van der Waals forces and the minimum of the collagen strands had opposing attractive and repulsive forces, there must be forces that balance the distance between cells, even if those processes have not yet been fully delineated.

The 20 nm gap between adjacent cells is shown in Figure 9.7. The electric field in the center region of the gap is shown. The areas near the membrane will have a much higher electric field, shown previously in Figure 9.6 above. The electric field will be symmetrical between the two cells. The dashed vertical lines cross the exponential electric field decay at 500 V/cm, an electric field sufficiently high to dissociate virtually all molecular complexes. The solid lines cross the electric field at 100 V/cm: electric fields below this value may not be able to dissociate a complex. This results in distinct zones between the cells, with a region of bound complexes in the center flanked first by transition zones of partial complexation then by regions of dissociation near the membrane. Complexes that predominate in bound form in solution, such as catecholamines and ascorbate (Dillon *et al.*, 2000), will separate near the membrane, removing the protection from oxidation that the binding provides, but now allowing the catecholamine to bind to its membrane receptor.

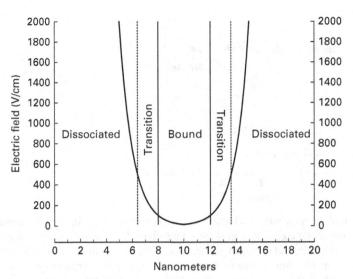

Figure 9.7   Electric field regions in the 20 nm gap between cells. The electric field will exponentially decay away from both cell membranes, producing a symmetrical electric field in the gap. In the center, the electric field is too small to cause complex dissociation. Between 8 nm and 6 nm from the membrane, the complex will dissociate as the electric field increases. In the regions adjacent to the membrane, the electric field will always be high enough to dissociate the complex.

This raises the immediate question that if high electric fields cause dissociation, how can an agonist bind to its receptor in the high electric field near a membrane? If the electric field were uniform along the membrane it would not be possible for any agonist to bind, a condition that clearly does not occur. Using a range of molecular pairs, including small molecule pairs NE–Asc and NE–morphine sulfate, small molecule–protein pair epinephrine–bovine serum albumin, and protein–protein pair insulin–glucagon (Dillon *et al.*, 2006), the effect of the electric field on complex dissociation was measured as a function of complex size. The slope of the log dissociation constants as a function of the electric field showed steep slopes for small molecule pairs, indicating a strong influence of the electric field. Pairs that included a protein, either with a small molecule or another protein, had a shallow slope, or a weak electric field dependence. Recalling that the dissociation constant can be converted into an association energy term (Chapter 2), the ratio of the slope/energy was plotted as a function of the sum of the inverse of the radii of the components of the molecular pair (Figure 9.8) using a constrained fit through the origin. The constrained fit is logical, as the inverse of the radii will approach zero as the radii approach infinity: a molecule with an infinite radius cannot have its binding site near the membrane, and so its dissociation constant will not be affected by the electric field. Both the slope and the inverse of the radii must be zero, so the line must go through the origin. The slope of the line in Figure 9.8 defines the molecular shielding constant, 7.04×10^{-8} cm^2/V, the cross-sectional projection of a molecule that shields its binding site from an electric field. The molecular shielding constant defines what is essentially an electrical shadow. For receptors with agonist

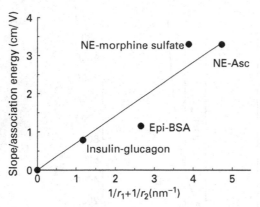

Figure 9.8   Determination of the molecular shielding constant. The slope of the log(Ke), the dissociation constant of a molecular complex in an electric field divided by the association energy of the complex at zero electric field is plotted as a function of the inverse of the radii of the components of the molecular pair. The slope of this relation over a range of pairs, constrained to go through the origin, yields the molecular shielding constant, 7.04×10^{-8} cm^2/V, the amount of electical shielding from an electric field produced by the cross-sectional projection, the electrical shadow, of a molecule. (Reproduced from Dillon *et al.*, 2006, with permission from Elsevier.)

binding sites buried within their peptide chains, such as the binding site for catechol-amines within the loops of a 7-transmembrane-spanning receptor, the protein will shield the agonist site from the electric field and allow binding. Thus, the dissociating effects of the electric field are circumvented.

The membrane electric field also appears to play a role in ion entry into channels. There is extensive research on channel selectivity that will be discussed below. But in addition to this, the structure of channels and their orientation in the electric field influences the state of the ions near the channel opening. When an ion in solution is subjected to an electrical force, it will reach a steady-state velocity v_i when the electric force is counterbalanced by its drag force (Kuhn and Hoffstetter-Kuhn, 1993), yielding a velocity

$$v_i = \frac{z_i e}{6\pi\eta r_i} E \qquad (9.34)$$

where z_i is the charge number of the ion, e is the elemental charge, η is the viscosity, r_i is the hydrated radius and E is the electric field. There is a linear relation between the velocity of the ion and the electric field with a constant slope of velocity/electric field. Note that this equation describes the movement of the hydrated radius. At a sufficiently high electric field, the water ions are stripped from the ion, and it will now have a smaller radius and a higher slope of velocity as a function of electric field (Barger and Dillon, 2010), shown for calcium ions in Figure 9.9.

A calcium ion has a high charge density (Chapter 2), and water molecules will form a hydrated shell around it in solution. At a sufficiently high electric field, the water molecules will be stripped from the ion, as the electrical energy driving off the water molecules will be greater than the bond energy holding the molecules in complex with the ion. Figure 9.9 shows that this stripping occurs between 300 and 500 V/cm, the same region

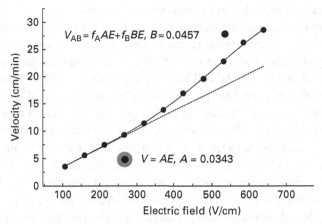

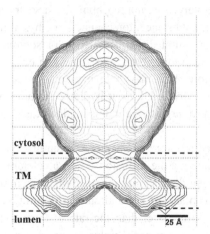

Figure 9.9   Electric field dependence of calcium velocity. The velocity and the hydration state are functions of the electric field. At low electric field, the velocity will increase with the electric field at a shallow slope reflecting the drag of the hydrated radius of the ion. At higher electric fields, the hydration shell is stripped and the smaller ionic radius has less drag. The equation models two separate slopes with a sigmoidal transition between them, with A the hydrated velocity and B the stripped velocity.

Figure 9.10   Three-dimensional structure of the type 1 inositol 1,4,5-trisphosphate receptor. This receptor extends only 1 nm into the extracellular lumen, well within the molecular dissociation range that extends to 6 nm. The receptor extends 11 nm into intracellular cytosol, well past the 8 nm limit of electric field dissociation. Calcium ions will enter without a hydration shell on the outside and be immediately rehydrated upon exiting, producing valve behavior. (Reprinted from Jiang *et al.*, 2002, with permission from Macmillan Publishers Ltd.)

where the dissociation constants for molecular complexes are increased. This stripping then will occur between 6.4 and 8 nm from the membrane, with ions stripped when nearer the membrane than 6.4 nm and hydrated when greater than 8 nm from the membrane.

Figure 9.10 shows a molecular model of the inositol triphosphate receptor calcium channel (Jiang *et al.*, 2002). This channel extends about 1 nm into the lumen of the extracellular space, and about 11 nm into the intracellular space, which will also have an exponential decay of its membrane electric field. Its structure is similar to that of other calcium channels: the dihydropyridine receptor calcium channel extends ~3 nm outside the cell and ~11 nm inside (Wang *et al.*, 2002); and the ryanodine receptor calcium channel extends < 2 nm outside the cell and ~15 nm inside the cell (Serysheva *et al.*, 2008). In all cases, the entry point for the channel is within the dissociated region with the water molecules stripped from the ion. At the exit point of the channel inside the cell, there is no opposing membrane 20 nm away as in the extracellular case, and the ions will emerge and immediately be rehydrated. This structure makes these calcium channels, even in the absence of any moving parts when they are open, behave as valves.

A valve has four components: a pore, a door, a conducted substance, and a driving force. For macroscopic valves, such as those in the heart, the door is connected to the pore, and the conducted substance blood is driven by a pressure gradient. This type of valve is thyroporetic, from the Greek θυρα for door, in which the valve door is attached to the pore. For open ion channels with asymmetric structures resulting in differential ion hydration, the channel is the pore and the stripped ion is the conducted substance. The door is the hydration shell removed by the electric field as the concentration gradient, not the electric field, moves the ion into the perimembrane region. Within this region, the electric field will then remove the hydration shell before entry into the channel. In this case, these ion channel valves are not thyroporetic, since the door is not attached to the pore, but to the conducted substance. Ion channels are thyrofluidic valves (Barger and Dillon, 2010). Similar structural differences exist for sodium channels, whose entry points are within the dissociation region, and potassium channels, whose entry points are in the hydrated region. This may help explain, in addition to the selectivity factors within the channels, why smaller but hydrated sodium ions do not enter channels for potassium, which has a larger ionic radius but is virtually unhydrated (Chapter 2).

Using the Einstein–Smoluchowski equation from Chapter 3 (assuming the interstitial fluid has the viscosity of water) to calculate the diffusion times, a calcium ion with a radius of 0.1 nm diffuses 1.6 nm, the total distance of the transition zone, in 0.5 ns. The lifetime of water clusters is 10^{-11} s. This places an approximate upper limit on the rate of calcium rehydration, a rate about 50 times faster than the rate of transition from the dissociated to the bound regions. Since calcium has a higher charge density than water dipoles, the rate of rehydration may be faster and the rate of dehydration slower. Hydration of a calcium ion by two layers of water molecules will increase its radius to about 0.3 nm. The change in velocity between the hydrated and stripped ions shows a velocity increase of 33%, less than predicted by the velocity equation for a threefold increase in radius. The drag force on the hydrated ion as it moves through the fluid may produce a smaller net radius as some weakly bound water molecules on the outside of the hydration shell are removed by friction. Also, the ionic charge is distributed over the outer surface of the stripped ion and the hydrated ion, with a lower charge density on the hydrated surface. Ionic interactions with water dipoles may contribute to a reduction in the velocity of the stripped ion. Further research in this field is needed to understand these phenomena more completely.

9.6 Membrane electrical conductance

The advent of molecular analysis has led to thousands of papers on different ion channels. Our understanding of these channels and their control of graded and action potentials in cells continues to grow. The starting point for this work is the mathematical analysis of the spread of electrical activity along a membrane by Hodgkin and Huxley (1952). The relative conductances for sodium and potassium that they measured in nerve cells are the basis for the later work on individual components of electrical activity.

The starting point of the Hodgkin–Huxley analysis of the action potential is the conductance of potassium g_K

$$g_K = n^4 \bar{g}_K \tag{9.35}$$

where $\bar{g}_K$ is the maximum potassium conductance and $n \leq 1$ is a factor that varies with membrane potential and time. Recognizing from experiments that the permeability of potassium varies with time, the change in n dn/dt is

$$\frac{dn}{dt} = \alpha_n(1-n) - \beta_n \cdot n \tag{9.36}$$

where α is the open state and β is the closed state for potassium crossing the membrane. From this, n is the proportion of molecules in the open state α and $1 - n$ is the proportion of molecules in the closed state β. The reaction rate constant for the change in state from closed to open is α_n and the reaction rate constant for the change from open to closed is β_n. The reaction rates for a resting membrane are voltage dependent, not time dependent. At rest, the β state dominates, and most of the potassium channels are closed. When depolarization occurs, α_n increases and β_n decreases in a voltage-dependent manner and, after a time, the α state dominates the β state. The numerical values of α_n and β_n can be measured from voltage-clamp experiments without any knowledge of the underlying molecular mechanisms. The fourth power of n in the conductance equation is needed to fit the delay in the rise in α_n after the application of voltage change. It implies cooperativity in the opening mechanism, again even if the mechanism is unknown.

Analysis of the sodium conductance is more complicated because the sodium conductance first rises and then falls after a membrane depolarization. This requires inclusion of both activation and inactivation terms. The sodium conductance g_{Na} is

$$g_{Na} = m^3 h \bar{g}_{Na} \tag{9.37}$$

where $\bar{g}_{Na}$ is the maximum sodium conductance, m is an activation factor similar to n for potassium, and h is an inactivation factor. The changes in m and h with time are

$$\frac{dm}{dt} = \alpha_m(1-m) - \beta_m \tag{9.38}$$

$$\frac{dh}{dt} = \alpha_h(1-h) - \beta_h \tag{9.39}$$

where the voltage-dependent rate constants are α_m, β_m, α_h, and β_h, analogous to those for potassium when the membrane is at rest. When depolarization occurs, the m state controls the opening of the sodium channels, reaching a new steady state with an increased sodium conductance, just as potassium did. The value of h is less than m, so that the process closing sodium channels occurs more slowly than the opening process, with the net result being a rise in sodium conductance. After a time, the sodium closing process h increases and dominates the opening process, and the sodium conductance is inactivated. Again, the values for the rate constants can be numerically determined using voltage-clamp experiments. These experiments showed that hyperpolarization of the membrane enhances the h process, keeping sodium channels closed, and if sufficiently large preventing the action potential from occurring. Also, the h process is prolonged after the end of the action potential. This analysis leads directly to the decrease in the occurrence of action potentials in neurons after IPSPs, inhibitory postsynaptic potentials, and the refractory period following an action potential. The large variability in the duration of the refractory period, up to 200 ms in cardiac muscle cells, indicates that the h process keeping sodium channels closed can be very different in different cells.

Combining the Hodgkin–Huxley equations yields the total membrane current I

$$I = C_m \frac{dV}{dt} + n^4 \bar{g}_K (V - V_K) + m^3 h \bar{g}_{Na} (V - V_{Na}) + g_1 (V - V_1) \qquad (9.40)$$

where C_m is the membrane capacitance, V_K and V_{Na} are equilibrium constants, g_1 and V_1 represent constants whose product is a leak current comprised of ions other than sodium and potassium. Stepwise integration of this equation yields the time course of the action potential. Adjustment of the different state values is made until the calculated time course closely matches the actual action potential in a given cell. The different stages of the action potential and the ion conductances are shown in Figure 9.11.

These changes in conductance and current occur at a specific place on a membrane. The physiological value of electrical change is the transmission of the signal along the membrane, rapidly conveying information from one end of the cell to the other. These changes are of two types, graded potential and action potentials. Graded potentials, also called local potentials or receptor potentials in appropriate settings, are not propagated to the end of a cell, but exponentially decrease over both time and distance. To have current through the membrane requires a complete circuit, shown in Figure 9.12.

The membrane current i_m is

$$i_m = \frac{1}{r_i} \frac{d_2 V}{dx^2} \qquad (9.41)$$

where r_i is the intracellular resistance including the membrane channels and dx is the incremental distance along the cell (Dudel, 1983). The extracellular resistance is so small compared with the intracellular resistance that it can be ignored. The longitudinal resistance of the membrane is so large compared with the intracellular fluid resistance that no current travels longitudinally through the membrane. The membrane current then represents only transverse current: that is, current normal to the membrane. The membrane current has both capacitive and ionic i_i components

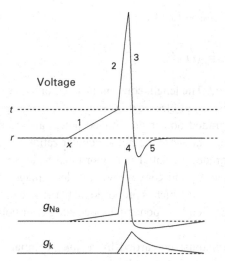

Figure 9.11 The action potential, and sodium and potassium conductances. The membrane potential has a resting voltage of r. A physical or chemical stimulus opens sodium channels at time x, causing an increase in sodium conductance g_{Na}. The membrane potential moves toward the sodium equilibrium potential (1) until reaching threshold voltage t. Voltage-gated sodium and potassium channels open at threshold, with the sodium conductance increasing faster than the potassium conductance, producing the action potential spike (2). The voltage-gated sodium channels close in a few milliseconds, decreasing the sodium conductance below the potassium conductance, moving the membrane potential toward the potassium equilibrium potential (3). Sodium channels enter a refractory period with a very low conductance. If the potassium conductance remains elevated above its value at rest, there is a hyperpolarization (4), which lasts until both the sodium and potassium conductances return to their resting values (5).

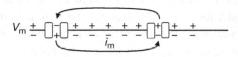

Figure 9.12 Local circuit in a membrane. The opening of sodium channels reverses the local polarity of the membrane, producing a small region with positive charge on the inside of the membrane and negative charge on the outside. Current i_m will flow driven by the voltage V_m across the membrane through the open sodium channel and a nearby open potassium channel, completing the circuit. The current is constant everywhere in the circuit.

$$C_m \frac{dV}{dt} + i_i = \frac{1}{r_i} \frac{d_2 V}{dx^2} \tag{9.42}$$

where C_m is the membrane capacitance. For small voltage-clamp depolarizations or hyperpolarizations, there is no action potential. Voltages measured at a distance from the clamp site will change over time, approaching a new steady-state voltage that is a function of the intracellular resistance, the membrane capacitance, and the applied voltage. When the applied voltage is removed, the new voltage will decay exponentially with a membrane time constant τ_m and a membrane length constant λ_m of

$$\tau_m = r_m \cdot C_m \tag{9.43}$$

$$\lambda_m = \sqrt{r_m(r_e + r_i)^{-1}} \tag{9.44}$$

where r_e is the extracellular resistance. The length constant in a neuron is on the order of millimeters, precluding the transmission of electrical information from the periphery to the central nervous system using graded potentials. Graded potentials are required to initiate action potentials, however, as shown in the initial depolarization to threshold in Figure 9.11. The magnitude of a graded potential is proportional to its stimulus. The image of a pebble dropped into a still pond conveys an excellent image of a graded potential, with the wave size, the height of which is proportional to the size of the pebble, spreading out across the pond and its height exponentially decaying over both time and distance.

Long-range electrical information must be carried by action potentials. Since the current in a circuit is the same everywhere, the conduction of an action potential along the membrane, solved by Hodgkin and Huxley, will be

$$C_m \frac{dV}{dt} + g_K(V - V_K) + g_{Na}(V - V_{Na}) + g_l(V - V_1) = \frac{1}{r_i} \frac{d_2 V}{dx^2}. \tag{9.45}$$

Thus, the propagation of the action potential will have a wave of increased g_{Na} preceding a restorative g_K increase as the g_{Na} returns to resting values. The return of g_K to resting values completes the wave. As the electrical activity spreads along the membrane, new voltage-gated channels are opened, allowing the action potential to spread undiminished to the end of the cell. At the end of the cell, the action potential stops, since the recently activated sodium channels are now in their refractory state and cannot be reopened. If the pebble in the pond is the model for a graded potential, a tsunami carrying a wave over great distances without decrement is the model for an action potential. The net flux of potassium from a cell during an action potential is many orders of magnitude below its concentration. Likewise, an action potential does not alter the sodium gradient across the membrane. Because the ion concentrations essentially remain constant, consecutive action potentials in the same cell are indistinguishable. For this reason, information conveyed using action potentials is based on the frequency of the signal, not the magnitude.

Myelin, the wrapping of exogenous cell membranes around neural axons, greatly increases the action potential conduction velocity. Looking at the right side of the above equation, for a given voltage any change in velocity has to involve a change in resistance. One way in which the velocity increases is by increasing the internal diameter of the cell. This decreases the resistance and increases the velocity in a myelin-independent manner. Myelin creates regions of very high transverse resistance across the membrane. Completion of the circuit involves longer distances between regions of low membrane resistance, the nodes of Ranvier. This would mean that for a given cell internal diameter, the intracellular resistance between regions of low transverse resistance is greater. For an increase in velocity to occur, rate of change of the local voltage must be greater and/or the resistance across the membrane must be lower. Thinking in extremes, if the number of open ion channels is very small, approaching zero, then the gradient driving current down the inside of the cell must also approach zero and the resistance will approach infinity. Therefore, if there are more channels open, the local intracellular voltage gradient will be larger, the resistance will be lower and the velocity will increase. The density of sodium channels at the nodes of Ranvier is 2000–12 000/μm^2, while in an unmyelinated neuron the sodium channel density is 110/μm^2, with similar differences in potassium channel numbers (Ganong, 2005). The high channel density at the nodes of Ranvier would produce large increases in g_{Na} and g_K, more than offsetting local increases in intracellular resistance and producing large increases in conduction velocity. It is a common experience to brush your hand across a potentially hot object like an iron and have to wait a noticeable time before knowing if you were burned or not. This difference arises because of conduction velocity differences. Myelinated touch neurons are 5–12 μm in diameter, with conduction velocities of 30–70 m/s, while unmyelinated pain receptors with neuron diameters of 0.4–1.2 μm have conduction velocities of 0.5–2 m/s (Ganong, 2005). For a distance of 1 meter from the hand to the brain, and using the fastest unmyelinated speed and the slowest myelinated speed, the time difference in the arrival of the burn and touch information would be $0.5\,s_{burn} - 0.033\,s_{touch}$ or 0.467 s, a discernible duration that matches personal experience.

9.7   The electrocardiogram

While individual cell action potentials are small and require special recording equipment to measure, there is one set of coordinated electrical responses that yields internal information at the body surface, the electrocardiogram. The ECG, or EKG as it is also called, detects the coordinated electrical activity of the heart. Its precise information is in contrast to the electroencephalogram (EEG), which detects net brain activity. While the lack of coordinated activity prevents the EEG from providing coordinated information like the ECG, since the 1950s the EEG has been the standard of life and death. A flat EEG is interpreted as brain death, and allows initiation of activities like organ donation without legal consequences. Quite differently, the ECG not only provides information that is spatially and temporally precise, the information often provides the basis for life-saving treatment.

The electrocardiogram was first developed by Einthoven using a string galvanometer, an advancement for which he was awarded the Nobel prize in 1924. The ECG is essentially the first derivative of the sum of the electrical activity of the heart. It is measured by comparing the voltage in separate leads that are different distances from the heart. When there is no signal coming from the heart there will be no voltage potential between the leads. When the heartbeat occurs, the coordinated electrical discharge will travel through the electrolyte solutions of the body and reach the skin. This signal will reach the closer lead first, producing a differential voltage compared with the other lead, and an ECG wave will appear. Refinement of the ECG has led to the use of 12 leads, three bipolar leads on the two arms and a leg, and nine unipolar leads, six on the chest and three

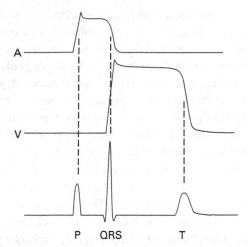

Figure 9.13   The electrocardiogram. The ECG detects coordinated activity in the heart by measuring differential voltages between leads from electrodes placed at different distances from the heart. The ECG is shown in the lower trace, with the atrial (A) and ventricular (V) action potentials (APs) shown above. The atrial AP precedes the ventricular AP, and the ventricular mass is greater than the atrial mass, aspects shown by the temporal offset of the action potentials and the difference in action potential sizes, respectively. Atrial depolarization is detected as the ECG P wave. The ECG QRS complex represents ventricular depolarization and atrial repolarization, with the larger ventricles dominating the signal from the smaller atria. Ventricular repolarization is shown in the ECG T wave. Temporal differences between the ECG waves also convey information, with the P − QRS difference indicating the duration of atrial systole, and the QRS − T difference indicating the duration of ventricular systole, for example.

on the limbs (Ganong, 2005). Standardization of the ECG produces consistent results and interpretations by clinicians.

During the cardiac cycle, the atria contract first, about 0.1 s ahead of the ventricles. The electrical activation of the heart is delayed as it passes through the atrioventricular (AV) node, a delay that allows the ventricles to complete their filling. The atria are smaller than the ventricles. The initial deflection, the P wave, represents atrial depolarization (Figure 9.13). The second deflection, the QRS complex, has components of the atrial repolarization and ventricular depolarization. Although occurring virtually simultaneously, slight temporal and positional differences result in the triphasic complex, with the small downward Q and S deflections flanking the large R peak. The T wave indicates ventricular repolarization.

In addition to accurate information about the timing of cardiac events, the ECG has considerable clinical applications. Not the least of these is the non-invasive nature of the ECG: the heart generates these signals every time it beats; all that is needed to see it is external amplification of the signal. One of the consequences of muscle and neural specialization is the loss of mitotic ability. When heart muscle cells die during myocardial infarction, a heart attack, they are not replaced with other muscle cells, but with scar tissue formed by fibroblasts and their collagen extrusions. Scar tissue does not conduct electricity nearly as well as muscle tissue. As a result, electrical activation moving through the cardiac walls has to detour around the scar tissue. Changes in electrical path show up as distortions of the normal ECG. Skilled technicians and physicians, using the differential information that comes from the different leads, can interpret the signal differences as specific abnormalities.

Among the simplest deviations that appear in the ECG are misshapen peaks. Because the scar tissue formed following a heart attack never goes away, ECGs show evidence of a heart attack even years after it has occurred. Damage in the atria would show up as a misshapen P wave, and possibly as a misshapen QRS, unless masked by the ventricular signal. Ventricular damage appears as misshapen T waves, and again as possibly misshapen QRS signals. Complete AV block, stopping the transmission of activation from the atria to the ventricles, separates the faster atrial rhythm generated by the sinoatrial node from the slower ventricular rhythm initiated somewhere in the AV node–Bundle of His–Purkinje fiber autorhythmic tissue of the ventricles. The P waves will not coordinate with the QRS and T wave, with many more P waves appearing than those of ventricular origin. Atrial fibrillation, a non-fatal condition, has complete discoordination of atrial cell activation, eliminating atrial pumping during the cardiac cycle. Because most ventricular filling is passive, sufficient blood is pumped by the ventricles to maintain the body's oxygen needs. The ECG detects atrial fibrillation by the absence of P waves. The primary danger of atrial fibrillation lies in the reduction of shear stress in the atrial cavities, increasing the possibility of clot formation. If coupled with non-laminar flow in the aorta due to atherosclerotic plaque formation, an embolus released from the left atrium will easily pass through the ventricle, with the danger of embolus entry into a carotid or coronary artery a real possibility. With luck, the embolus goes to the foot, far preferable to a stroke or heart attack. Ventricular fibrillation, and its attendant loss of blood pumping, is an immediate life-threatening condition. The ECG during ventricular fibrillation will exhibit no regular peaks, just noise. Electrical or manual shock to the heart to recoordinate its electrical activity is required within minutes before serious damage or death results.

More complex ECG readings than those listed here are also possible. It is said that learning to interpret the ECG is like learning golf: you can learn the basics in a few minutes, but it takes years to get really good. The best interpreters of ECG recordings can not only detect the problem, but where in the heart the problem is located, a terrific advantage in targeting treatment. And it is all non-invasive.

9.8 Channel ion selectivity

The conductance of the sodium and potassium ions leading to an action potential requires a specific sequence of ionic events: local depolarization through ligand- or mechanical-gated sodium channels to threshold; rapid opening then closing of voltage-gated sodium channels producing the action potential spike; slower opening and closing of voltage-gated potassium channels producing repolarization; return to resting levels of sodium and potassium permeability. In addition to being driven by their chemical and electrical gradients, as well as the electric field effects on ion hydration and the entry sites of

Table 9.2 Relative ion conductance in different ion channels

	Sodium	Potassium	Calcium
Sodium channels	1	0.048–0.086	<0.093–<0.11
Potassium channels	<0.004–0.07	1	0.011*
Calcium channels	0.00085	0.00033	1
Non-specific cation channels	1	1–1.11	0.2–87

Sources: *Inoue, 1981; all others Hille, 2001.

channels, many ion channels also selectively filter ions. Table 9.2 shows the relative permeabilities of sodium, potassium and calcium through different channel subtypes. Regardless of the mechanisms that produce the relative differences in conductance, this table summarizes the key fact, that each physiologically important ion-specific channel is not highly permeable to the other important ions. Multiple factors may be needed to produce this effect.

Different types of channels have different minimum pore size channel dimensions (Hille, 2001). Sodium channels have 0.31×0.51 nm rectangular pores. Potassium pores are circular with a diameter of 0.33 nm. The non-selective nAChR pore is a 0.65×0.65 nm square with beveled corners. Each serves as a selectivity filter, not allowing ions or molecules to pass that exceed these dimensions. There is more flexibility to the sodium channel than to the potassium channel. The sodium channel forms an oxygen-bounded pore through which sodium and larger organic ions can pass. From Table 9.2 the conductance of potassium through the sodium channel is less than 10% of the sodium conductance, despite a pore size that is larger than the 0.276 nm potassium diameter. Water hydrogen bonded to an oxygen in the channel narrows the space, allowing sodium to pass between the water and a carboxyl group, while potassium has too great an energy barrier to overcome to squeeze between the water and carboxyl group (Hille, 1975). Calcium with an even larger charge density than potassium would also bind so tightly on the entry side of the carboxyl group that it could not traverse the narrowest part of the pore and would leave the channel on the entry side. Sodium and calcium do not pass through potassium channels as the charges at the entrance to the potassium channel do not have sufficient electric field strength to dissociate the bound water, while potassium can enter as water is not strongly bound to potassium (Hille, 2001). Recognition that the electric field of the membrane can dissociate the water from sodium requires that the potassium channel entrance be in the bound region of the intercellular gap. This does appear to be the case.

Sodium and potassium do not pass through calcium channels to any significant degree. Potassium is much larger than calcium (diameter 0.2 nm), and can be excluded based on size. Sodium ions are virtually identical in size to calcium ions and cannot be selectively excluded on that basis. Figure 9.14 shows a general model for ion permeation. Ions are localized near the channel opening by the anomalous mole fraction effect, in which the attraction of an ion to the entrance of a channel produces lateral depletion of similar ions along the membrane, repelled by the presence of the same charged ions near the channel

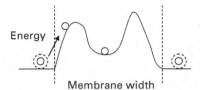

Figure 9.14   General model of ion channel permeation. Ions must overcome an energy barrier to enter the channel. This often involves removal of a hydration shell, shown as the dashed line around the ion. Within the channel the ion reaches a metastable state with a depth specific to a given ion in a given channel. There may be multiple metastable states within a single channel. The ion has to have sufficient energy to cross the exit barrier followed by rehydration as it exits the channel.

mouth (Nonner *et al.*, 1998). While this does not require that ions enter the channel in single file, it does increase that probability.

Ions have metastable states within a calcium channel. Statistically, the higher sodium concentration in the extracellular fluid means that more sodium ions will enter the channel than calcium. An anomalous mole fraction effect between sodium and calcium results in preferential calcium permeation of the calcium channel (Hille, 2001). The energy well for the calcium metastable state is much deeper than the sodium metastable state. Note that the depth of an energy well is not a physical depth, but the strength of the local energetic binding. The calcium energy depth overcomes the concentration differences between the ions, and statistically there will be a much higher probability that calcium will be in the channel. The entry of a second calcium into the channel creates a repulsion that drives the first calcium over the exit barrier, with the second calcium now replacing the first in the metastable state. Sodium entry does not have sufficient energy to drive calcium away from the metastable state, so that sodium effectively never crosses the entire channel. Only calcium ions can complete the transition process.

Figure 9.15 shows two important aspects of ion function, the activation gate and ion selectivity for a potassium channel (Kuo *et al.*, 2005). X-ray crystallography of the channel in the closed and open states shows that this channel consists of four subunits, a tetramer formed by the dimerization of two dimers. Bending of the inner helices produces significant movement of the outer helices, opening the activation gate region of the channel. The selectivity filter at the top of the channel is not significantly altered by the opening of the activation gate, which allows easy access to the selectivity filter by ions. Other types of channels have different numbers of subunits. Ligand-gated channels fall into several super families: pentameric cys-loop receptors that include GABA, glycine, serotonin, and nicotinic acetylcholine receptors; tetrameric glutamate receptors; and trimeric ATP receptors. Voltage-gated ion channels form tetramers, while gap junctions are hexameric. In general, the more subunits a channel has, the larger the open pore it forms as a consequence of packing limitations (Hille, 2001). Tetramers pass single ions, pentamers distinguish anions from cations, while hexameric gap junctions not only pass ions but many small metabolites as well.

One of the most complex aspects of ion permeation is the binding and dissociation of water. Structure-making ions, sodium and calcium, and the structure-breaking ion,

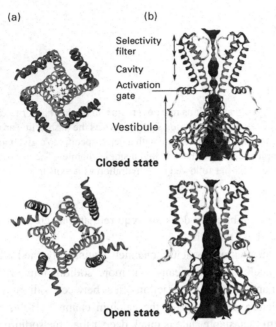

(a)   **(b)**

Selectivity filter

Cavity

Activation gate

Vestibule

Closed state

Open state

Figure 9.15   Gating mechanism of potassium channel KirBac3.1. (a) Top view in the closed and open configurations showing the rearrangement of the helices as the channel opens. (b) Side view showing two monomers in the open and closed configurations. The activation gate opens from a diameter too narrow for a single water molecule to pass to a diameter that allows many water molecules to pass. (Reproduced from Kuo *et al.*, 2005, with permission from Elsevier.)

potassium, were discussed in Chapter 2, along with the hydrogen bonds formed between water molecules. Sodium and calcium have a sufficiently high charge density to structure the water around them, decreasing the rate of exchange of bound water with the surrounding bulk water. Potassium does not extensively structure the water around it. Those ions that have structured water associated with the ion will have to have significant energy input in order for the dehydrated ion to enter a portion of an ion channel too narrow to accommodate bound water.

There are five ways in which water can be removed from an ion-bound state near a channel: (1) random thermal energy input from its surroundings in bulk solution, i.e., kT; (2) separation of the water and ion by the electric field within 6 nm of the membrane, discussed above; (3) frictional drag on the hydrated ion as it moves through the solution, converting a spherical structure into an ovoid one, reducing the Stokes radius; (4) association of the water with areas of high charge density within the channel; and (5) statistical separation of water and ion in restricted volume within the channel.

The turnover of water–water hydrogen bonds within bulk occurs on the timescale of 10^{-11} s, driven by ambient energy. Separation of water from ions on a similar timescale can be incorrectly used to infer that all ions approach channels stripped of hydration shells, an inference made by considering only the dissociation rate constant. As discussed in Chapter 2, the rate constant and the dissociation constant are different, with the bound fraction a function of both the forward and backward rate constants and the concentration. Given the almost 55 M water concentration in physiological bulk water, the thermodynamics produce a strong driving force for hydration, even given rapid dissociation and association rate constants.

Hydrated ions will diffuse in bulk solution at a rate calculated above of about 1 nm/ns. In addition, attraction or repulsion from local membrane-associated charges will increase the rate of movement in solution. At steady-state velocity, the driving force is exactly balanced by the frictional drag force. Water molecules bound to the ion normal to the direction of movement have an increased probability of dissociation, producing an oval cross-section of the complex normal to the direction of flow and a reduced circular profile in the direction of flow.

The association of water with the protein of a channel was described above, reducing the dielectric constant of the surface-bound water. It may not occur that protein surface charge density is greater than the surface charge density on an ion, but the important variable is the degree to which the surface charge density is different from the charge density of bulk water. If the charge density is greater, as in areas with hydrogen-dissociated amino acid groups, the statistical rate of water dissociation from the ion will increase. Conversely, if channel areas had hydrophobic side groups with low charge density, water would be less likely to dissociate. For channels in which the water must be stripped from the ion in order for the ion to pass, binding of dissociated water to a site in the path of the ion would decrease conductance, while binding behind the ion's path would increase the possibility of transiting the channel.

The separation of water and ion in the restricted space of a channel is a special case of ambient energy induced dissociation. The Boltzmann distribution of molecules assumes an isotropic distribution of an ensemble of molecules. Within the channel, a water–ion complex is not an ensemble, and is not in an isotropic environment. There will always be a statistical probability of dissociation or binding, but each new position of the complex will produce a different probability.

Considering the difficulty in assessing the input from each of the five elements for an ion in the vicinity of a channel, producing an accurate model of water dissociation is daunting. The effect of magnesium on smooth muscle contraction is an example. Magnesium sulfate is used to lower blood pressure in pregant women with pre-eclampsia (McLean, 1994). Magnesium ions have about twice the charge density of calcium ions. The mechanism of action is presumed to be the blocking of calcium channels by magnesium, with the external charges insufficient to dehydrate the magnesiuim ion so that it will block the channel but not enter (Hille, 2001). The proximity of the calcium entry port to the membrane means the ions must first traverse the dissociation zone as they approach the membrane, may or may not enter the unstirred layer, and have very high charge densities that will strongly bind to local negative charges at the channel mouth. This would imply that the magnesium effect may be due to strong magnesium binding at the mouth of the channel, having its effect in blocking calcium while at the same time not entering as it has too deep an energy well to be displaced. Given all the factors involved, a definitive answer has not been found. Similar cases can arise for all ion transport. In addition to ion dehydration, having to further consider the binding of the ion to the protein, local charges, changes in protein structure and membrane potential, experiments and models that focus on limited aspects of ion transport become the norm.

Given the diversity of elements in ion channel function, from structural differences, to water–ion interactions, local and membrane electric fields, ion competition and energetic considerations, it is not surprising that this is one of the most vibrant fields of biophysical research.

The membrane potential is based on the differential permeability of potassium and sodium. At rest, potassium permeability dominates, with non-permeable protein inside the cell providing the negative charge that counters the positive charge on potassium. Membrane potential is measured using microelectrodes injected into cells. Membrane potentials at which no current passes define the equilibrium potential for each ion, which can also be calculated by the Nernst equation. The potassium equilibrium potential is $-90\,mV$ and the sodium equilibrium potential is $+60\,mV$. Actual membrane potentials reflect the relative permeability of potassium and sodium, and vary in different cells. The Goldman equation uses the membrane potential measurements and the ion concentrations inside and outside cells to calculate the relative permeabilities of sodium and potassium. Patch-clamp experiments study the conductance of individual channels in membrane patches and whole cells. Membrane proteins structure the water near their surface. The membrane potential decreases exponentially with distance from the membrane yielding a high electric field that can dissociate molecular complexes and water–ion complexes within 6.4 nm of the cell membrane. Asymmetrical extension of calcium channels produces valve behavior based on differential ion hydration. Graded and action potentials occur based on changes in potassium and sodium conductance. Coordinated electrical activity in the heart produces the electrocardiogram that allows non-invasive measurement of cardiac electrical activity. Experimental determination of membrane ion channel structure, local electrical conditions, ion and water activity have provided a wealth of data on the control of membrane conductance and channel selectivity. This area of biophysics has generated numerous Nobel prizes, and continues to be an area of active research.

References

Aguilella-Arzo M, Andrio A, Aguilella V M and Alcaraz A. *Phys Chem Chem Phys*. **11**:358–65, 2009.

Barger J P and Dillon P F. Biophysical Society Meeting, Biophysics of Ion Permeation, Poster 1727, 2010.

Chapman D L. *Phil Mag*. **25**:475, 1913.

Debye P and Hückel E. *Phys Z*. **24**:185–206, 1923.

Delahay P. *Double Layer and Electrode Kinetics*. New York: John Wiley and Sons, pp. 33–44, 1966.

Dillon P F, Root-Bernstein R S, Sears P R and Olson L K. *Biophys J*. **79**:370–6, 2000.

Dillon P F, Root-Bernstein R S and Lieder C M. *Am J Physiol Heart Circ Physiol*. **286**:H2353–60, 2004.

Dillon P F, Root-Bernstein R S, Lieder C M. *Biophys J.* **90**:1432–38, 2006.

Dudel J. In *Biophysics*, ed. Hoppe W, Lohmann W, Markl H and Ziegler H, 2nd edn. New York: Springer-Verlag, pp. 641–56, 1983.

Ganong W. *Review of Medical Physiology.* New York: Lange, 2005.

Glaser R. *Biophysics.* Berlin: Springer-Verlag, 2001.

Goldman D E. *J Gen Physiol* **27**:37–60, 1943.

Gouy G. *J Phys.* **4**:457, 1910.

Hamill O P, Marty A, Neher E, Sakmann B and Sigworth F J. *Pflugers Arch.* **391**:85–100, 1981.

Helmholtz H. *Pogg Ann.* **89**:211, 1853.

Hille B. *Ion Channels of Excitable Membranes*, 3rd edn. Sunderland, MA: Sinauer Associates, 2001.

Hille B J. *Gen Physiol.* **66**:535–60, 1975.

Hodgkin A L and Huxley A F. *Nature.* **144**:710–12, 1939.

Hodgkin A L and Huxley A F. *J Physiol.* **117**:500–44, 1952.

Hubel D H and Wiesel T N. *J Physiol.* **148**:574–91, 1959.

Inoue I. *J Gen Physiol.* **78**:43–61, 1981.

Jiang Q X, Thrower E C, Chester D W, Ehrlich B E and Sigworth F J. *EMBO J.* **21**:3575–81, 2002.

Kuhn R and Hoffstetter-Kuhn S. *Capillary Electrophoresis: Principles and Practice.* Berlin: Springer-Verlag, 1993.

Kuo A, Domene C, Johnson L N, Doyle D A and Venien-Bryan C. *Structure.* **13**:1463–72, 2005.

McLean R M. *Am J Med.* **96**:63–76, 1994.

Moore, W. *Physical Chemistry*, 4th edn. Englewood Cliffs, NJ: Prentice Hall, 1972.

Neher E, Sakmann B and Steinbach J H. *Pflugers Arch.* **375**:219–28, 1978.

Nonner W, Chen D P and Eisenberg B. *Biophys J.* **74**:2327–34, 1998.

Putintsev N M and Putintsev D N. *Russ J Phys Chem A.* **80**(12), 1949–52, 2006.

Root-Bernstein R S and Dillon P F. *J Theor Biol.* **188**:447–79, 1997.

Rüppel H. In *Biophysics*, ed. Hoppe W, Lohmann W, Markl H and Ziegler H, 2nd edn. New York: Springer-Verlag, pp. 163–78, 1982.

Selkurt E E. *Physiology*, 3rd edn. Boston, MA: Little Brown, 1971.

Serysheva I, Ludtke S, Baker M, *et al. Proc Natl Acad Sci U S A.* **105**: 9610–15, 2008.

Stern O. *Z. Electrochem*, **30**: 508, 1924.

Tien H T. In *Bilayer Lipid Membranes (BLM): Theory and Practice.* New York: Marcel Dekker, pp. 134–42, 1974.

Wang M, Velarde G, Ford R C, *et al. J Mol Biol.* **323**:85–98, 2002.

10 Agonist activation and analysis

This chapter covers the quantitative aspect of agonist activation: how the blood, and by extension the interstitial fluid, presents a pharmaceutical agent to the cell; quantification of drug concentration and tissue response; diffusion and elimination of an intracellular agent; statistical analysis of the tissue response; and the difficult problems associated with drug development. Because of the necessary role played by agonist receptors in this process, the background of some of the major receptor types is also included.

10.1 Membrane receptor proteins

Agonists initiate a cell response by binding to a receptor. Without trying to give a complete list of every kind of receptor, three of the major categories of receptors are the G-protein coupled receptors (GPCRs), the tyrosine kinase receptors, and nuclear receptors. GPCRs dominate cell activation using the temporally limited G-protein system. There are more than 800 GPCR proteins, and the complexity of their activation scheme is daunting. Tyrosine kinase receptors are activated by insulin and a variety of growth factors, and activate processes associated with the structural growth of cells and tissues. Nuclear receptors bind to hydrophobic hormones, steroids, thyroid hormones, etc., that do their work by initiating gene transcription. This section gives a brief overview of these receptor types, with the understanding that all of them have far more complexity than can be covered here. Recognizing the pivotal role these molecules play in linking the quantitative aspects of cell activation, pharmacokinetics, dose–response relations, intracellular diffusion, and data analysis of biological responses, it is appropriate that a general discussion of their functions is included.

There are hundreds of different GPCRs. Structurally, all have seven membrane-spanning domains dominated by alpha helices through the hydrophobic membrane region. The N-terminal is external and the C-terminal internal. The seven helices form a barrel structure with a hydrophilic core. All are linked to G-proteins and thus are part of activation schemes that have only limited temporal activity. The barrel structure is stabilized by conserved cysteine disulfide bonds. Palmitoylation of cysteine residues in the extracellular loops increases binding to cholesterol and sphingomyelin, localizing GPCRs into lipid rafts. Following activation by the binding of an agonist, or in the case of opsins after the absorption of light energy, the protein is phosphorylated, leading to the association of the GPCR with β-arrestin at the intracellular membrane surface, G-protein

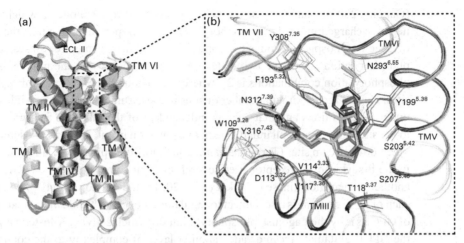

Figure 10.1 Structure of GPCR protein beta 2-adrenergic receptor. The orientation in (a) has the extracellular loops on the top. The transmembrane helical regions are labeled TM and numbered from the N-terminal. Extracellular loop II is also labeled. The inset (b) shows the agonist binding site with three different agonists superimposed in the space. There is considerable flexibility and space in the agonist pocket, with binding requiring an amine group and an aromatic ring on the agonist. (Reproduced from Wacker *et al.*, 2010, with permission from American Chemical Society.)

decoupling and the internalization of the GPCR (Tobin, 2008). This leads to desensitization of the cell to the agonist. The β-arrestin molecules also form complexes with Src tyrosine kinases and the ERK and JNK3 MAP kinase cascades, thereby associating agonist-bound GPCRs with major intracellular signaling pathways (Luttrell and Lefkowitz, 2002). There is considerable conservation of amino acids in all GPCRs, especially in the transmembrane regions. Conversely, there is significant variability in the extracellular loops and agonist-specific structures at the binding sites.

Among the most studied GPCR molecules are the opsins of phototransduction and the beta 2 adrenergic receptor (β2AR) (Ganong, 2005). GPCR proteins are differentiated by significant variations in the structure of loop 2. The binding of agonists within the barrel transmembrane region and the covering of the barrel opening by the unique structure of loop 2 creates the differential binding of agonists. These agonists include odorants, neurotransmitters, and a wide range of peptide and protein hormones. In all cases, binding of the agonist leads to GDP–GTP exchange by the GPCR-associated G-protein, activating the G-protein function.

Figure 10.1(a) shows the structure of a β2AR receptor with the extracellular domain at the top (Wacker *et al.*, 2010). The highlighted region is the agonist binding site. Figure 10.1(b) shows three superimposed agonists within the binding site. There is considerable flexibility within the agonist binding region, allowing different molecules access to the site. Binding requires hydrogen binding with an agonist amine group and hydrophobic interaction with an aromatic ring. The easy adaptability of the agonist binding site to molecules possessing the minimal structural binding requirement implies that the regulatory sites must be at different locations. In fact, there are multiple phosphorylation sites on the β2AR receptor, all in the intracellular C-terminal tail

(Ganong, 2005). Conflicting schemes differentiate between a model in which it is the net negative charge imparted by phosphorylation that is important, rather than the alternative of specific phosphorylation sites controlling specific receptor responses. Given that phosphorylation sites have two states, phosphorylated or not, the number of possible phosphorylation combinations is 2^n, where n is the number of different phosphorylation sites. Having the maximum number of sites in use assumes each is accessible to a kinase under all conditions and that each is independent of the other. In fact, while there are many sites, they are not all independent, as a mutation in one prevents phosphorylation of other, downstream sites (Torrecilla et al., 2007). Multiple phosphorylation site activation in the bradykinin B2 receptor uses several different signal pathways, GRKs, PKC and another unidentified kinase (Blaukat et al., 2001), adding more layers of complexity, as each of these systems is also subject to regulatory processes. While the primary activation of GPCRs is by their agonist, phosphorylation stops their activity. Allosteric regulation of the GPCR glutamate receptor adds another layer of complexity to the control of these receptors, with the allosteric sites within the membrane-spanning domain (Gregory et al., 2011). Further, binding of dihydroxyl molecules such as ascorbate to extracellular loop 1 of the β2AR receptor allosterically increases the sensitivity of the receptor without decreasing efficacy, and effectively blocks desensitization produced by phosphorylation, while at the same time acting as a redox enzyme, converting dehydroascorbate to ascorbate (Dillon et al., 2010). In summary, a range of different agonists containing a common motif can bind to the active site of a GPCR, triggering G-protein and related activities including multiple receptor phosphorylations through a number of kinase cascades, with β-arrestin binding to the GPCR and kinase cascade elements producing further cell effects, among them inactivation of the GPCR and the G-protein except in those circumstances where allosteric factors prevent desensitization of the agonist. It is difficult to imagine a more complex, interacting control system, but one in which understanding the key control elements can lead to better pharmaceutical interventions. Given the breadth of GPCR activation, the dynamism of this field is understandable.

Tyrosine kinases are primarily associated with growth. Insulin, epithelial growth factor and platelet-derived growth factor all activate protein kinases. Most membrane protein kinases are monomers with a single transmembrane domain. The binding of an agonist causes dimerization of two monomers, conveying active tyrosine kinase activity on the complex. If no ATP is involved in the dimerization, the energy gained by the agonist binding will be less than that of a single ATP hydrolysis, so the energy between the monomer and dimer states cannot be high. After dimerization, the kinase autophosphorylates, and has increased kinase activity. The autophosphorylation seems to increase the structural stability of the complex. This may be a case of retention time/reaction time that was discussed in Chapter 2: binding of the agonist links the two monomers together long enough so that the autophosphorylation creates an energy well deep enough to hold them together. In the case of the insulin receptor, an $\alpha_2\beta_2$ tetramer, binding of insulin does not cause complex formation with other insulin tetramers (Ulrich and Schlessinger, 1990). Tyrosine kinase receptors activate kinase cascades that result in increased gene expression leading to growth. The cascading factors and/or the final transcription factor must traverse the distance from the membrane to the nucleus to have this effect on growth.

Not surprisingly, tyrosine kinase activity is also linked to the uncontrolled growth in cancers (Zwick *et al.*, 2001), with pharmaceutical intervention in tyrosine kinase activity a potential treatment tool.

Hydrophobic receptors have a different consideration. Steroid hormones and thyroid hormone can penetrate the hydrophobic core of the membrane, but their movements through a hydrophilic milieu often involve complexing with proteins. The hormones are only active in their free form, so the dissociation from their transport proteins is an important element in quantifying their mechanisms. If the initiation of transcription is the final step in the hydrophobic hormone activation path, binding to the nuclear receptor is the penultimate step. Whether these receptors are in the cytoplasm or the nucleus, the binding of the hormone to the receptor must confer the ability to bind to the target gene and start transcription (Mangelsdorf *et al.*, 1995). Each hormone binds to a specific nuclear receptor, the hormone displacing a heat shock protein. The complex will activate RNA polymerase, the receptors being part of the nuclear receptor superfamily. The central part of a nuclear receptor has a highly conserved DNA binding domain with two zinc fingers and a highly conserved C-terminal ligand domain. These receptors are quite specific for the genes they will activate and the agonists that switch them on. They are not mixed with other transcription factors. Class I steroid receptors are homodimers that bind to DNA half-sites organized as inverted repeats. Class II nuclear receptors include thyroid hormone and all the other non-steroid hydrophobic ligand dependent receptors.

10.2   Pharmacokinetics

The introduction of pharmaceutical drugs into the body uses several different mechanisms: continuous intravenous infusion; oral ingestion; aerosol inhalation; and transdermal absorption. Each of these can be modeled mathematically, calculating the rate and final concentration of drugs in the bloodstream. The physical nature of the drug plays an important role in its free concentration and elimination. Hydrophilic drugs do not generally bind to circulating proteins in the blood, and are subject to filtration from the renal glomeruli into the tubular system and subsequent excretion. In addition, many drugs are organic acids or organic bases and are secreted into the tubules causing additional elimination. Hydrophilic drugs are easily distributed through the interstitial fluid of the body and may be eliminated by enzymatic activity in the liver or other tissues. The blood is a very oxidizing environment, and drugs may be oxidized into nonfunctional forms. Because of these processes, hydrophilic drugs have to be taken at regular intervals or their concentration falls below effective levels. Hydrophilic B and C vitamins must be taken regularly for the same reasons.

Hydrophobic drugs (and vitamins) generally cross membranes easily but, as discussed earlier, getting to these membranes through the water-based blood and interstitial fluid is difficult. In the blood many hydrophobic drugs circulate bound to protein, particularly albumin. Since only free drugs have activity, this binding forms a significant buffer capacity for hydrophobic drugs. Binding within adipose tissue forms an additional buffer for these drugs. Binding to proteins, not filtered by the kidneys, greatly reduces the rate of excretion. Binding also reduces the rate of oxidation of these drugs. Conversely, the rapid partition of these compounds into any available membrane will increase their rate of removal from the interstitial fluid. When the free concentration is decreased by elimination or transport into a cell, hydrophobic drugs dissociate from bound or storage sites until the partition constant ratio is restored. The ratios of bound-to-free hydrophobic molecules can reach 1000:1 or more in the case of some drugs, such as artificial thyroid hormone given to Hashimoto's syndrome patients. Because the rate of elimination of these drugs is so slow, only low doses are needed to maintain desired concentrations. The turnover rate, unlike hours for hydrophilic drugs, has a time course of days or weeks in some cases.

The rapid turnover of hydrophilic drugs makes them subject to mathematical modeling of intake and elimination. Figure 10.2(a) shows the sequence of regular intravenous drug injection concentration changes; it shows the alternating rise and fall in concentration of a drug injected at time intervals that match the half-time for elimination of the drug from the body. For exponential elimination, either of two time constants are used for analysis: the half-time time constant $T_{0.5}$ and the natural logarithm constant τ, which is the time at which $1/e$ of the concentration remains. Since both are used, $T_{0.5}$ is shown in the individual injection model and τ is shown in the continuous infusion model below. The term rate constant is also used; a rate constant κ is the inverse of the time constant

$$\tau = \frac{1}{\kappa}. \tag{10.1}$$

Whichever term is most convenient for a particular analysis can be used. The units of $T_{0.5}$ and τ are time and κ are time^{-1}.

When a hydrophilic drug is injected the drug will be eliminated from the system exponentially, with the rate of loss proportional to the extracellular concentration. Each method for eliminating a drug, renal excretion, enzymatic degradation and oxidation will have an exponential time constant for elimination. If one of the time constants is much smaller than the others, that process goes the fastest, and will dominate. If two rate constants are near one another and smaller than any others, the mathematics of elimination will have a net rate constant that is the sum of the two, and still fit a single exponential elimination. If a drug partitions into separate, slowly exchanging pools, the net elimination will then be the sum of the exponential decay from each separate pool.

Returning to Figure 10.2(a), the upward arrows represent the dose injections, each identical in the increase in drug concentration they produce. The curving downward lines show the exponential decrease in concentration, reaching one-half the peak concentration when the next injection is made. The boundary lines show the high and low values that the drug will have, with the regimen eventually showing the asymptotes approached. The low D_{low} and high D_{high} concentrations reached will be

$$D_{\text{low}} = \frac{D_{\text{o}}}{2^{T_{\text{D}}/T_{0.5}} - 1} \tag{10.2}$$

$$D_{\text{high}} = D_{\text{low}} + D_{\text{o}} = \frac{D_{\text{o}}}{2^{T_{\text{D}}/T_{0.5}} - 1} + D_{\text{o}} \tag{10.3}$$

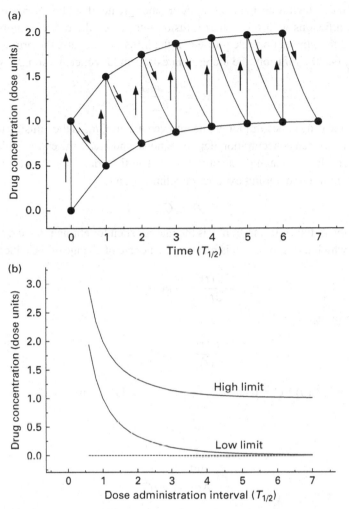

Figure 10.2 Drug concentration during interval drug injections. (a) The changes in drug concentration during interval injections. The upward arrows indicate the injection of a constant dose. The downward arrows indicate the exponential elimination of the drug. The injections are given at the $T_{0.5}$ interval, when one-half of the drug is eliminated. (b) The upper and lower limits of the drug concentration relative to the dose when the interval is varied relative to $T_{0.5}$.

where D_o is the injected dose, T_D is the injection interval, and $T_{0.5}$ is the half-time time constant. For this regimen to be functional, D_{low} must be above the effective drug concentration, while D_{high} must be below toxic levels. Figure 10.2(b) shows the upper and lower limits when the injection interval varies relative to the time constant. For injection intervals longer than six half-times, the drug concentration will fall to less than 1% of the injected dose. This is not the most common method for drug delivery.

For an intravenous infusion regimen, often used in hospitals, the extracellular concentration will eventually reach a steady state. The medicine will be infused at a constant rate. If we again consider a hydrophilic drug, it will be distributed in the plasma and the interstitial fluid, which is normally about four times greater than the plasma volume. For intravenous infusions with a constant infusion rate, when the concentration reaches a certain value the elimination rate will equal the infusion rate, and the drug will be at a steady state. For the infusion, the increase in extracellular concentration ΔC will be

$$\Delta C = \frac{C_{iv} R_{iv} t}{V_{ext}} \tag{10.4}$$

where C_{iv} is the drug concentration in the intravenous drip, R_{iv} is the infusion rate, t is the time interval between concentration determinations, and V_{ext} is the extracellular volume, including both the plasma and interstitial fluid. The time dependent concentration $C(t)$ for a concentration undergoing exponential elimination is

$$C(t) = C_o e^{-\kappa t} \tag{10.5}$$

where C_o is the initial concentration, κ is the rate constant for elimination and t is the time period over which the elimination is measured. The rate of change of the concentration is

$$\frac{dC(t)}{dt} = -\kappa C(t) e^{-\kappa t} \tag{10.6}$$

which is rearranged to

$$\frac{dC(t)}{C(t)} = -\kappa e^{-\kappa t} dt. \tag{10.7}$$

Initially, at $t = 0$, $C(0) = C_o = 0$ and $e^{-\kappa t}$ equals one, reducing the equation to

$$\frac{\Delta C}{C} = -\kappa t \tag{10.8}$$

and

$$\Delta C = -\kappa t C \tag{10.9}$$

where C is the concentration at time t. Setting the two ΔC equations equal yields

$$\frac{C_{iv} R_{iv} t}{V_{ext}} = -\kappa t C \tag{10.10}$$

where the equality occurs when the infusion and elimination rates are the same, and the concentration will be at a steady state, C_{ss}. Solving for C_{ss} yields the steady-state concentration during an intravenous infusion

$$C_{ss} = \frac{C_{iv} R_{iv}}{\kappa V_{ext}}. \tag{10.11}$$

The time course for the change in concentration $C(t)$ is

$$C(t) = C_o + \frac{C_{iv} R_{iv} t}{V_{ext}} - \kappa t C = C_o + t\left(\frac{C_{iv} R_{iv}}{V_{ext}} - \kappa C\right). \tag{10.12}$$

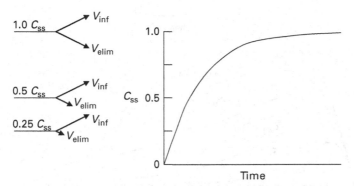

Figure 10.3   Approach to a steady-state concentration during drug intravenous infusion. For a new infusion, the starting concentration is zero. The steady state is reached when the entry rate for the drug for infusion, shown as the vector V_{inf} on the left, is balanced by elimination rate of the drug, shown as V_{elim}. The elimination rate is concentration dependent, and is zero at the start of the infusion. As the concentration rises, the constant rate of infusion is approached by the increasing rate of elimination, until at $1.0\ C_{ss}$ the infusion and elimination rates are equal, and the concentration remains constant. The higher the rate constant for elimination, the lower will be the steady-state concentration.

As noted above, when starting an infusion, C_o will be zero. C_{iv} and R_{iv} are known quantities. During cases of severe edema or dehydration, V_{ext} will increase or decrease respectively: the change must be taken into consideration when calculating the drug dose. The body's fat-free mass is often determined by measuring the current conducted through the body, as adipose tissue does not conduct electricity well. Using the conductance measurement and the total body weight the fat-free mass can be determined, from which the extracellular fluid volume can be estimated. Under most circumstances V_{ext} will be a constant and can be estimated with confidence. In this case, the steady-state concentration will be inversely related to the elimination rate constant. For cases in which the elimination process may be in question, such as renal or liver dysfunction, C_{ss} can significantly increase. The vectors for infusion and elimination at different fractions of C_{ss} and the time course of approach to C_{ss} are shown in Figure 10.3.

Injection or infusion of drugs is not the norm for administration. Oral ingestion, aerosol inhalation or transdermal absorption are much more common. The mathematical models for all of these are similar. In all cases, there will be a rate constant for transfer of the drug into the body: that is, out of its administration sites. Given their different chemical composition, every drug would have a specific rate of absorption from the GI tract, the alveoli or the skin. The rate constant for entry into the body τ_{in} will be the negative of the rate constant for removal from the administration site. The concentration of drug D in the extracellular fluid is

$$D = D_o\left(1 - e^{-t/\tau_{in}}\right) \cdot e^{-t/\tau_{out}} \qquad (10.13)$$

where D_o is the dose taken, t is the time since administration, and τ_{out} is the rate constant for elimination from the body. Different temporal profiles as a function of the ratios of the

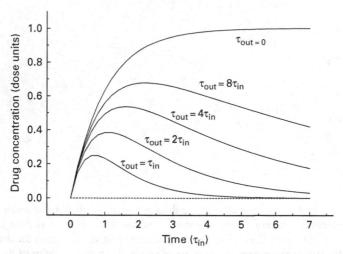

Figure 10.4   Time courses of drug entry into the extracellular volume using drug administration by ingestion, lung aerosol inhalation, or skin patch absorption. The dose has a set amount of drug, and the rate of entry matches the rate of removal from the GI tract, the lungs of the skin. The different lines show the time course of entry, peak value, and removal from the extracellular fluid by any means, entry into cells, renal excretion, oxidation or enzymatic metabolism, relative to the rate of entry.

entry and elimination time constants are shown in Figure 10.4. Since each dose has a limited amount of drug, the steady state is only reached if τ_{out} is zero, a condition that will not occur. In all real cases, the drug concentration will rise and then fall as the drug first enters and then is eliminated from the extracellular fluid. As with other routes of administration, elimination can include entry into cells, excretion, oxidation and enzymatic degradation.

In looking at the three models of drug administration, the time courses of extracellular fluid drug concentration show significant differences. Only by using an infusion can a steady-state concentration be reached. The need to precisely treat very sick patients shows why hospitals routinely use infusions, and set patients up for them even if they are never needed. The rise in the use of infusion pumps for diabetics further emphasizes the precision of infusion. Interval injections show the greatest rate of change in drug concentration, not surprising given the sharp increase at each administration. Ingestion, inhalation and transdermal patch administration show an intermediate buffering of the drug concentration, with the greatest buffering occurring when the rate of elimination is much slower than the rate of administration. Perhaps the most rapid route of administration not requiring needle injection is the sublingual administration of nitroglycerin to cardiac patients. The perception of symptoms indicating coronary artery vasospasm leads to the self-administration of this drug under the tongue, sending the hydrophobic compound through the skin and into the blood in seconds, treating a condition that could not tolerate the tens of minutes or more that oral ingestion could require. The variability of drug administration can be tailored to the necessary clinical application.

10.3   The dose–response curve and the Hill equation

The dissociation constant for a particular agonist and receptor will determine the fraction of bound receptors at a given agonist concentration. This is the basis for the dose–response curve. The dissociation constant K_D for the binding

$$A + R \leftrightarrow AR \tag{10.14}$$

is the ratio of the free concentrations over the bound concentration

$$K_D = \frac{[A][R]}{[AR]} \tag{10.15}$$

where A is the free agonist, R is the unoccupied receptor and AR is the agonist-bound receptor. The fraction R_{Bound} of the receptors that are agonist bound is

$$R_{Bound} = \frac{[AR]}{[AR] + [R]} = \frac{[AR]}{[R_{tot}]} \tag{10.16}$$

where R_{tot} is the total number of receptors. Combining the two equations yields

$$R_{Bound} = \frac{[AR]}{[A][R]/K_D + [R]} = \frac{[AR]K_D}{[A][R] + [R]K_D} = \frac{[A][R]}{[A][R] + [R]K_D} = \frac{[A]}{[A] + K_D} \tag{10.17}$$

which can be rearranged to

$$\frac{R_{Bound}}{1 - R_{Bound}} = \frac{A}{K_D} \tag{10.18}$$

where $(1 - R_{Bound})$ is the fraction of receptors that are free of agonist, or R_{Free}. Taking the logarithm of both sides yields

$$\log\left(\frac{R_{Bound}}{R_{Free}}\right) = \log[A] - \log K_D. \tag{10.19}$$

This equation describes the binding of a single agonist to a single receptor. When the ratio of bound-to-free is 1 (i.e., when they are equal), the agonist concentration is the K_D. This equation is often used to calculate K_D values, and can be used for agonist-receptor binding, or the binding of molecules in solution, such as oxygen and myoglobin, when the binding is one-to-one. When the binding is greater than one agonist to one receptor, the equation is modified to

$$\log\left(\frac{R_{Bound}}{R_{Free}}\right) = n\log[A] - n\log K_D \tag{10.20}$$

where n is the number of agonists bound to the receptor. This equation is known as the Hill equation from its derivation for oxygen-hemoglobin binding (Hill, 1910). When data is plotted as the log of the ratio against the log of the agonist, the slope at $\log_r(1)$ is the value of n. It is the effective binding constant that is determined by the Hill equation, which yields a value of $n = 2.8$ for oxygen-hemoglobin binding (Chang, 1977). It also applies to multisite binding to membrane proteins, such as the multiple binding of calcium ions to the calcium ATPase.

Often it is easier to see the response to an agonist, rather than direct binding by the agonist. The formula for measuring the percent response of an agonist is

$$\frac{R}{R_{\max}} \times 100\% = \frac{A^n}{\text{ED}_{50} + A^n} \tag{10.21}$$

where R is the response, $R_{\max}$ is the maximum response, A is the agonist concentration, n is the number of molecules of A binding to the receptor and ED_{50} is the concentration of A that gives one-half of the maximum response. The binding of an agonist to a receptor or a protein in solution is shown in the semi-log plots of the response in Figure 10.5(a). The curves are semi-logarithmic because the response is plotted linearly. The n values for the two plots in Figure 10.5 are $n = 1$ for the left plot and $n = 2.8$ for the right plot in both graphs, corresponding to 1:1 binding and O_2 binding to hemoglobin. For a drug response, the potency of the response is the ED_{50}, and the efficacy of the response is the maximum response. Dose–response data are often plotted as a percent response, removing the quantitative measurement of the response efficacy. For example, two different smooth muscle preparations, often used to test drugs, may have very different forces/cross-sectional area or forces/tissue mass, indicating differences in the health of the tissues. If both have their response to a drug plotted as a percent response, information on the vitality of the preparations is lost. Responses are best given as measured values relative to some standard such as tissue mass, not percents or fractions of the maximum. Reporting percent responses is only acceptable if the maximum efficacy as a function of tissue size is reported separately. This is also the only way to measure the relative efficacies of multiple drugs.

Note that in the top panel in Figure 10.5 the curve is much steeper for the right hand plot than the left hand plot. When the slope at the ED_{50} equals 1 on a semi-log graph, it indicates one-to-one binding. When the slope is steeper, it indicates multiple agonist binding and/or cooperativity of binding between different subunits, as in the case of hemoglobin. When the data are plotted in a linear–linear graph, as in Figure 10.5(b), one-on-one binding exhibits a rectilinear curve, with the differences in the low dose responses lost with the change in axis. Cooperative curves will still show sigmoidal behavior on a linear–linear plot. The values on the curves in Figure 10.5 were designed to match the responses of CO and oxygen to hemoglobin, with CO having a 200-fold tighter (i.e., lower) K_D for binding hemoglobin than oxygen. Carbon monoxide binding to hemoglobin lacks the cooperative nature of oxygen binding. It is easy to see the mathematical similarity between K_D measurements and ED_{50} measurements, even though the former is a physical constant and the latter a tissue response.

10.4   Intracellular molecular diffusion and elimination

The production of second messengers at the cell membrane and the entry of hydrophobic hormones into the cell are the two primary ways in which cells are chemically activated.

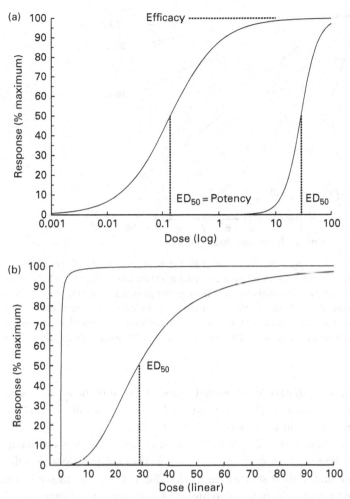

Figure 10.5   Dose–response curves. (a) Semi-log curves plot the percentage maximum response against the log of the dose. The drug potency is the dose that gives 50% of the response, also called the ED_{50}. The drug efficacy is the maximum response. When the percent response is plotted, the quantitative response is not shown. (b) When the same data are plotted on a linear–linear graph, the left curve is rectilinear, while the right is sigmoidal. The ED_{50} values vary by a factor of 200.

The production and entry sites are seldom in the same location as the processes these molecules activate. Most hydrophobic hormones, for example, work by activating transcription in the nucleus. Intracellular transport of hydrophobic molecules has the same difficulties associated with extracellular transport. The hydrophobic molecules are not very soluble in the cytosol and will bind to proteins to minimize contact with water. If these proteins are transported by dynein toward the nucleus, the movement into the nucleus will be enhanced. The details of how hydrophobic molecules dissociate from their transporters and bind to the genome are still being ascertained. If they do dissociate from the transport molecules without being directly passed to their binding sites on the

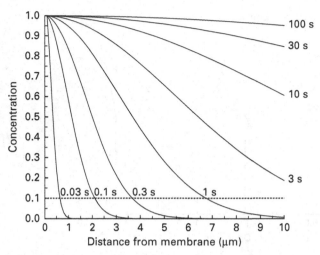

Figure 10.6   Intracellular diffusion profiles. For an agent that is either produced at the membrane or enters through the membrane, if its membrane concentration remains constant it will eventually permeate the entire cells. The exponential decrease in concentration with distance from the membrane is eventually overcome by the temporal saturation of the cytosol. For processes that require a certain concentration of an agent to be activated, such as that indicated by the dashed line, the distance of the process from the membrane and the rate constant for diffusion will determine when activation will begin.

genome, they will still have to diffuse to those sites. Hydrophilic second messengers such as cAMP will diffuse throughout the cytosol, binding to various proteins. The diffusion of molecules in the cytosol is modeled in Figure 10.6.

For cells of average size, 20 μm across, a membrane will always be within 10 μm, in the case of a spherical cell. As most cells are not spherical, the distance to the membrane would be one-half of their shortest dimension. Some cells such as skeletal muscle cells are much larger, with a diameter of 50 μm across and many centimeters long. Figure 10.6 shows diffusion profiles over 10 μm. When a cell is activated by a drug or hormone, it will be surrounded on all sides by the agonist. This occurs because in reaching the cell by way of the blood, the filtration/reabsorption of fluid around a capillary will continually flush the interstitial fluid around all cells. If the receptors for hydrophilic agents are evenly distributed around the cell, all sides of the cell will be activated simultaneously. Although hydrophobic agents in general do not use cell membrane receptors, these agonists will also surround the cell, entering through the membrane on all sides. If the agonist concentrations remain elevated for a period of time, the processes of second messenger production and hormone entry will occur at the membrane, with molecules diffusing inward from there.

Assume that the extracellular agents of activation have a constant concentration. Whether a second messenger is produced at the membrane or a hormone is entering, the concentration at the membrane will remain constant as long as the cell activation continues. For systems with G proteins the activation process is temporally limited, as discussed below. If the intracellular agent concentration remains constant at the membrane, eventually the entire cell will have that concentration as diffusion occurs. Modeling of this process has two considerations. First, the concentration of the agent will be highest at the membrane, with an inward diffusion gradient from the site of agent production or entry at the membrane. Second, over time the concentration everywhere will increase. The concentration of the agent C across the cell then will be

$$C = C_\mathrm{m} e^{-(l/\lambda * \tau/t)} \qquad (10.22)$$

where C_m is the concentration of the agent, drug or hormone at the membrane, l is the distance from the membrane, λ is the length constant describing the concentration profile over distance, τ is the time constant describing the change in concentration over time, and t is the time. The diffusion profiles of a relatively small molecule are shown in Figure 10.6. The times would increase by roughly tenfold for protein diffusion. The profiles show a progression of distributions across the length of the cell as a functions of different times. The dashed line indicates the concentration that a local process would need in order to activate some process, such as transcription. If the genome were 10 μm from the membrane in this cell and needed 10% of the membrane concentration to activate transcription, more than 1 second but less than 3 seconds would be needed to start the process. Note that initiation of the process does not stop the activation process immediately. Processes controlled by negative feedback will decrease activation, but will always require some time to do this. The concentration will rise toward the limiting membrane concentration, temporally buffering this system. Even if negative feedback systems initiated by protein synthesis decrease the membrane concentration of this agent, it will be some time before the agent diffuses away from the genome.

The agent profiles in Figure 10.6 apply when there is continuous production or entry of an agent increasing its total intracellular content, and the agent is not eliminated inside the cell. This could be the case when a hydrophobic agent with a steady-state extracellular concentration enters the cell, if its removal inside the cell is very slow. For cellular activation by hydrophilic agonists that trigger a G-protein-controlled system, the production of the intracellular second messenger will only have a limited duration. The distribution profiles of agents produced for only a limited time are shown in Figure 10.7.

Figure 10.7(a) shows the distribution of a limited production agent that is not eliminated from the cell. The total content of the agent will remain constant in the cell, and will eventually be evenly distributed across the cell. In the model in Figure 10.7(a), for agent production that stops in 0.1 seconds, the agent is uniformly distributed by 3 seconds. Of course, all intracellular signaling agents have to be removed eventually. For systems in which activation is constant, the removal process will result in a steady-state concentration like that shown above in Figure 10.6. The steady-state agent concentration is inversely related to the removal rate: the higher the rate of removal, the lower will be the steady-state agent concentration. For a G-protein-activated production system coupled with an exponential elimination process, the distribution of the agent in the cell is a function of the ratio of the duration of production to the rate of elimination. The rate of elimination yields a local concentration C

$$C = C_\mathrm{tot} e^{-(l/\lambda * \tau_\mathrm{dif}/t)} \bullet e^{t_\delta/\tau_\mathrm{Elim}} \qquad (10.23)$$

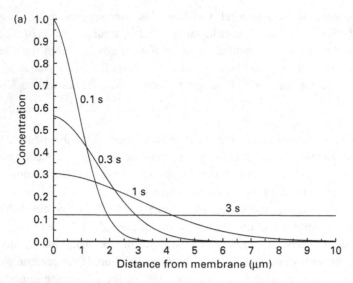

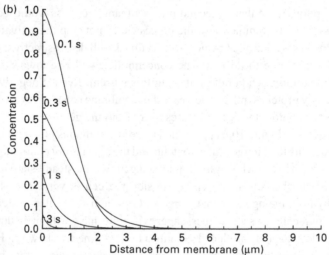

Figure 10.7   G-protein effect and intracellular elimination. (a) G-proteins have limited temporal activity. Distribution profiles are shown when the activation stops at 0.1 s after initiation. The agent reaches a steady-state concentration spread across the cell. (b) If the cell removes the agent with an exponential time constant of 5 seconds, five times the diffusion time constant, the concentration will fall as shown by the profiles. The agent will not reach the center of the cell unless the elimination time constant increases, slowing the rate of elimination and allowing the agent to diffuse farther.

where C_{tot} is the total concentration of the agent in the cell, l and λ are the distance and length constant defined above, τ_{dif} is the diffusion time constant, t is the diffusion time, τ_{elim} is the elimination time constant, and t_δ is the time since the agent production stopped. Figure 10.7(b) shows the temporal change in a system in which the elimination time constant is five times larger than the diffusion time constant, with the elimination processes uniformly distributed throughout the cell. Increasing the elimination time constant will slow the rate of removal, allowing the agent to penetrate deeper into the cell.

G-proteins function as timing elements in biology. Gilman and Rodbell were given the Nobel prize for their discovery. Heteromeric G-proteins are a category of G-proteins that link cell surface receptors to a wide range of cellular functions. The G-protein receptors are termed serpentine or seven-transmembrane (7 TM) spanning receptors based on their structure. More than 1000 different molecules use these receptors, from small molecules to proteins (Ganong, 2005). A G-protein is a trimeric protein on the cytoplasmic side of the membrane. The α subunit binds GTP and dissociates from the βγ complex. Both the separated α subunit and the βγ complex activate multiple cellular functions, including vision, taste, smell, many neurotransmitters and hormones such as TSH, FSH, LH, VIP and PTH. Activation in G-protein systems is temporally limited. The α subunit activity lasts until the GTP is hydrolyzed to GDP, which remains attached to the α subunit, stopping its activity. The GDP-bound α subunit then reassociates with the βγ complex which then has its activity stop as well. The system remains at rest until the GDP dissociates from the α subunit and is replaced by GTP, causing α subunit dissociation and reactivation of the system. One of the first systems in which G-proteins were discovered was in visual transduction, allowing the cycling of our visual images.

There are multiple isoforms of all the G-protein subunits, producing the diversity of intracellular activity agents, including the second messengers, cAMP, cGMP, and IP3.

10.5   Statistical analysis

In general, statistical analysis is not a major problem in biophysical systems. The changes measured during experiments are so dramatic that statistics are seldom needed to see that something significant has happened. Still, it is ethically correct to properly analyze the experiments. A good starting point is the number of experiments needed to demonstrate a phenomenon. This varies across the academic spectrum. In mathematics, statistics are not needed when a new theorem is proven. Likewise, in physics the demonstration of a new subatomic particle, difficult as it may be to generate, does not require statistical verification at the $p < 0.05$ level. Just showing that the particle exists is enough. Perusal of chemistry experiments shows that an N of 3 is often used, with the experiments usually so reproducible that no or little statistical analysis is provided. Many biophysical experiments are done at this level: the fluorescence of a molecule or the osmotic pressure of a solution often do not need advanced statistics. When more complex systems are considered, and especially when cells and tissues enter the picture, the results have enough variability that statistics are required. The N for many cellular and tissue experiments is 4–6, with rejection of the null hypothesis at the $p < 0.05$ level most common. In population studies in other fields, the N value is dozens or hundreds in order to obtain statistical significance.

In biophysical experiments where statistics are necessary, why is the $p < 0.05$ value most commonly used? If one looks at a normal distribution (Figure 10.8), the probability of falling within two standard deviations is 95.44%, and outside of two standard

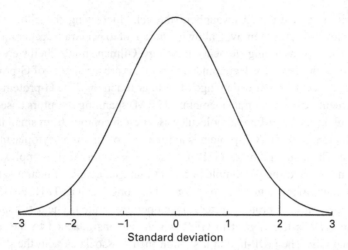

Figure 10.8 The normal distribution. More than 95% of the members of a normal distribution will have values that fall within two standard deviations of the mean. Only 0.26% of the members of a normal distribution will fall outside three standard deviations.

deviations is 4.56%. If a p value is less than 0.05, it is near to or more than two standard deviations from the mean. There is only a one in 20 chance of that element being a part of that distribution with the normal distribution mean and standard deviation and thus you can reject the null hypothesis that the element is part of this distribution. There is still the possibility that the element is a member of the group, as the statistics can only provide a probability, not certainty. For $p < 0.0026$, the element must be more than three standard deviations away from the mean to reject the null hypothesis.

The normal distribution is useful in another matter: data rejection. Every experiment at one time or another generates a piece of data that appears to be very different from all the other measurements. A practical example is the one student who gets 99% on a test and everyone else gets below 60%. When can a data measurement be discarded? Well, it can't be done after the fact; that is, you can't do an experiment with no advance standard and decide what to reject afterward. Instead, based on prior experience, set a standard before the experiment is done. A common standard is rejecting data more than two standard deviations from the mean, calculating the mean with all the data, including the questionable point. If it falls outside the range, discard it. You must also discard any other points that also fall outside. When reporting this experiment, you report what your standard was and how many points were rejected. This is not dishonest: you have established a reasonable standard and acted upon it. Selective data rejection, leaving out data that does not buttress your hypothesis without an advance standard, is dishonest and should not be done.

Sometimes an experimental system is changing over time, and the baseline is different for some measurements than for others. When data is reported as a difference above baseline, as a percent or fractional change from baseline, or as a peak response, the analysis assumes that the system analyzed operates in a specific molecular way. As seen in Figure 10.9, the same data set can be reported as reflecting no difference, the first smaller

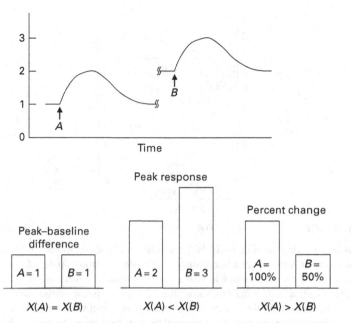

Figure 10.9   Data analysis of changing baseline experiments. The A stimulus causes a response from baseline 1 to peak 2. The B stimulus causes a response from baseline 2 to peak 3. When the peak response is subtracted from the baseline, the differences are the same for mechanism X ($X(A) = X(B)$). When the peak response is chosen, the change in X is smaller for A than for B ($X(A) < X(B)$). When the percentage change from baseline, or the fraction of response above baseline, is chosen, the change in mechanism X is larger in A than in B ($X(A) > X(B)$). The choice of analysis presupposes how the mechanism works.

than the second, or the first greater than the second. This clearly presents a problem, since each kind of analysis routinely appears in journals, especially endocrinology journals. What is the right way to analyze data in the face of a changing baseline? The different analyses have the potential for unethical conclusions (Dillon, 1990).

Each analytical method presumes a different mechanism X (Figure 10.10). When equal changes in stimulus produce equal changes in the mechanism that controls the response, the mechanism has to be linear. For system changes near equilibrium, this will be true. But over a sufficiently large change in stimulus this mechanism would still have to be linear. There are very few linear mechanisms in biology. Perhaps one of the few is the linear increase in muscle force with an increase in the number of crossbridges. Analyzing data by just subtracting the baseline without knowing that the relation is linear is usually not scientifically valid.

Measuring the peak response in the face of a changing baseline is not intuitively obvious. The mechanism X must have progressively greater changes in order to cause the same external change in response. The response is approaching an upper limit. This is what many biological systems do as an increasingly high concentration of an agonist is needed for a step change in response. The upper ranges of sigmoidal dose–response curves fit this model, and are often the appropriate way to analyze data.

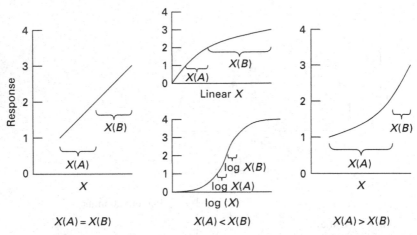

Figure 10.10　Mechanisms for changing baseline analysis. Concluding that equal responses indicate an equal change in mechanism X presumes a linear stimulus–response relation, $X(A) = X(B)$. Concluding that the change from the lower baseline indicates a smaller change in mechanism X than the change from the higher baseline $X(A) < X(B)$ presumes the stimulus–response relation that approaches an upper limit for mechanism X; this is a sigmoidal relation on a log(X) curve. Concluding that the change from the lower baseline indicates a larger change in mechanism X than the change from the higher baseline $X(A) > X(B)$ presumes that this part of the relation describes a cooperative system.

When the percent or fractional change from baseline is measured, the change in X needed to produce the same external response will be progressively smaller. Having an ever greater response for a step change in stimulus will occur in cooperative systems, such as the loading of oxygen onto hemoglobin. At some point, of course, these systems will ultimately be limited by some other factor, recalling the multiple inputs resulting in inflection points that were discussed previously. This additional factor does not negate the range over which cooperativity can occur, but there must be some reason to assume that fractional changes are the right way to report data when the baseline is changing.

When can the baseline be ignored? That is, when can the change from the baseline alone be used for analysis? Consider two situations, muscle contraction and hormonal activation. If a muscle has significant passive tension when the contractile system is activated, the passive tension can be subtracted from the total tension to give the active tension. Why? The passive tension in the connective tissue is not calcium or ATP dependent, but the contractile system is both calcium and ATP dependent. When the baseline can be shown to be independent of the stimulated response system in some way, the baseline can be ignored and its value subtracted from the response. When a hormone response is measured, a baseline of hormonal concentration cannot be independently separated from the stimulated hormone concentration, and the baseline cannot be ignored. Any reported analysis that does not include the original data will be assuming one of these three models. The best way of avoiding the data analysis problem is to report the raw data in addition to any analysis relative to the baseline. When the raw data is reported, other people can use it for other, perhaps different, analyses and conclusions.

Besides measuring the p value in the distribution of your data, there is another aspect of data analysis, the power. While this will seldom be a factor when the control and experimental results are very different, it does come up when population studies are done, as in assessing medical tests. The effects of statistical differences measured by low p values and the power of the test can have multiple outcomes (Bhardwaj *et al.*, 2004). The power of a test is the probability of detecting a difference between two groups when a true difference exists. The power of a test generates a confidence interval and is strongly dependent on the number of members of each tested group. For the test of whether there is a significant difference between a drug and a placebo response, where the placebo response has a mean of 1.0, a drug confidence interval of 0.9–1.1 would indicate no significant difference. A drug confidence interval of 1.002–1.003 would indicate a significant but clinically irrelevant difference. A confidence interval of 3–10 would indicate both statistical difference and clinical importance, while a confidence interval of 0.8–10 would indicate no significant difference, but that the power of the test was insufficient to rule out a large difference between the drug and placebo. The last test should be repeated with a larger test size. The significant but irrelevant group, 1.002–1.003, may have had so large an N that a difference was found that had no practical value. It is possible to calculate in advance how big the N should be for a particular difference between means and the power desired to show a difference. This pre-experimental calculation needs the difference between the control and test means, the desired p value, and the power, where power values will range between 0 and 1, with higher number indicating group differences but also needing a larger N. Granting agencies sometimes ask for calculation of N in advance of funding, as too large an N is impractical to fund. The power analysis calculation can be done by a number of software analysis packages. When medical personnel talk about evidence-based medicine, it is the combination of statistical difference and power that they are talking about. Power calculations will eventually become as common as p values in many aspects of research.

When a pathological condition is rare in the general population, a problem can develop over the use of diagnostic tests. Calculation of false positives in healthy individuals when a disease is rare in the population can be made using Bayesian statistics. In this case, the fraction of the population with the disease is P(A). The probability of getting a positive test for the disease is P(B). The probability of having the disease if you have a positive test is P(AifB)

$$P(AifB) = \frac{P(A) \cdot P(BifA)}{P(A) \cdot P(BifA) + P(NotA) \cdot P(BNotA)} \tag{10.24}$$

$$P(NotAifB) = 1 - P(AifB) \tag{10.25}$$

where P(BifA) is the probability of having a positive test if you have the disease, P(NotA) is the probability of not having the disease, P(BNotA) is the probability of having a positive test if you don't have the disease, and P(NotAifB) is the probability of not having the disease if you have a positive test. Note that these last two are not the same thing. P(BNotA) is false positive rate. In general this will be a low number, but no test is perfect,

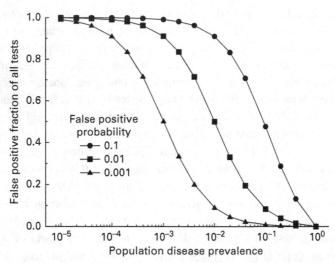

Figure 10.11   Probability of a false positive test in a healthy individual. Bayesian analysis calculates the probability of a false positive diagnostic test in a healthy individual as a function of the prevalence of the disease in the population and the probability that the test can yield a false positive. The test has a 0.999 possibility of yielding a positive result when the disease is present. When the disease is rare in the population, most of the positive tests will occur in healthy individuals.

and some false positives will occur. When the occurrence of the disease is rare, most people will not have the disease, but they can still get a false positive and be thought to have the disease. The probability of this in the entire population is P(NotAifB), and is plotted in Figure 10.11. The rarer the disease, the more likely that any tested individual who has a positive test will have a false positive. As the prevalence of the disease increases in the population, the fraction of false positives in the population will decrease. Estimates of the HIV infection rate in the general population range from 1 in 300 to 1 in 3000, a statistically rare probability. The test for HIV detects the virus in infected individuals with an accuracy of 0.99 to 0.999. The variation in this factor has little effect on the false positive rate. The rate of false positive testing for an individual is 1–10%. Using these parameters, the fraction of false positives in the general population is 0.750–0.997.

For surgical personnel cut during surgery on HIV positive patients, the infection rate is about 1 in 3000. Even with a test that only yields a 1% false positive rate, more than 96% of the surgical personnel tested for HIV infection following a surgical cut will be false positives. This is not to say that testing these individuals would be futile, for 24 out of 25 tested individuals would test negative and be negative. For the other healthy individual with the positive test, the test could be positive because of a testing error, or because of a cross-reaction in the person. A re-test would correct the former, but not the latter. Sometimes the results of a false positive can lead to treatments with severe consequences, consequences the person acquires inappropriately. The major technical advance that limits false positives in the population is a reduction in the false positive rate of the test.

10.6   Drug development and orphan diseases

The actual ED_{50} of a drug is not specifically important: regardless of a drug's potency, it is the efficacy, the maximum response to the drug, that must be known. Because there are only a limited number of modes of cellular activation, every drug produces side effects. It is easy to find a drug that will do what you want, but to find one that does what you want but does not have serious side effects is very difficult. This represents one of the most challenging aspects of drug development. Given the possibility of side effects from a drug, the ED_{50} is important relative to the drug concentrations that produce deleterious effects. Just as the ED_{50} is the drug concentration that produces 50% of the pharmacological response, the LD_{50} is the median lethal dose. The therapeutic index TI is

$$TI = \frac{LD_{50}}{ED_{50}} \qquad (10.26)$$

with high values corresponding to relatively safe drugs, while low TI values have effective doses near to potentially lethal ones. Just as the dose–response curve encompasses several orders of dose magnitude between initial and maximal responses, so too will some lethal doses occur at lower concentrations than the LD_{50}. In this regard, other standards are also used. The LCt_{50} is the lethal concentration multipled by time of exposure, recognizing that both elements can be important in producing the lethal event. There may commonly be toxic events that occur at concentrations far lower than fatal doses, yielding the protective index PI

$$PI = \frac{TD_{50}}{ED_{50}} \qquad (10.27)$$

where TD_{50} is the toxic dose, used in place of the LD_{50}.

Whichever standard is used, the concept of the range over which a drug will have both beneficial effects and toxic effects, while well known to the medical community, is one of the least understood concepts by the general public. Many people regard pharmaceutical therapy as digital: give a certain amount of drug, and the symptoms will be relieved with no side effects. The extensive warnings associated with every drug are just so much "noise", never applicable in one's own case. In fact, the statistical nature of pharmaceutical responses, both therapeutic and toxic, means that every individual falls somewhere on that statistical distribution, a position that cannot be known in advance. This is dramatically demonstrated when the drug development process is considered.

There are multiple phases during the development of a new drug. In the pre-clinical phase, the drug's function is discovered and its characteristics are determined in tissue and animal studies. Following this, if the drug is deemed to have possible therapeutic benefits without significant deleterious side effects, the clinical trial phases begin using people who have consented to be included in the tests. Phase 0 uses doses believed to be too small to have any therapeutic effects. During this phase the pharmacokinetic parameters can be established in humans, beyond those previously made in different animal models. These studies establish the safety floor of the drug, setting the doses and dosing regimens to be used in subsequent phases. Phase I uses a small number (usually < 100) of healthy volunteers to test the tolerance of humans to different doses of the drug. Some groups get a single dose, others get escalating doses of the drug, measuring how much can be given before side effects appear. In some cases, patients with very serious diseases such a incurable cancers may also be included in Phase I, an informed decision in cases where there may be no currently available treatment. Phase II uses larger groups of volunteers and patients to determine the dose regimen for therapeutic responses and how efficacious different doses are. Failure of the drug to be efficacious or the appearance of significant side effects can derail a drug's development at this point. Phase III involves randomized controlled trials at multiple sites, often over an extended period of time using several thousand patients. These tests are extensive and expensive, comparing the new drug to the current standard of treatment in the field. Particular attention is given to negative responses that require withdrawal of the drug from use. Successful completion of Phase III leads to governmental approval and bringing the drug to market. Phase IV covers post-marketing studies of the drug over time, assessing its long-term safety and effectiveness. For every drug that enters Phase I, 71% enter Phase II and 31% enter Phase III (DiMasi *et al.*, 2003). When considering the cost of drug development, note that most drugs never get to Phase III. All the pre-clinical and clinical costs of failed drugs must be borne by those projects that are successful.

Separate studies of the cost of drug development (DiMasi *et al.*, 2003; Adams and Brantner, 2006), both found that preclinical costs for a new drug to be just under $400 million, and just over that for the clinical studies. The total cost to develop a new drug is $800–900 million, with the range going from $500–2000 million (Adams and Brantner, 2006). When the entire process is complete, there will still be a range of responses from the population. Consider drugs for diseases like asthma, diabetes or arthritis, where tens of millions of patients may take the drug. For a drug in which 3000 people were tested before the drug went to market, a single negative result represents 1/3000. When 30 million people take the drug, this would mean 10 000 negative results. Of course, that one bad result out of 3000 has no real statistical significance, and may not have kept the drug from the market. But 10 000 bad results would both pull the drug from the market and result in many lawsuits. This potential has to be considered when companies develop a new drug, and has led to the concept of the orphan disease and the orphan drug.

Orphan diseases are those diseases that have either too few people to warrant the cost of drug development, little prospect of improved treatment, or are only present in significant numbers in underdeveloped areas where the prospect of cost recovery is minimal. The U.S. National Institutes of Health lists 5954 orphan diseases. While most of these would go unrecognized even by health professionals, some are well known: amyotrophic lateral sclerosis (ALS), also known as Lou Gehrig's disease; Guillain-Barre syndrome; Marfan syndrome, the connective tissue disease that Abraham Lincoln may have had; and tuberculosis, not prevalent and easily treatable in the developed world, but growing in prevalence in developing countries. The NIH grants special circumstances to encourage scientists and pharmaceutical firms to work in these fields, recognizing that there is little chance of profit. Progress in these fields may be piggy-backed on work in more profitable fields. The development of orphan drugs to treat these diseases must still meet the same requirements for safety, dosing and efficacy, but governments may also provide tax incentives, extended patent protection and financial support for clinical trials. A numerical example demonstrates the difficulty of inducing companies to develop orphan drugs. Suppose it takes $500 million to develop a drug that a company will have patent rights to for 10 years, the patent clock starting much earlier than when a drug comes to market. Just breaking even would take $50 million a year. In the United States, an orphan disease is one that affects less than 200 000 people. Using this number, each of these patients or their insurers would pay $250 per year, just about what people would expect to pay for a medicine to treat a serious disease. If the disease though had only 20 000 patients, the yearly cost would go to $2500, a lot of money to an individual. And that assumes that the drug actually works, without significant side effects, and just to break even. Conversely, 30 million patients spending $100 per year is three billion dollars every year. It is not a surprise that drug companies go after major diseases: they both help many people and are very profitable. The support of orphan disease research and treatment may take a combined government/industry/academia effort that is more formalized than at present.

Not all drug development involves new chemical compounds. Sometimes naturally occurring toxic agents can be used for medical purposes, such as digitalis in the treatment of heart ailments. Other uses go beyond medical treatments. The most toxic of all known substances, blocking the release of the neuromuscular junction neurotransmitter acetylcholine, is botulinum toxin, which has an LD_{50} of 1 ng/kg body weight intravenously and 3 ng/kg when inhaled (Arnon *et al.*, 2001). If the average person weighs 70 kg, 490 g (just over 1 pound) of toxin would be enough to give everyone in the world an LD_{50} dose, and 10 kg would be enough to kill every person in the world by stopping diaphragm contractions and respiration. It has been used therapeutically to treat medical conditions in which blocking local muscle contraction is desirable. It is not surprising that security agencies are concerned about the use of botulinum toxin as a weapon. Conversely, after what had to be the most convincing FDA request ever submitted, this most powerful poison, marketed as botox, has been approved and is widely used for cosmetic treatment of wrinkles, despite its potential for side effects.

Drug companies also take into account potential lawsuits. As noted above, both therapeutic and toxic responses have a range of doses. Even the safest drug has that rare instance when a negative result occurs. It cannot be avoided. People expect that a tested drug is always safe, but given the physics of pharmacokinetics and the statistical response to a drug, that safety can never be guaranteed. When the cost, time, range of responses, side effects and patient numbers are considered, it is not surprising that the system can seem so broken. There is no obvious solution to the difficulties in this area.

Agonist activation of cells uses both hydrophilic and hydrophobic molecules. G-proteins are activated when hydrophilic agonists bind to GPCRs. There are hundreds of GPCR proteins. Desensitization of GPCRs occurs when the receptors are phosphorylated and bind to β-arrestins. GPCRs activate several kinase cascades and have several allosteric regulatory sites. Receptor tyrosine kinases bind insulin and other activation factors associated with growth. There are two major classes of nuclear receptors, those that bind steroids and those that bind all other hydrophobic hormones including thyroid hormone. Agonist binding to nuclear receptors triggers mRNA production when zinc fingers on the receptor bind to DNA. The pattern of drug activation of tissues is dependent on the method of dose delivery: intermittent intravenous injection, continuous intravenous perfusion, or delivery methods that depend on epithelial transport: ingestion, inhalation, or transdermal patch. The dose–response curve defines a drug's potency and efficacy, and the slope of the curve at the ED_{50} concentration defines the effective ratio of molecular number to receptor using the Hill equation. The rate of diffusion by a second messenger or hydrophobic hormone defines how long it takes to saturate the cell when the production or inflow is constant and the diffusion rate greatly exceeds drug elimination. For temporally limited G-protein systems, the average concentration is a function of the duration of second messenger production, the rate of diffusion and the rate of elimination. When the rate of elimination is on the order of the rate of diffusion, the second messenger will not penetrate a significant distance into the cell. Analysis of differences in agonist response should use the appropriate statistical test, understanding that statistics offer probabilities, not proof. Temporally changing baselines can create presumption of a particular physical model depending on how the data are analyzed. For diseases that are rare in the population, Bayesian statistics show that most positive tests will actually be false positives. All drugs have the potential for toxicity: treatments come with a probability of negative outcomes, although the probability will vary between different drugs. Drug development is costly, and most drugs do not make it through the phases of clinical testing. The high cost of drug development limits work on orphan diseases, diseases that affect a small fraction of the population. Governments may need to partner with pharmaceutical firms to foster orphan drug development.

References

Adams C P and Brantner V V. *Health Aff.* **25**:420–428, 2006.

Arnon S S, Schechter R, Inglesby T V, *et al. J Amer Med Assoc.* **285**:1059–70, 2001.

Bhardwaj S S, Camacho F, Derrow A, Fleischer A B Jr, Feldman S R. *Arch Dermatol.* **140**:1520–3, 2004.

Blaukat A, Pizard A, Breit A, *et al. J Biol Chem.* **276**:40431–40, 2001.

Chang R. *Physical Chemistry with Applications to Biiological Systems*, New York: MacMillan, 1977.

Dillon P F. *Persp Biol Med.* **33**:231–6, 1990.

Dillon P F, Root-Bernstein R, Robinson N E, Abraham W M and Berney C. *PLoS One.* **5**:e15130, 2010.

DiMasi J A, Hansen R W and Grabowski H G. *J Health Econ.* **22**:151–85, 2003.

Ganong, W. *Review of Medical Physiology.* New York: Lange, 2005.

Gregory K J, Dong E N, Meiler J and Conn P J. *Neuropharmacology.* **60**:66–81, 2011.

Hill A V. *J Physiol.* **40**(Suppl):iv–vii, 1910.

Luttrell L M and Lefkowitz R J. *J Cell Sci.* **115**:455–65, 2002.

Mangelsdorf D J, Thummel C, Beato M, *et al. Cell.* **83**:835–9, 1995.

Tobin A B. *Br J Pharmacol*. **153** Suppl 1:S167–76, 2008.
Torrecilla I, Spragg E J, Poulin B, *et al. J Cell Biol*. **177**:127–37, 2007.
Ullrich A and Schlessinger J. *J Cell*. **61**:203–12, 1990.
Wacker D, Fenalti G, Brown M A, *et al. J Am Chem Soc*. **132**:11 443–5, 2010.
Zwick E, Bange J and Ullrich A. *Endocr Relat Cancer*. **8**:161–73, 2001.

11 Stability, complexity and non-linear systems

One of the hallmarks of physiological systems is homeostasis. Homeostatic systems maintain a steady-state set point over time using negative feedback. Negative feedback activates recovery processes when there is deviation from the set point. While some states are always tightly maintained, some changes can occur during the lifetime of the individual. These changes take two forms: adjustment of the homeostatic set point, or a step change to a new set point by a positive feedback mechanism. Positive feedback systems do not use a recovery process to return to the original set point. New negative feedback processes now keep the organism at the new set point. A set point must have an energy minimum, such that small variations from the minimum create a driving force returning the system to the set point. Energy minima occur at multiple levels, from atoms to organisms. For systems in transition the energy profile may exhibit metastable states between two large minima. Metastable states offer temporal buffering of state changes. In enzymes and ion channels the energy barriers between the external states and internal metastable states and the depth of the metastable state control the rate of the metabolic reaction or ionic current. Complex analytical methods such as catastrophe theory provide insight into state transitions. Catastrophe theory models changes in space, not time, with state transitions occurring whenever the state variable reaches particular values in parameter space. The feedback systems that maintain homeostasis are deterministic far from equilibrium, but become chaotic when transitions are near equilibrium. Not all catastrophic transitions are fatal, but for living systems catastrophes can lead to death in two ways, homeostatic instability or global thermal equilibrium. These two modes of death correspond to having too much ambient energy or too little ambient energy. This leads to the Goldilocks problem, in which ambient energy will be too high, too low or just right. Since life must be a continuum, a system cannot leave the viable energy range without extinction, leading to restrictions in modeling evolution. All living systems will have feedback systems, both negative and positive, that keep them away from either of the life–death transitions. Fractal systems regularly transition between deterministic and chaotic states. Conditions with multiple fractal inputs can have significant variability around a set point using allometric regulation. Narrowing of the fractal distribution has been associated with pathological states and increased risk of death. Apoptosis is programmed cell death, in which processes that lead to self-destruction are activated. Ultimately, all individuals reach global thermal equilibrium after death.

11.1   System control

Systems analysis uses differential equations to assess changes over time. The order of a system is its highest derivative in the differential equation. The degree of a differential equation is the highest power to which one of the variables or its derivative is raised. In many cases the differential equations are too complex for exact solutions, but in other cases exact solutions are possible. Even in those systems that cannot be solved exactly the qualitative properties of the system can sometimes be determined.

Physiological systems have both mechanical and chemical controls. Mechanical systems have elements that control position, as well as its first and second derivatives, velocity and acceleration. If a mechanical system is at physical equilibrium, as an object held in place by a spring, any distending force F_L that changes the position of the object will return it with a given Hooke's law spring constant κ,

$$F_L = -\kappa L \tag{11.1}$$

where L is the length of the spring. The negative sign indicates that the force is restorative, returning the object to its original position. This is the same principle as we saw in Chapter 1, where an element at the bottom of an energy well would entropically return to the bottom of the well after temporary energy input. The energy well equivalent to the spring constant would be the width of the well, with higher values of κ represented by narrower wells. If there is a viscous damping element resisting movement, the damping force F_v will be proportional to the velocity v at which the object moves, with

$$F_v = -h\frac{dL}{dt} = -hv \tag{11.2}$$

where h is the damping constant. For a system at equilibrium, the total force F_{tot} is defined by Newton's second law, with

$$F_{tot} = m\frac{d^2 L}{dt^2} = ma \tag{11.3}$$

where m is the mass of the object and a is the acceleration. Since

$$F_{tot} = F_L + F_v \tag{11.4}$$

the combined equation will be

$$ma = -\kappa L - hv \tag{11.5}$$

or

$$\kappa L + hv + ma = 0 \tag{11.6}$$

where the 0 indicates that the system is at equilibrium. For oscillations of this system, the natural frequency ω_o of the system will be

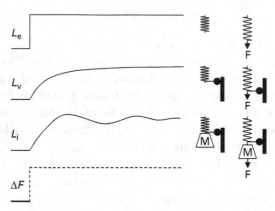

Figure 11.1   Mechanical models of elastic, viscous and inertial responses to a force-induced step change in length. From top to bottom, the systems shown are elastic (L_e), elastic-viscous (L_v), elastic-viscous-inertial (L_i) and the step change in force (ΔF).

$$\omega_0 = \sqrt{\frac{\kappa}{m}}. \tag{11.7}$$

If the system were entirely undamped (note that h is not present in this equation), ω_0 would be the frequency of the system oscillation. If there is damping in the system, the rate of damping will be determined by the dimensionless damping ratio ζ:

$$\zeta = \frac{h}{2\sqrt{m\kappa}}. \tag{11.8}$$

When $\zeta = 1$, the system is critically damped; that is, it will return to rest in the fastest time without further oscillations. When $\zeta < 1$, the value of h is small for the system, and will not rapidly damp the system, such that there will be further oscillations around the equilibrium position before the system comes to rest. When $\zeta > 1$, the system will take longer to come to rest than when critically damped, the time a function of the damping constant h. The approach to rest will be exponential. Mechanical systems can be overdamped, critically damped, or underdamped, while chemical systems behave as overdamped systems, with virtually all changes approaching equilibrium exponentially.

When an external force F is applied to a spring, the system is not at equilibrium. The system will eventually reach a new equilibrium at a different length, longer for a stretching force and shorter for a compressive force. The force equation will be

$$F = \kappa L + hv + ma \tag{11.9}$$

with the three terms representing elastic (κL), viscous (hv) and inertial (ma) elements. The successive application of each of these elements is shown in Figure 11.1 for a given change in force ΔF. In the top panel, when a force is applied to a massless spring without either viscous or inertial elements, the length will exhibit a step change, with the relation between L_e and ΔF determined by the spring constant κ. The second panel shows the length response L_v with the inclusion of a viscous damping element, friction slowing the approach to equilibrium, along with the elastic element. This makes h an exponential time constant. The length response L_i includes all three elements, elastic, viscous and inertial. The inertial term is proportional to the mass of the system. This particular system is underdamped, so it oscillates as it approaches the new equilibrium. It is important to note that the contributions of the viscous and inertial elements will vanish as both the velocity and acceleration approach zero as the spring approaches its new equilibrium. The final length response to a step change in force is determined by the elastic term alone.

There are circumstances in which the inertial term is insignificant because the mass is so small. One site in the body where sinusoidal mechanical oscillations occur is in the cochlea. The mechanical characteristics of the tectorial membrane were measured in both the radial and longitudinal directions during forced sinusoidal oscillations of the membrane surface (Gu et al., 2008). Varying the oscillation frequency is the equivalent to changing the velocity applied to the membrane. Impedance measurements were used to assess the physical properties of the membrane. Mechanical impedance Z of a material is the ratio of an applied force to the resulting velocity v:

$$Z = \frac{F}{v}.$$ (11.10)

Thus, the impedance measurement tests the elastic, viscous and inertial components of the force displacement equation, dividing each term of the force equation by the velocity. This mathematical transformation is appropriate when the material's mechanical properties are linear and the force is the integral of sinusoidal components at different frequencies. The resulting equation for impedance $Z(\omega)$ as a function of the angular frequency ω is

$$Z(\omega) = \frac{\kappa}{j\omega} + h + j\omega m$$ (11.11)

where k is the elastic spring constant, h is the viscous damping constant, m is the mass and $j = \sqrt{-1}$, representing a 90° phase shift. The real and imaginary components of impedance form a vector the length of which is the impedance magnitude and angle of which is the impedance phase. All of the terms in the impedance equation must have the same units, and from the middle term must be in units of mass/time or resistance. The spring constant κ has units of mass/time^2, while the angular frequency ω has units of time^{-1}, while j is dimensionless. Thus each term has its unit in the force equation divided by velocity. In a system in which oscillations are applied to a material, the impedance will be lowest at the resonant frequency of the material; that is, the lowest force is needed to produce a given velocity. This is what happens in the cochlea, where the specific oscillations of the basilar membrane, much more flexible than the tectorial membrane, will have their greatest amplitude of oscillation to a given sound pressure at a specific distance from the oval window. The graded potentials produced at this site will be larger than at any other site for that frequency of sound. The brain will interpret signals coming from this site as representing that frequency.

In the measurements of the impedance of the tectorial membrane, taking into account the real and imaginary components, Gu et al. (2008) showed that the phase of a purely

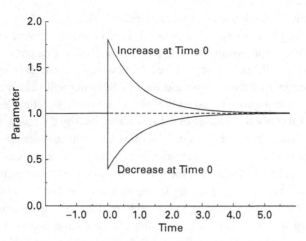

Figure 11.2   Recovery from perturbations of negative feedback systems. Changes in the control parameter away from the set point result in recovery processes that return the system to its original set point. The recovery systems can accommodate changes that have either increases or decreases from the set point.

elastic membrane would be at $-90°$ with a negative slope of magnitude vs. frequency, a purely viscous membrane would have a phase of $0°$ and zero slope, and a purely inertial membrane would have a phase of $+90°$ and a positive slope. Experiments over more than two orders of magnitude of angular frequency showed that the tectorial membrane behaves as a combined elastic and viscous system, with no measurable inertial component: the mass of the tectorial membrane is so insignificant that the inertial term can be ignored. This is radically different from whole body mechanics, where inertial forces due to acceleration can have profound consequences, such as the potentially fatal effects of whiplash caused by automobile accidents. The important terms of the force equation will be tissue dependent.

11.2   Negative feedback and metabolic control

While mechanical systems can have different damping behavior, chemical systems, including those of intermediary metabolism, behave as overdamped systems for a single variable change. Metabolic systems that oscillate must have multiple control variables. Step changes in a controlled variable will return the system to its original state using negative feedback. Figure 11.2 shows graphs of negative feedback after a step increase and a step decrease in the control parameter. In chemical systems this will be a concentration, but mechanical parameters like muscle position can also be the control parameter. A negative feedback system is stable over time, maintaining a steady state. Perturbations from this state, either positive or negative, result in recovery reactions that return the system to its original state at set point C_0. The value C of the control parameter will be

$$C = C_0 \pm (C_0 - C_t)e^{-\alpha t} = C_0 \pm (\Delta C)e^{-\alpha t} \qquad (11.12)$$

where C_t is the time-dependent value of C, α is the negative feedback time constant, t is the time since the perturbation from the set point and ΔC is the difference between the set point and the current parameter value. The $\pm$ indicates the direction of the change away from the set point, $+$ being an increase and $-$ being a decrease. Note that the exponential factor will approach zero at long times, minimizing the difference and returning the parameter back to the set point. As illustrated in Figure 11.2 both increases and decreases in the parameter will return to the set point in a negative feedback system.

In Chapter 1 it was noted that organisms are not at equilibrium, but in a steady state removed from equilibrium. The entropic cost of each reaction moves the system closer to equilibrium, so that a continuous influx of energy is needed to maintain life. It is not the case, however, that every reaction in the body is removed from equilibrium. Indeed, most metabolic reactions are at equilibrium between their substrates and products. The maintenance of metabolic concentrations requires negative feedback of regulatory enzymes. The body as a whole is kept from global thermal equilibrium by the presence of regulatory enzymes that are not at equilibrium (Newsholme and Start, 1973). It is not the case though that all non-equilibrium reactions are regulatory. A non-equilibrium reaction will have a large drop in free energy, driving the reaction forward. In Chapter 4 we found that three enzymes associated with glycolysis had large free energy drops, hexokinase (HK), phosphofructokinase (PFK) and pyruvate kinase (PK). Of these, PFK is substantially more regulated than the other two. The introduction of glucose into most cells, hepatocytes, the mucosal epithelium and renal tubular cells excepted, results in its phosphorylation by HK. Phosphorylated glucose is not transported by either class of hexose transporters, thus trapping the glucose in the cell. The equilibrium for HK is such that the phosphorylated form dominates. In skeletal muscle there is no detectable free glucose, so that even if HK can be regulated, such as inhibition by high concentrations of its product glucose-6-phosphate, the absence of detectable substrate indicates that there is no functional regulation. In the liver hepatocytes an alternative enzyme, glucokinase (GK), can phosphorylate glucose. The K_m for HK is 0.01–0.1 mM glucose, while for GK it is 10 mM (Newsholme and Start, 1973). An increase in plasma glucose after a meal will have little effect on HK, as the normal plasma concentration of 5 mM glucose would saturate the enzyme. The low K_m indicates a relatively slow enzyme, so the absence of free glucose in tissues like skeletal muscle indicates that the rate-limiting step in phosphorylation must be glucose entry. For the liver, with its higher K_m, the increase in plasma glucose leads to rapid phosphorylation of glucose in the liver hepatocytes without saturating GK. This will lead to a preferential accumulation of ingested glucose into the liver, storing it for future use by the brain when plasma glucose levels decrease between meals. Pyruvate kinase, the last enzyme of glycolysis, also has a large free energy drop. The absence of changes in its substrate PEP during large variations in energy flux indicates this enzyme is also not regulated, despite being a non-equilibrium enzyme.

All enzymes, both equilibrium and non-equilibrium, can have an increase in flux through the enzyme when there is an increase in the concentration of substrate. The identification of a regulatory enzyme, in addition to being non-equilibrium, depends on whether that enzyme can have an increase in flux in the presence of a decreased substrate

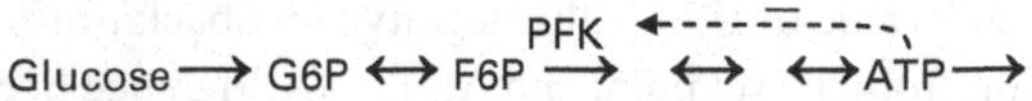

$$\text{Glucose} \longrightarrow \text{G6P} \longleftrightarrow \text{F6P} \xrightarrow{\text{PFK}} \longleftrightarrow \longleftrightarrow \text{ATP} \longrightarrow$$

Figure 11.3   Allosteric control of glycolysis. PFK activity is regulated by the concentration of ATP. ATP is the product of the glycolytic pathway, and can bind to PFK and reduce its activity and thus the net flux through the glycolytic path.

concentration, a condition known as the crossover theorem (Newsholme and Start, 1973). The identification of a regulatory enzyme can therefore be made through measurement of substrate and product under different conditions, even if the regulating substance is not known. PFK is a regulated enzyme. ATP is a substrate of PFK, but at high ATP levels PFK activity is inhibited. Since in many cells the concentration of ATP is highly buffered by creatine kinase, increased flux through PFK during increased energy consumption occurs through increases in AMP and ADP, which allosterically increase PFK activity at physiological concentrations. When small decreases in ATP occur, the concentration of ADP will increase significantly. The adenylate kinase reaction

$$2\text{ADP} \leftrightarrow \text{ATP} + \text{AMP} \tag{11.13}$$

causes an even fractionally larger increase in AMP, as the skeletal muscle ATP concentration (5 mM) is so much larger than ADP (~0.1 mM) or AMP (~0.001 mM). As PFK responds to the AMP/ATP ratio, the large relative change in AMP causes a large increase in PFK activity.

Product inhibition is a common enzyme control mechanism. Conceptually, this control manifests itself in two ways. In its simplest form, the buildup of product decreases the activity of the enzyme. We have already seen this in the inhibition of the myosin ATPase by phosphate, an important mechanism that limits muscle contraction. A product must be released from an enzyme to complete the reaction. The product will have a reverse binding rate, returning to the enzyme that produced it. The reverse reaction rate, as will all reaction rates not under allosteric control, will be dependent on the concentration of the binding substance, as long as the concentration is near or below the K_m. When the free concentration of the product rises, the statistical dissociation of the product from the enzyme decreases. The phosphate effect on myosin reduces the rate of the force-generating step, resulting in lower force. At the organism level, increased inorganic phosphate in skeletal muscle decreases the ATPase rate and muscle contraction speed.

The second method of product inhibition uses the product of a reaction pathway to limit the regulatory enzyme in the pathway. In glycolysis, PFK is the regulated enzyme. The product of glycolysis is ATP. Increased amounts of ATP decrease the activity of PFK, even though ATP is a substrate for the reaction. As no glycolysis would occur without some ATP, there must be an ATP concentration at which the glycolytic rate is maximized. Figure 11.3 shows the glycolysis pathway, with single-headed arrows indicating the non-equilibrium reactions HK, PFK and PK. ATP allosterically inhibits PFK. The allosteric theory (Monod et al., 1963) proposed that an enzyme can have both a catalytic site and a regulatory site. ATP binding to the PFK regulatory site decreases the activity of the enzyme. Decreasing the concentration of ATP will allow ATP to leave the regulatory site and return the enzyme to its uninhibited state. Allosteric regulation does not always involve the product of an enzyme or

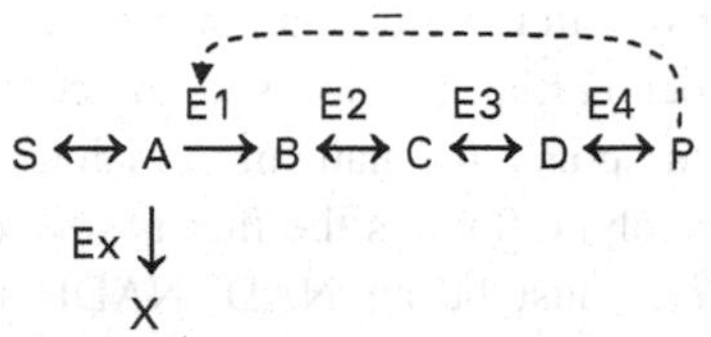

Figure 11.4   Feedback control at metabolic branch points. The product P of the pathway allosterically inhibits the first enzyme after the branch point that leads to its production. This process is a negative feedback mechanism that controls the activity of the pathway independent of other pathways that also use molecule A, such as the X pathway.

a pathway. As noted above, AMP increases the activity of PFK. This activation is also allosteric, and although AMP concentrations are controlled by the substrate (ADP) and product (ATP) of glycolysis, AMP is not strictly a part of glycolysis.

The feedback of pathway products often occurs at branch point enzymes (Figure 11.4). Branch points occur when a molecule can be used as the substrate for multiple reactions. At least one of the pathways has to be allosterically controlled for differential pathway control. Otherwise, an increase in molecule A would increase the flux through both pathways only dependent on the relative activities of the two branch point enzymes and their K_m values. Allosteric control can limit the activity in one pathway using pathway product inhibition, as shown in Figure 11.4, or conversely by allosteric activation. The branch point enzymes are shown as non-equilibrium reactions where control can be independent of substrate concentration.

As covered in every basic biochemistry course, the Michaelis–Menten constant K_m is the substrate concentration that produces one-half of the maximum velocity of an enzymatic reaction (Chang, 1977). For substrate concentrations below K_m, there is an approximate linear increase in activity with concentration. For substrate concentrations above K_m, the velocity approaches a maximum. The reaction is independent of substrate concentration changes far above K_m. In most metabolic pathways, the concentrations are well below K_m: indeed, it is often difficult to measure the free concentrations of many metabolites. The direct transfer of metabolites between enzymes further complicates assessment of metabolic concentrations. While there are conditions in which flux is regulated either by the K_m, such as the phosphorylation of glucose in skeletal muscle cited above, or by an allosteric factor, such as AMP's effect on glycolysis, in most cases an increase in substrate concentration will result in an increase in flux.

Flux regulation can also be limited by cofactor availability. Many B vitamins are enzyme cofactors such that dietary deficiencies can result in decreased metabolic flux. Perhaps the most striking example of substrate control occurs in glycolysis, where NAD^+ is a substrate for the glyceraldehyde-3-phosphate dehydrogenase (G3PDH) reaction. The equilibrium reaction limits glycolytic flux when the available NAD^+ has been converted to NADH. NADH is returned to the NAD^+ by two pathways, anaerobically using lactate dehydrogenase

$$\text{Pyruvate} + \text{NADH} \overset{\text{LDH}}{\longleftrightarrow} \text{Lactate} + \text{NAD}^+ \tag{11.14}$$

and aerobically using the electron transport system in mitochondria. Under high ATPase rate conditions in cells with minimal mitochondrial activity, such as contracting fast glycolytic skeletal muscle, lactate will build up as a mechanism to maintain glycolytic flux. Although PFK normally regulates glycolytic flux, as the flux is severely reduced when NAD^+ is not readily available, there must be an $NAD^+/NADH$ ratio where glycolytic flux regulation passes from PFK to G3PDH.

While the extensive studies of negative feedback involving cellular metabolic regulation have demonstrated substrate, allosteric and cofactor control mechanisms, negative feedback also occurs at tissue and organ levels. The control of muscle position and motion involves the muscle spindle nuclear chain and nuclear bag fibers feeding back on alpha motor neurons. In the knee-jerk reaction, stretching the patellar tendon elicits a contraction of the quadriceps. This is a negative feedback reaction: the muscle is extended by the tap, and reflexly returns to its original length. The knee-jerk reaction is a monosynaptic reflex: the afferent neuron synapses directly onto the quadriceps alpha motor neuron, generating an action potential that activates the motor endplate on the muscle. At the same time, interneurons are activated that inhibit the contraction of the hamstring, the muscle that closes the knee joint. Since different numbers of synapses are involved, the time constants for the stretch activation and paired muscle inhibition have to be different. In a complex activity such as sprinting, alternating contraction/relaxation cycles ideally sequence quadriceps/hamstring activity so that when one muscle is contracting the other is relaxed, preventing the stretching of an activated muscle that can produce structural damage, as previously discussed. The differences in activation and inhibition time constants are unimportant during slow movements, but may play a role when a high caliber athlete pulls a muscle when sprinting, potentially having the paired muscle contraction cycles out of phase, leading to simultaneous contraction and injury. It is these muscle groups that have the highest changes in biomechanical load during sprinting (Schache *et al.*, 2011), changes in which phase differential could be damaging.

11.3   Positive feedback

Positive feedback occurs when a system moves from one set point to another. In physiological systems this can occur in two distinct ways. In the first a molecular reaction has such a large drop in free energy that the reaction is essentially irreversible. Initiation of the reaction by enzyme E_{pos} is the trigger for a state transition from A to B:

$$A \xrightarrow{E_{pos}} B. \tag{11.15}$$

The concentration of A is normally controlled by conventional negative feedback. The state change reaction is not part of the normal negative feedback system. The state change is not limited by the production of B, but the disappearance of A, which vanishes exponentially:

$$A = A_0 e^{-\beta t} \tag{11.16}$$

where A_o is the concentration of A in the vicinity of E_{pos}, β is the rate constant of the reaction and t is the time since the activation of E_{pos}. The body will eventually make more of A, using a mechanism independent of the state change reaction. Blood clot formation is an example of this type of positive feedback.

A positive feedback system converts liquid blood to a solid, an obvious state change. In this case, the control parameter is the free concentration of fibrinogen. The conversion of fibrinogen to fibrin results in the formation of the fibrin mesh through the end-to-end binding of free fibrin. Activation of the enzyme thrombin converts fibrinogen to fibrin. This reaction is strongly weighted toward the hydrolysis of fibrinogen at equilibrium. The thrombin reaction is essentially irreversible. Once the mesh forms, removal of thrombin does not reverse the reaction, with the solid mesh remaining until plasmin slowly breaks down a clot over several weeks. When this occurs free fibrin or fibrinogen is not the result, but the component amino acids. Fibrinogen is manufactured in the liver and released into the blood, where it has a steady-state concentration, indicating a match of production and elimination. In the equation above, the conversion to fibrin would be a function of the time constant during thrombin activity. Since free fibrinogen is the measured parameter, its concentration is the A_o value that would approach zero as the thrombin activity rises.

The second type of positive feedback moves the control parameter to a value outside that normally maintained by negative feedback. In this case, the previous negative feedback is still present, but must be sufficiently exceeded by the positive feedback system that it cannot maintain the original homeostatic condition. The initial state of the system is C_1. Activation of the positive feedback mechanism will send the system toward a new set point C_2. The value of the parameter C at the onset of a positive feedback system will be

$$C = C_2 - C_2 e^{-\beta t} + C_1 e^{-\beta t} \qquad (11.17)$$

where β is the time constant for the onset of the new control subsystem driving the parameter to the new set point as well as the time constant for the suppression of the original control subsystem keeping the control parameter at C_1. The positive feedback equation shows that the parameter will both activate the control system approaching C_2 and suppress the control subsystem maintaining C_1. The presence of the suppression term does not imply anything about its mechanism. If the rate constant maintaining C_1 is much smaller than β, no active mechanism is necessary. If the rate constants are similar, an active suppression mechanism would be needed. The values of the C_2 onset time constant and the C_1 suppression time constant need not necessarily be the same, as they are in the equation above and in Figure 11.5, but as they must be in response to some stimulus, they must be similar. If the onset were much slower than the suppression, the parameter would approach zero for a time. If the onset were much faster than the suppression, the parameter would go through a rapid increase before settling at the new set point. The rate of approach to the new set point will be proportional to the ratio of the time constants maintaining initial parameter C_1 (α time constant) and the new set point (β time constant), as shown in Figure 11.5. Anaphylactic bronchiole constriction, gastric ulcers and ovulation all have this type of positive feedback.

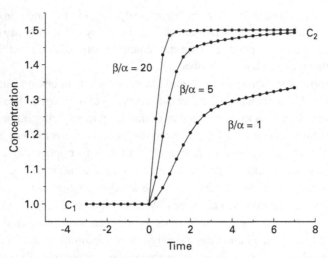

Figure 11.5   Positive feedback changes from state C_1 to state C_2. Activation of a positive feedback system moves the control parameter, in this case the concentration, to a new set point. The rate of approach to the new set point is determined by the ratio of the rate constant β for suppression of the original system negative feedback system to the rate constant α for maintenance of the original negative feedback system.

The lungs have a reserve that is about tenfold greater than the tidal volume of 500 ml. The ventilation–perfusion ratio is about 0.8, meaning that most of the blood in the lungs goes to those areas that are inflated, with minimal blood going to uninflated areas. Constriction of the bronchioles minimizes the lung volume at rest, matching the tidal volume to the needs of gas exchange. Some allergens significantly increase the production of histamine and leukotrienes C4 and D4, C4 and D4 together formerly called slow reactive substance of anaphylaxis, causing an alteration of the set point for bronchoconstriction. Leukotriene D4 has been shown to significantly disrupt the normal ventilation–perfusion ratio, constricting bronchiole beds that receive blood (Casas *et al*, 2005). This resulted in a decrease in the oxygen saturation from 100 to 75 mmHg. This change in set point during allergen exposure is a positive feedback reaction because the new set point is outside the normal concentration range of these bronchoconstrictors. There are many substances that contract or relax the bronchioles, so the normal set point is a combination of the concentration, binding and intracellular transduction of all the stimuli at the bronchioles. The treatment of anaphylaxis requires a concentration of much higher than normal epinephrine, countering the effects of histamine and the leukotrienes and returning the system to the appropriate level of inflation. In the absence of epinephrine or similar treatment, the person could die, the ultimate positive feedback state change.

In some cases, positive feedback systems carry the mechanism for their own destruction without being fatal to the organism, but to a part of the organism. In the stomach, there is a positive feedback system regenerating ulcerations. Excess stomach hydrochloric acid and/or bacterial infection cause ulcers when the mucus layer separating the acid from the epithelium is breached. The acid or infection causes tissue damage, which in turn triggers the release of histamine as a non-specific defense mechanism. Histamine then triggers more acid release, leading to the positive feedback cycle of more ulcers, more histamine, more acid, more ulcers, etc. Medical treatments break the cycle by either acid neutralization or suppression of acid secretion. Left unchecked, in the extreme case the acid would kill all the acid secreting cells and thus end the positive feedback. Fortunately, even for people with excess acid secretion, the rate of cellular reproduction matches the rate of cellular destruction, and they can live, even if in an unfortunately painful steady state.

A similar case, but without pathological consequences, occurs in the female reproductive tract. For many years the trigger for ovulation was thought to be the stimulation of luteinizing hormone (LH) by very high levels of estrogen, which normally inhibits LH release in a negative feedback manner. The reversal of the estrogen effect was troublesome, as the reversal of effect did not have a strong theoretical basis. The discovery of the hypothalamic hormone kisspeptin showed that very high levels of estrogen trigger the release of kisspeptin, estrogen levels higher than those that are normally needed to suppress LH release. Kisspeptin in turn causes release of gonadotropin-releasing hormone (GnRH) in sufficient concentration to overcome the suppressive effects of estrogen on LH, leading to the LH surge that causes ovulation. The rupturing of the ovarian follicle causes a drop in estrogen, breaking the positive feedback mechanism that caused ovulation. In a mechanistic sense, the change in the tissue (i.e. follicular rupture) parallels the destruction of the stomach epithelium above. In those transitions that do not result in death, new negative feedback processes will keep the system at its new set point. All long-term steady-state systems must have energetic stability.

11.4   Models of state stability

The stability of a physical system requires that it be at the lowest part of an energy well, its equilibrium position, or have continuous energy input to maintain a steady state removed from equilibrium. The force equation that includes time-dependent derivatives will change its value as a given system approaches equilibrium. In a system with one parameter, the state of the system can be defined by a two-dimensional graph of the parameter and time. Systems with multiple parameters require multidimensional graphs. For a system with n parameters, those parameters will all have a given value at a point in time, and will describe a particular value in n-dimensional phase space (Glaser, 2001). A two-parameter system will have two-dimensional phase space and three-parameter systems will have three-dimensional phase space, like those in Figure 11.6.

Figure 11.6 shows the surface of different three-parameter spaces. Two of the three parameters are defined by the two sets of mutually perpendicular ribs describing an XY plane, with the third parameter the Z distance above or below the crossing point. If energy is plotted in the Z direction, all the systems will move toward the lowest energy state. Importantly, time is not part of phase space diagrams, so that the absolute rate of movement cannot be inferred from the diagrams. In the upper left graph, both non-energy parameters have a common energy minimum, so this system is stable and will

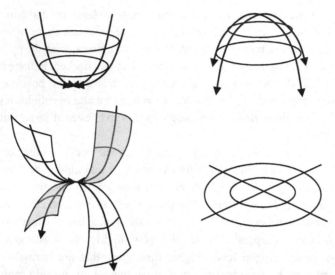

Figure 11.6 Models of energetic stability. The upper left shows a three-dimensional stable system with all vectors directed toward a lowest energy state. The upper right shows an unstable system with all vectors directed away from the center state. The lower left shows a saddle system with some vectors directed toward the center and others directed away. A saddle system will have a transfer of variables where the vectors cross. The lower right shows an indeterminate system with no vectors. All parts of an indeterminate system are equally accessible.

move toward that minimum. The upper right system is unstable: there is no minimum, so any element in this system will migrate away from the common crossing point that is an energy maximum. The lower left shows a saddle diagram, where elements above the crossing point will move toward their minimum by changing one parameter, but when the minimum for that parameter is reached, the system is at an unstable point for the second parameter, so now the second parameter will change, carrying the system away from the crossing point. In the lower right graph there is no change in the Z direction. This system is indeterminate. In this system, the energy is independent of the parameters. The entire system is at an energetic steady state, so that the parameters can be changed without altering the energy of the system. This would occur when a series of enzymes, like those in the middle of glycolysis, has virtually no net change in free energy.

When defining the surfaces of parameter space it is not immediately apparent that all the surface in that space is not equally accessible. In any n-dimensional space, the accessibility of any parameter phase is a function of its independence from the other parameters. When the parameters that occur at the same time are plotted together, they will define a line in phase space, as shown in Figure 11.7. This three-dimensional space includes the data from Figure 5.12, in which smooth muscle force, velocity and myosin light chain phosphorylation were plotted as a function of time. The data defines a particular pathway in this phase space. If other data were included, more lines would be defined. For example, if the velocity data for different loads were included, the defined surface would curve, since the force–velocity relation is curved. As we can only perceive

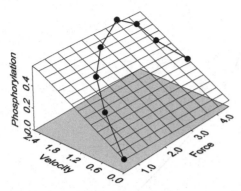

Figure 11.7   Phase space of Figure 5.12. The time course of smooth muscle force, velocity and light chain phosphorylation is plotted in parameter phase space, independent of time. The phosphorylation is plotted as the fraction of light chain phosphorylation. Velocity is plotted as the $0.12\,F_o$ velocity in units of mL_o/s. The force axis is plotted as the load-bearing capacity in units of 10^5N/m^2.

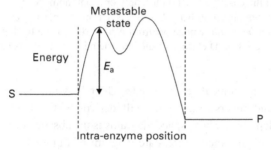

Figure 11.8   Enzyme metastable state. The transition from substrate molecule S to product molecule P occurs at an intra-enzyme binding site. The transition molecule will occupy a metastable state. The depth of the metastable state will determine the transition rates to the product or back to the substrate.

three-dimensional space, adding further parameters such as tissue length or calcium concentration would be impossible on one graph, but could be part of a family of graphs more completely defining the phase space of muscle contraction.

11.5   State transitions

An important, and relatively simple, example of phase space is shown in Figure 11.8. Here the metastable state of an enzyme intermediate is defined. The enzyme is converting molecule S into molecule P. It is common for enzymes to bind to their substrate, putting strain on particular bonds that leads to the molecular rearrangement that is the catalyzed reaction. The substrate S will need to overcome a certain activation energy to enter the active site of the enzyme. Here there will be an energy well that defines the metastable state. The system is metastable in that the energy barriers keeping the molecule within the well are smaller than those needed to enter the well. Over any given time, the molecule will exit the well, either in the forward direction completing the conversion to P, or in the

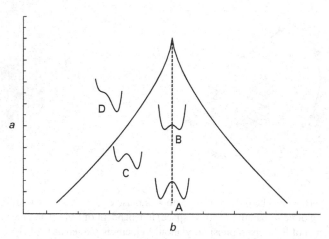

Figure 11.9   State transitions in the A+3 cusp catastrophe. Within the cusp this function has two minima. The dashed line is at $b=0$. A smooth change of parameter crossing this line changes which minimum is lower. Outside of the cusp the function has only one minimum. At A, $b=0$. Increasing parameter a changes the function from A to B, decreasing the barrier separating between the minima. Parameter b gets more negative from A to C, decreasing the left side minimum. Crossing the cusp from C to D eliminates the left side minimum, a state catastrophe.

reverse direction returning the molecule to its S state. The relative height of the energy barriers in the forward and reverse directions will determine the fraction of each transition. An important element of phase space diagrams is the absence of their relation to physical space. In the enzyme metastable state diagram of Figure 11.8, a different but related molecule would have a different activation energy E_a and depth of the metastable state, even if the two molecules had virtually the same position within the enzyme. The enzyme reaction rate would reflect these differences, as molecules with the lowest activation energy and shallowest energy well will have the fastest rate constants.

One of the consequences of energetic state transition is the limitation on the phase space occupied by living systems. Two essential properties of living systems are the availability of free energy and homeostatic stability. If either of these were lost in a living being, that being would die. This dichotomy can be modeled using the dual minimum function of the cusp catastrophe (Figure 11.9). Catastrophe theory is a canonical mathematical program that models state transitions that occur due to a small change in a control parameter. A canonical system in one in which a smooth conversion of variables into a standard format is possible. Catastrophe theory is a branch of bifurcation theory, developed by Thom (1970) and Zeeman (1972). There are several different catastrophe models, each with a different set of control parameters (Gilmore, 1985). The simplest catastrophe is the fold, which uses a cubic equation. The next simplest is the cusp, which uses a quartic equation with two minima. The A+3 cusp catastrophe fits the control parameters a and b to the equation

$$f = 0.25x^4 + 0.5ax^2 + bx. \tag{11.18}$$

Transitions of this equation occur when the *a* parameter changes in such a way that the function crosses a critical or semi-critical line, shown in Figure 11.9. The solid cusp lines are the critical lines, where a minimum vanishes (or appears, depending on the direction of the parameter change), when the line is crossed. The dashed line in the middle of the cusp is a semi-critical line: when the function crosses the center line, the lower minimum changes from one side to the other. It is important to note that the wells in the cusp catastrophe function are not the same as the energy wells that have been frequently discussed. The catastrophe function is qualitative. The system will cross a critical line at a particular parameter value, creating or abolishing a local minimum in the system. There is a qualitative change in the system, but the change cannot be quantified without additional information about the local conditions. Catastrophe theory is also time independent. Consider, for example, moving toward the edge of a cliff. When you approach the edge, nothing will happen until you take that one step off the edge. Whether you stroll toward the edge, run toward it, or drive a car toward it, the catastrophe occurs when the edge is crossed, regardless of how one gets there. Catastrophe theory has been used to model energy transitions in living systems (Dillon and Root-Bernstein, 1997).

The living state must exist at a higher free energy than the global free energy minimum. Entropy will drive all systems toward the global free energy minimum, and after death all life forms will reach this state. Prior to death energy production and acquisition will keep living cells at an energy state above this minimum. At the same time, the living state must remain in a sufficiently deep energy well to prevent spontaneous transition to the global minimum. Death will occur if the energy well is so deep that it reaches the same energy as the surrounding environment, shown in Figure 11.10 as the loss of free energy death. For example, in the case of cyanide poisoning, cyanide binds to the cytochrome oxidase, blocking ATP production (Ganong, 2005). This decreases the free energy of the cyanide-exposed cell, leading to death. Conversely, if the homeostatic energy well is too shallow, ambient energy will be able to overcome the local energy barrier and the death will result as the system moves into the global energy well. Death then occurs due to loss of homeostasis. The complement system, natural killer cells, and cytotoxic T cells all kill cells by injecting pore-forming proteins into the cell membrane. The cell has pumps, transporters and channels that homeostatically maintain ion concentrations across the membrane. Pore formation destroys this homeostasis, allowing sodium and potassium to rapidly flow down their gradients, resulting in osmotic lysis through the loss of volume control. Thermal excess would have the same effect on cellular homeostasis.

The net effect of having the requirements for free energy and homeostasis will restrict living systems to one slice of parameter space in the cusp catastrophe. When energy is one of the parameters, the parameter space of living system includes the C function in Figure 11.9, bounded by a critical and semi-critical line. Crossing the critical line is consistent with the loss of homeostasis, and crossing the semi-critical line is consistent with the loss of free energy. For other analytical conditions, the state transitions will depend on other parameters. If one is using the intracellular calcium concentration as a control parameter, under normal conditions it would be outside of the cusp. Normal

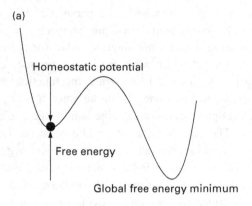

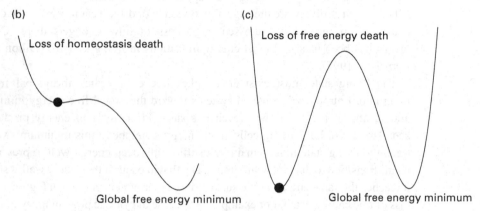

Figure 11.10   The two types of energetic death. Living systems exist in an energetic balance between loss of homeostasis and loss of free energy. (a) The dot indicates a system which has sufficient free energy to drive reactions but also a deep enough energy well to remain removed from the global energy minimum of the environment. (b) A system with a shallow energy well. Small ambient energy variations would be sufficient to drive this system to the ambient energy minimum. (c) A system with so little free energy that it cannot drive any reactions.

variations would change the depth of the well, but no state transition would occur. In people with muscular dystrophy, much of the damage done when the dystrophin protein is missing is caused by a calcium leak, increasing the resting calcium level in the cell, leading to the activation of calpain and proteolysis (Culligan and Ohlendieck, 2002). Stretch-induced muscle activation that leads to calcium-induced contraction also leads to calcium-induced damage (Allen *et al.*, 2010). When muscular dystrophy is modeled, there is a calcium concentration that causes a state change leading to damage and death: there will be a new minimum appearing in the A+3 function as the calcium reaches the concentration corresponding to the activation of pathological processes that are not normally present.

When catastrophe theory first appeared, it was applied to transitions in many fields, before subsequently being found to be less predictive than desired. Catastrophe theory

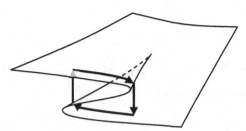

Figure 11.11   Hysteresis cycle on a cusp catastrophe.

continues to be studied in mathematics, and makes important contributions in physics and engineering, such as in modeling of black holes. One of its difficulties is that catastrophe theory does not have time dependence. Being able to define a value where a transition occurs may not help if the rate of approach to that value cannot be determined. Also, the transition values must dominate all other parameters in order to be useful. All the examples above in cellular physiological systems, cyanide poisoning, pore formation and calcium concentration, lead to such important changes that all other parameters are dwarfed in comparison. At more complex physiological levels, catastrophic models are still possible, but may have so many other factors that transition analysis is more difficult. An area of immediate applicability is in systems with hysteresis (Gilmore, 1985). For the fold of a cusp catastrophe, moving across the top of the fold reaches a critical point of transition to the lower surface, followed by movement across the lower fold surface to a critical point transition back to the upper surface (Figure 11.11). The entire cycle describes a hysteresis curve. Multiple methods of analysis combining differential equations of the changes in variables with identification of variable values that result in state transitions are needed to fully describe dynamic systems.

In addition to the effects of specific parameters on state changes in any system, the ambient energy always has to be considered. Figure 11.12(a) shows the three cases of energy relative to kT, the ambient molecular energy. A static system in which the depth of the energy well is much deeper than kT is shown on the left: the energy of the environment is too small to cause any transition out of that state. In physiological systems, covalent bonds fit this criteria. As discussed in Chapter 1, the Boltzmann distribution predicts the probability of a molecule having a particular energy, from which it can be estimated that it would take hundreds of years for a covalent bond to spontaneously break. Only by using enzymes to lower the energy barrier between the energy well shown and adjacent wells can a system leave this type of system. These systems will not be dynamic.

The middle panel of Figure 11.12(a) shows a system in which the ambient energy is greater than the depth of the energy well. This type of system will be unstable. Ambient energy will drive the system out of this energy well easily. Thermal denaturing of proteins during cooking fits this model. Cooking temperatures are not high enough to break covalent bonds, but the secondary structure maintained by hydrogen and hydrophobic bonds will be entirely disrupted and irreversible.

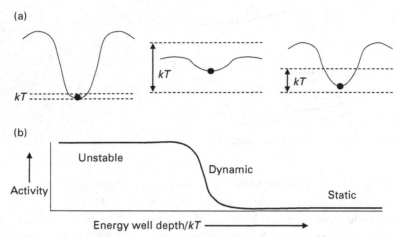

Figure 11.12   Effect of ambient energy on system stability. (a) The ambient energy is kT. On the left, the energy well is so deep that this system is essentially static when buffeted by ambient energy. While covalent bonds have this energy, living systems need enzymes to lower the energy barrier to make or break covalent bonds. The system in the middle panel has ambient energy greater than the depth of the energy well. This system would be unstable. The right panel shows a system whose energy wells are deeper than but near ambient energy. This system will be dynamic. (Reproduced from Dillon and Root-Bernstein, 1997, with permission from Elsevier.) (b) The transition from unstable to dynamic to static as a function of the energy well depth/ambient energy ratio.

The right side of Figure 11.12(a) shows a system in which the depth of the energy well is of the same magnitude as the ambient energy. This system will be dynamic, with rapid transitions between different states. Small changes in this system will significantly alter the rates of transition. A plot of system activity as a function of the ratio of energy well depth to ambient energy, shown in Figure 11.12(b), will have a sigmoidal transition between high activity when the ratio of depth/kT is small to low activity when the ratio is high. Living systems have to exist, at least in part, in the dynamic range where the ratio is in the neighborhood of unity to 3–5 kT. Even a cursory look at this diagram shows that the middle mimics the dose–response curve of many physiological systems. The extensions into regions where there are no changes is intentional: reports of dose-dependent phenomena routinely do not include concentrations where nothing is changing. Interesting things happen in regions where there is change. This is one of the hallmarks of biophysics, that it studies systems that are dynamic. That unstable, static and dynamic parts of a system are sometimes colloquially referred to by the Goldilocks description, respectively too hot, too cold, and just right, is not a coincidence.

11.6   Non-linear systems: fractals and chaos

Systems that exhibit self-similarity at different dimensions are fractal. When fractal analysis first entered physiology and biophysics, both structural and time varying systems were discovered that exhibited fractal behavior. Fractal systems involve scaling relationships, such that there is a constant ratio of a function L at scale r to that function at scale ar (Bassingthwaighte et al., 1994):

$$\frac{L(r)}{L(ar)} = k \quad \text{for } a < 1 \tag{11.19}$$

where k is a constant. When there is a power law such that

$$L(r) = Ar^\alpha \tag{11.20}$$

the equations can be rearranged to yield

$$L(ar) = L(r)a^{1-D} \tag{11.21}$$

where D is the fractal dimension of the system. D can also be determined by examining the self-similarity between dimensions. When the scale is changed by a factor of F and the number of pieces similar to the original is N, the fractal dimension is

$$N = F^D \quad \text{and } D = \frac{\log N}{\log F}. \tag{11.22}$$

Both power law scaling and self-similarity yield the same value of D for a given system. The term fractal arises because many systems are characterized by a non-integer fractional dimension, in contrast to Euclidian space in which the dimensions are integers. When examining a fractal system, it will look the same at any scale. One of the first applications of fractals in physiology came in the examination of the bronchial tree. There are about 20 generations of the bronchioles before the tree terminates in the alveoli. The relationship between the log of the diameter of the bronchioles and their generation number had shown a logarithmic relation down to about the tenth generation, but a deviation from a log relation for generation occurs between 10 and 20 (Weibel and Gomez, 1962). When the log of the diameter (N) was plotted against the log of the generation (F), the relation was linear, with a slope of D, down to the twentieth generation (West et al., 1986).

Fractals show up in kinetic analysis as well. We saw the stochastic nature of ion channel opening and closing in Figure 9.3. When the log of the number of ion channel durations per millisecond was plotted against the log of closed time durations, the relation was linear (i.e., fractal) over more than three orders of magnitude (McGee et al, 1988). This behavior means that the channel kinetics are identical whether the duration of observation is very long or very short. The physical explanation for this implies a large number of similar conformational states with shallow local minima (Bassingthwaighte et al., 1994). This structural model is also consistent with a dynamic model in which the energy barrier between open and closed states varies in time. This model is in contrast to the Markov model in which the energy barriers are constant and occur only at discrete energies. Time variations or a wide distribution of energy barriers make a channel's behavior less discrete and more fractal. The dichotomy led to a controversy over the nature of ion channels. If small minima are independent of large minima, then the channel behavior will be discrete; if the small minima and large minima are functionally dependent the channel will exhibit fractal behavior. The physical limitation on a channel's size means that even conditions that exhibit fractal behavior have a physical limit, unlike mathematical scaling, which is infinite.

The physical limit of fractal systems poses interesting questions. Consider the fractal bronchiole tree. With every smaller generation, the diameter of the bronchiole decreases in a predictable fashion. Why does it stop at an alveolus? Fractals require a kind of system memory, or self-examination. The system can only take the next step by knowing how big the previous step was: there has to be some kind of control system that dominates the process. As the system gets smaller and smaller, other forces become relatively more important. Water bound to a surface will have minimal energetic influence on a surface when the surface is bending over centimeters, but will have a greater effect when the bending occurs over micrometers. Likewise, the angle between bending units will increase as the number of units needed to complete a circle decreases. Since molecules are assumed to be at their lowest energy position when unconstrained, the greater the angle, the more energy must be put into the structure, until the system cannot put enough energy into the structure to bend it. Bronchioles will have an energetically minimum diameter.

Part of the difficulty in the leap from a mathematical analysis of a fractal structure to the actual enzymatic processes that produce that structure are implicit in discussions over findings like the fractal nature of channel behavior. These considerations were addressed in studies of the structure of glycogen (Meléndez *et al.*, 1999). Starting with a core protein glycogenin on which glycogen is constructed, they modeled the glycogen molecule as the number of $N(k)$ boxes of ε diameter needed to re-cover the surface of the growing structure:

$$D_f = \lim_{k \to \infty} \frac{\log N(k)}{-\log \varepsilon} \tag{11.23}$$

where D_f is the fractal dimension. In this model, k(iterations) $= t$(tiers), the number of layers as the complex grows. From this ε can be calculated from the tier number t

$$\varepsilon = q \cdot \frac{t}{2^{(t-1)/2}} \tag{11.24}$$

and the fractal dimension is

$$D_f = \lim_{t \to \infty} \frac{\log 2^{t-1}}{-\log(q \cdot \frac{t}{2^{(t-1)/2}})} = 2 \tag{11.25}$$

where q is a constant comprising all the other parameters besides t. The glycosidic chain ends would form a compact surface for digestion by phosphorylase. The structure normally reaches 12 tiers, with only the outer four exchanged by the enzyme. This compact structure allows the enzyme to slide across the surface liberating glucose residues. This model can address the two central tenets of a fractal system, self-similarity and self-reference. The entire structure is fractal at every tier, including the number of available glucose units and the time energy metabolism can be supported by each tier. But, as noted above, the mathematical description of a fractal does not supply information about its construction. In this case, the behavior of the three enzymes involved, the synthase, the branching enzyme

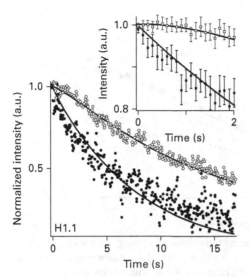

Figure 11.13   Time courses of chromatin repositioning. The lower curve shows the repositioning of euchromatin, and the upper curve shows the repositioning of heterochromatin. The inset shows the early time points of the curves. (Reproduced from Bancaud *et al.*, 2009), by permission from Macmillan Publishers Ltd.)

(which must be much faster than the synthase), and the phosphorylase, which will remove the ends of chains dangling from the surface, results in the fractal glycogen structure. Known differences in enzyme activity and differential activity in open vs. confined environments gives further support to this model (Meléndez *et al.*, 1999). Given the relatively simple structure of glycogen, that more complex systems like bronchioles have yet to yield workable models of fractal construction is not surprising, but no less desirable.

In Chapter 2 we discussed the importance of retention-reaction times in gene activation, and the digital nature of transcription factor dissociation. The process of molecular association with chromatin may be equally complex. Euchromatin is the more lightly packed form of chromatin, containing many active genes. Heterochromatin is more tightly packed, and may assist in the activation of nearby genes or contain genes inactivated by histone methylation. The kinetics of these two types of chromatin are quite different (Bancaud *et al.*, 2009). As shown in Figure 11.13, the repositioning of markers in the euchromatin region, kinetically between compact and fully open regimes, is much faster than those in the heterochromatin region. While each can be fit to a fractal, the fractal for euchromatin space is much larger than that for heterochromatin space. The activity of transcription keeps the euchromatin space more open, and transcription factors not currently involved in transcription are able to scan larger volumes, facilitating the search for rare or distant binding sites. In contrast, highly packed heterochromatin will limit the diffusion of a binding protein away from its current site, illustrated as the plateau in the initial time course of repositioning in the inset of Figure 11.13, resulting in a positive-feedback system with a prolonged duration of closely packed structures. The mechanisms controlling the regimes and the transitions between them are as yet undetermined, awaiting further experiments.

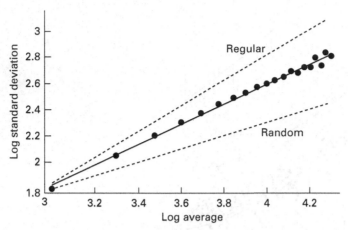

Figure 11.14 Fractal dimension of heart beat interval time series of a young adult male. The plot of log interval SD as a function of log interval mean has a slope of $D = 1.24$, intermediate between a rigidly regular beat interval with a slope of 1.0 and a random interval with a slope of 1.5 (Redrawn from West BJ. Front Physiol. 1: 1–17, Figure 2, 2010, used with permission.)

Among the most common areas susceptible to fractal analysis is time series variation, including heart beat, respiratory and walking gait intervals. Observations across many experiments and models have led to the allometric mechanism of system control (West, 2010). The mathematical basis for this mechanism is the development of fractional calculus, with both the fractional integral and fraction derivative defined. The mathematics of this process is beyond the scope of this book, but in its application to biological systems it defines allometric control. In contrast to homeostatic control which is local and rapid, allometric control retains long-time system memory, with correlations that are inverse power law in time. The net effect is a multifractal system that has multiple feedback of different subsystem outputs back into the input of the system. The system can vary in time, but with multiple inputs it is unlikely that the system as a whole will exhibit a state change. The overall system will have a fractal dimension D, which is related to the Hurst exponent H by the equation

$$D = 2 - H. \tag{11.26}$$

The Hurst exponent is the tendency of time series to regress to the mean. Its exact solution requires an infinite number of interactions, which is physically impossible. As such, the Hurst exponent is estimated, not exactly determined. Its utility is in defining specific limiting cases. Varying between 0 and 1, $0 < H < 0.5$ has negative autocorrelation – that is, a decrease in the interval between time values has a high probability of being followed by an increase; $0.5 < H < 1$ has positive autocorrelation, with an increase in time interval most probably followed by another increase; and for $H = 0.5$, the funcation has random behavior. In the calculation then of a fractal dimension, $D = 1.5 = 2 - 0.5$ indicates a random process. A value of $D = 1 = 2 - 1$ indicates a rigid, regular interval between time points. Where a time series fractal dimension falls defines its degree of regularity/randomness. The data plot that determines D is the log of the standard deviation of the data in a time series as a function of the log of the mean. The slope is D, as shown in Figure 11.14 for heartbeat interval data.

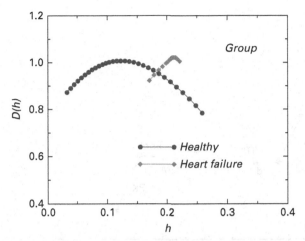

Figure 11.15 Relation of fractal dimensions distribution as a function of the Hölder exponent for heartbeat interval variation in healthy and heart failure individuals. The broad distribution of the healthy individuals indicates multifractal input. The narrow distribution of the heart failure individuals indicates monofractal input. (Redrawn from Stanley *et al.*, 1999, with permission from Elsevier.)

Allometric systems have multiple fractal inputs. The strength of allometric systems lies in these multiple inputs. Variations in the overall fractal dimension are measured using the Hölder exponent h, which varies from 0 to 1 for finite time series, and is related to the fractal dimension D by

$$D = 2 - h. \tag{11.27}$$

The Hurst and Hölder exponents are identical for infinitely long time series, but may be different for finite series. When the fractal dimension D is plotted as a function of h after a Legendre transform, the variations in D will form an arc with values of D falling for both positive and negative values of h relative to the value of h that yields the maximum D (Figure 11.15). It is this presentation that indicates the state of an allometric system. With many fractal inputs, the fractal dimension arc in h will be broad. The loss of complexity will narrow the arc. Comparisons of the middle cerebral artery blood flow velocity time series in control and migraine groups showed a narrower fractal dimension distribution in the migraine group (West, 2010). A similar Legendre plot is shown in Figure 11.15 for heartbeat interval fractal dimension variation for healthy and heart failure individuals. The healthy group had a broad, multifractal distribution, while the heart failure group had a narrow, monofractal distribution. Proponents of allometric regulation see that systems rely on multifractal inputs to maintain health, even being critical of regular, non-fractal ventilation and blood pumps as negating the health benefits of interval variation. This type of control is in sharp contrast to homeostatic control that strongly regulates a single variable. It will take further investigation to determine if one approach or the other dominates, or if different regulatory systems are specifically allometric or homeostatic.

The association of monofractal behavior with disease states implies that when only a single factor controls a system, failure of that factor can lead to a pathological state change.

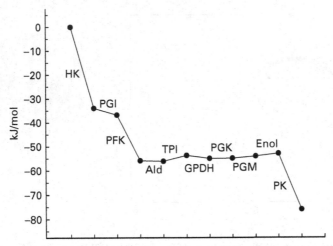

Figure 11.16 Free energy changes in glycolysis. Changes in free energy near ambient energy indicate enzymes near equilibrium. The sequence of enzymes from aldolase to enolase are near equilibrium. This sequence can exhibit chaotic behavior during sinusoidal glucose input.

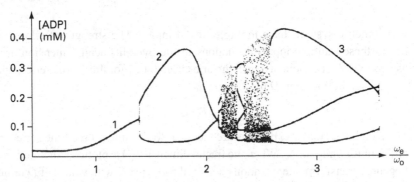

Figure 11.17 Variations in ADP as a function of sinusoidal glucose input. Increasing the frequency (ω_e/ω_o) of glucose input causes fractal transitions in ADP concentrations, with chaotic transitions occurring at intermediate frequencies. The chaotic states revert to deterministic behavior as the frequency further increases. (Redrawn from Markus *et al.*, 1985, with permission from Elsevier.)

An aspect of fractal systems that results in temporal state change is the tendency of some fractal states to produce chaotic behavior, particularly when the system has sinusoidal input. Chaotic behavior states sample more parameter space than a deterministic system. If there are particular parameter spaces that lead to pathological conditions, this would increase the probability of pathology. Sinusoidal glucose input can produce chaotic fluctuations in the equilibrium range of glycolysis. Figure 11.16 shows that for the enzymes between aldolase and enolase, there is no significant change in free energy relative to ambient energy. When the concentration of NADH, a function of glyceraldehyde-phosphate dehydrogenase activity, is used to estimate ADP concentrations, change in the rate of sinusoidal glucose input results in periodic chaotic behavior (Markus *et al.*, 1985), shown in Figure 11.17.

Given its name, its is easy to assume that chaotic systems will behave randomly. In chaotic systems however, the plotting of t vs. $t+1$ data reveals not random, but ordered, complex and non-linear behavior. Non-equilibrium reactions like PFK and PK will have large free energy decreases (Figure 11.16) and are not sensitive to initial conditions. In contrast enzymes with energy changes near ambient energy will be very susceptible to changes in initial conditions, an aspect that can produce chaos (Bassingthwaighte et al., 1994). But unlike truly random systems, chaotic systems are limited in their response, drawn to specific parameters within their system called strange attractors. Thus, chaotic systems will not sample all of parameter space, but only selected sections and are eventually returned to deterministic space. Both deterministic-to-chaotic and chaotic-to-deterministic transitions are shown in Figure 11.17. Different input conditions lead to chaotic activity of glycolytic ADP, which then reverts to deterministic activity (Markus et al., 1985). This would be statistically less likely to happen if a system transitioned from deterministic to random behavior.

Chaotic behavior may protect cells from catastrophic transitions. Since life is a continuum, living systems cannot reversibly undergo life–death state transitions. Apoptosis, programmed death of selected cells in a larger system, is the exception. Since cells contain the genetic ability to enter pathways leading to death, under most circumstances these pathways must be avoided. First considering all non-apoptotic subsystems, deterministic, non-equilibrium subsystems cannot lead into apoptotic pathways, as this would rapidly bring an end to life. During system transitions through near equilibrium sequences, such as the middle reactions of glycolysis, the elements in the sequence do not have strong driving forces. The possibility of crossing a catastrophic transition exists to the extent that equilibrium systems can sample all of parameter space. If these sequences are chaotic, they will not sample all parameter space, but will have a trajectory in parameter space around a strange attractor, eventually returning the subsystem to deterministic space, which as noted above will not be apoptotic. Chaotic behavior protects against catastrophic life–death transitions.

11.7   Apoptosis

Apoptosis is programmed cell death. As noted above, the continuum of life cannot occur if the processes that trigger a cell's own death are not only regulated, but strongly inhibited under almost all circumstances. Apoptosis, once activated, leads to a series of events that kill the cell. Cells have genomic sequences of death domains and death effector domains, domains which are normally kept inactive by antagonistic decoy receptors that bind activators of the death domains without activating any apoptotic processes (Danial and Korsmeyer, 2004). Under the right circumstances, these processes are activated, with abnormally high intracellular calcium and unfolded proteins – two factors that are often associated with apoptosis. Elevated calcium leads to caspase activation.

Caspases are a family of cysteine-dependent aspartyl proteases that are part of the apoptotic cascade. A wide range of factors can precipitate caspase activation, in which initiator caspases activate effector caspases, which then proteolytically attack many cellular proteins, leading to cell death. The caspase cascade is similar to the activation of proteolytic enzymes in the small intestine, where enterokinase converts trypsinogen to trypsin, which then activates many other proteases leading to the complete digestion of all ingested and secreted proteins. Caspase digestion of cellular cytoskeletal structures causes rounding and shrinkage of cells as they lose protein polymer support. The transport of proteins within the cell will also be affected by the caspase-induced alterations in actin filaments which in turn leads to mitochondrial dysfunction (Martin and Baehrecke, 2004). Cytoskeletal changes may account for the dense packing of organelles during apoptosis. Chromatin condenses against the nuclear membrane, which also fragments. The cell membrane forms blebs, further indicating cytoskeletal damage. Given the wide range of these and other insults, the death of the cell is not surprising. How do most cells prevent this from happening most of the time?

Among the key players in apoptosis is p53, an activator of apoptosis genes, which functions to apoptotically kill transformed cells, a major defense against cancer. The protein is kept in low concentrations inside cells through degradation, minimizing the probability of binding to apoptotic control sites. Signals that inhibit p53 degradation can induce apoptosis. Also, activation of the *BCL-2* gene plays a prominent role in preventing apoptosis by blocking the plasma membrane blebbing, volume contraction, nuclear condensation, and endonucleolytic cleavage of DNA associated with apoptosis (Hockenbery *et al.*, 1990), making this a branch point gene in the control of apoptosis (Danial and Korsmeyer, 2004). These and other regulatory sites have to be sufficiently strong to prevent inappropriate cell death, and still function in areas like cancer control to eliminate potential deadly cells.

Apoptosis is not the only process that can lead to cell death. We know that cells can survive a wide range of environmental conditions using multiple intracellular mechanisms, but given sufficient insults the systems can be overwhelmed and unable to cope. Unlike the well-programmed apoptotic processes, necrotic cell death uses enzymes with normal metabolic roles that essentially go rogue, destroying the cell they are in (Syntichaki and Tavernarakis, 2002). Activation of calpain and the deregulation of lysosome degradation leading to necrotic death appear to be common elements across species. The necrotic process is triggered by factors such as hyperactive ion channels or G-proteins, or by protein aggregation, that lead to increased intracellular free calcium either from the endoplasmic reticulum or the extracellular space through calcium channels. This in turn activates calpains, proteolytic enzymes that will both destroy intracellular structures independent of apoptosis and initiate the caspase cascade and apoptosis. This is consistent with the idea of a common mechanism that can be activated by a wide range of stimuli leading to cell death (Syntichaki and Tavernarakis, 2002). Other proposed mechanisms, such as poorly liganded iron leading to a number of apoptotically linked disease states (Kell, 2010), also reflect the theme of multiple factors leading into a common path that causes death.

The elements of allometric regulation discussed in the previous section and the activation of common elements leading to apoptotic and necrotic death overlap considerably. Cells clearly survive under a wide range of physical and chemical insults, yet there are conditions they cannot survive. This commonality sees multiple control systems, whether fractal or not, that weakly influence the state of a cell without rigidly holding it to a specific homeostatic point, allowing parameters to vary in time without the cell dying. This would be consistent with the axiom that the tree that does not bend will break. Moving to a condition where one factor has a great deal of influence puts the cell at risk, for if that subsystem fails, the cell will fail and undergo a life–death state transition.

Interestingly, at the extracellular level, strong homeostatic regulation of two ions, potassium and calcium, appears to be essential for life. Altered potassium leads to cardiac arrhythmias, which may in turn cause fatal ventricular fibrillation. Hypocalcemia can lead to death through muscular spasms of the respiratory system. Other extracellular components, while leading to pathological conditions that increase morbidity and mortality, may not lead as inexorably to sudden death. Increased sodium causes hypertension, but if a single hypertensive event always led to death who would watch their children compete in sports? Increased glucose causes diabetes, but do people die from a single episode of sugar overload? Increased hydrogen ions cause acidosis, but if a single acidotic event caused death, no one would exercise. Yet, over prolonged periods of time, hypertension, type II diabetes, and renal failure induced acidosis do cause death. The control of systems at the cellular and supercellular levels can be different, and the conditions that maintain states or lead to rare but important state transitions that alter life, or to the final state transition from life to death, are not always easily established, either through failure of a single important system, or the widespread failure associated with apoptosis.

Controversy of a mild sort arose from the inception of studies of programmed cell death. Apoptosis has two pronunciations, one with the accent on the penultimate syllable and a silent second "p", and the other with the accent on the second syllable and both p's pronounced. In the very first paper on apoptosis (Kerr *et al.*, 1972), where the word was first used in this context, a footnote (p. 241) addresses this point. The authors, after consulting with a professor of Greek, proposed that the pronunciation with the silent p be used. Letting the originators of a term determine its use, especially after consulting with an expert, seems the proper thing to do. In the grand scheme of things, this may be a minor point, although the suggested pronunciation does have a more soothing sound. No matter the regard for our fellow academicians, popular usage will ultimately determine the accepted pronunciation.

It is in the nature of scientists to try to minimize the variables in their experiments so that only one factor is being tested. This point was made when the global Gibbs energy equation was discussed in Chapter 1. Looking at multiple inputs in the regulation of cell activity or cell death is not intuitive for many scientists. Focusing on a single aspect of negative or positive feedback is the most common, though not the only approach to systems control. Research looking for links between different elements is growing. Projecting the models of catastrophe theory or apoptosis to supercellular levels will always involve assumptions that are difficult to prove. Likewise, if there are multiple feedback processes supporting systems with a fluctuating, average state, does studying them one at a time really tell us anything? Synthesis of the multiple experimental results and the ideas they generate will be necessary to advance concepts of systems control.

References

Allen D G, Gervasio O L, Yeung E W and Whitehead N P. *Can J Physiol Pharmacol.* **88**:83–91, 2010.

Bancaud A, Huet S, Daigle N, *et al. EMBO J.* **28**:3785–98, 2009.

Bassinthwaighte J B, Liebovitch L S and West B J. *Fractal Physiology.* New York: Oxford University Press, 1994.

Casas A, Gómez F P, Dahlén B, *et al. Eur Respir J.* **26**:442–8, 2005.

Chang R. *Physical Chemistry with Application to Biological Systems.* New York: MacMillan, 1977.

Culligan K G and Ohlendieck K. *Basic Appl Myol.* **12**:147–57, 2002.

Danial N N and Korsmeyer S J. *Cell.* **116**:205–19, 2004.

Dillon P F and Root-Bernstein R S. *J Theor Biol.* **188**:481–93, 1997.

Ganong, W. *Review of Medical Physiology.* New York: Lange, 2005.

Gilmore R. In *Mathematical Analysis of Physical Systems,* ed. Mickens R E. New York: Van Nostrand Reinhold, pp. 299–356, 1985.

Glaser, R. *Biophysics.* Berlin: Springer-Verlag, 2001.

Gu J W, Hemmert W, Freeman D M and Aranyosi A J. *Biophys J.* **95**:2529–38, 2008.

Hockenbery D, Nunez G, Milliman C, Schreiber R D and Korsmeyer S J. *Nature.* **348**:334–6, 1990.

Kell D B. *Arch Toxicol.* **84**:825–89, 2010.

Kerr J F R, Wyllie A H, Currie A R. *Br J Cancer.* **26**:239–57, 1972.

Markus M, Kuschmitz D and Hess B. *Biophys Chem.* **22**:95–105, 1985.

Martin D N and Baehrecke E H. *Development.* **131**:275–84, 2004.

McGee Jr R, Sansom M S P and Usherwood P N R. *J Membr Biol.* **102**:21–34, 1988.

Meléndez R, Meléndez-Hevia E and Canela E I. *Biophys J.* **77**:1327–32, 1999.

Monod J, Changeux J-P and Jacob F. *J Mol Biol.* **6**:306–29, 1963.

Newsholme E A and Start C. *Regulation in Metabolism.* Chichester, UK: John Wiley & Sons, 1973.

Schache A G, Blanch P D, Dorn T W, *et al. Med Sci Sports Exercise.* **43**:1260–71, 2011.

Stanley H E, Amaral L A N, Goldberger A L, *et al. Physica A: Stat Mech Appl.* **270**:309–24, 1999.

Syntichaki P and Tavernarakis N. *EMBO Rep.* **3**:604–9, 2002.

Thom R. In *Towards a Theoretical Biology 3: Drafts,* ed. Waddington C C Edinburgh, UK: Edinburgh University Press, pp. 89–116, 1970.

Weibel E R and Gomez D M. *Science.* **137**:577–85, 1962.

West B J. *Front Physiol.* **1**:1–17, 2010.

West B J, Bhargava V and Goldberger A L. *J. Appl. Physiol.* **60**:1089–97, 1986.

Zeeman E C. In *Towards a Theoretical Biology 4: Essays,* ed. Waddington C C Edinburgh, UK: Edinburgh University Press, pp. 8–67, 1972.

12 Concluding remarks

Thank you for the time and effort you have made to read this book. I hope you have found it profitable. Like every science textbook, it reflects the work of thousands, as every idea in this book was discovered by somebody. Discovering something new, even if it is just new to you, is very satisfying. When I am working on a new idea, I see it as a many-sided physical object in my mind, changing in shape, color and texture, rotating in all directions, until I can see the edges are parallel. When they are, I have the answer. I don't know why my mind works like this, but it does. Sometimes the problem percolates for days or weeks, but sometimes it can take on-and-off thinking for years. Looking for confirmation of an idea means thinking up original experiments and making them work. When you do a novel experiment, you have to have so much trust in and understanding of your equipment that if you get an unusual result, you can believe the result is correct because you know the equipment is working properly, and therefore the hypothesis is true. This moment of discovery of something really original is thrilling: you know that you know something new, you know that it is important, and you know that no one else in the whole world knows it. It is a great gift to have that flash of insight when an experiment confirms an original idea. In over 35 years in science, I have been lucky enough to have this happen exactly four times. Each time, it is a golden moment, a glimpse of nature never seen before. In the spirit of discovery, I have climbed a hill to the west, away from a settled valley, and seen the dawn illuminate a new, unknown valley before me. Many scientists make that same climb, find a new valley and settle there, plowing the fields, building the towns, and leading others to the benefits of their discovery. Others, and I count myself lucky to be one of them, go into that new valley, stay for a while, and whether others come behind or not, set their restless mind on the new, undiscovered land over that next hill. Whether you have gone often, or just once, or if you are new to this wonderful world, when you get that urge to travel on, ignore the naysayer words that there is nothing left to find, that no one would want to go there, that it's not important or that it's just too hard. Follow your instincts and climb that hill.

Go.

Travel light.

Index

Printed in the United States
By Bookmasters